ADVANCES IN
METAL-ORGANIC CHEMISTRY

Volume 6 • 1998

ADVANCES IN METAL-ORGANIC CHEMISTRY

Editor: LANNY S. LIEBESKIND
Department of Chemistry
Emory University
Atlanta, Georgia

VOLUME 6 • 1998

 JAI PRESS INC.

Stamford, Connecticut   *London, England*

Copyright © 1998 JAI PRESS INC.
100 Prospect Street
Stamford, Connecticut 06904

JAI PRESS LTD.
38 Tavistock Street
Covent Garden
London WC2E 7PB
England

ISBN: 0-7623-0206-2

ISSN: 1045-0688

Manufactured in the United States of America

QD
410
A38
v.6
1998
CHEM

CONTENTS

LIST OF CONTRIBUTORS

Ana M. Castaño

Departamento de Química Orgánica
Universidad Autónoma de Madrid
Madrid, Spain

Antonio M. Echavarren

Departamento de Química Orgánica
Universidad Autónoma de Madrid
Madrid, Spain

Mark Lautens

Department of Chemistry
University of Toronto
Toronto, Ontario, Canada

Rai-Shung Liu

Department of Chemistry
National Tsinghua University
Hsinchu, Taiwan

Norio Miyaura

Division of Molecular Chemistry
Hokkaido University
Sapporo, Japan

Michel Pfeffer

Laboratoire de Synthèses-Métallo-induites
CNRS
Strasbourg, France

John Spencer

Laboratoire de Synthèses-Métallo-induites
CNRS
Strasbourg, France

William Tam

Department of Chemistry
University of Toronto
Toronto, Ontario, Canada

PREFACE

Welcome to Volume 6 of *Advances in Metal-Organic Chemistry*. You will find an international offering of topics spanning both metal-catalyzed and stoichiometric metal-mediated transformations. In Chapter 1, "Oxa- and Azametallacycles of Nickel: Fundamental Aspects and Synthetic Applications," Professor Antonio M. Echavarren and his coworker, Dr. Ana Castaño, of the Universidad Autónoma de Madrid introduce us to the mechanistically fascinating and synthetically versatile reactions of nickelacycles bearing either oxygen or nitrogen within the ring. Although stoichiometric in the metal, these easily prepared metallacycles are shown to be very useful reagents that undergo unusual and functionally rich transformations.

Metallacycles are also featured in Chapters 2 and 3 of this volume. Cobalt- and nickel-based metallacycles are key catalytic intermediates in "Transition-Metal-Catalyzed Cycloaddition Reactions of Bicyclo[2.2.1]heptadienes" by Professor Mark Lautens and Dr. William Tam of the University of Toronto (Chapter 2). The high strain energy of norbornadiene makes possible many of the interesting processes covered in this informative chapter. It is followed by a comprehensive review of the "State of the Art in Selective Hetero- and Carbocyclic Synthesis Mediated by Cyclometallated Complexes"

(Chapter 3) offered by Professor Michel Pfeffer and Dr. John Spencer of the Universite Louis Pasteur. Here, Spencer and Pfeffer explore some of the many unusual organic ring systems that can be obtained when cyclometallated complexes are treated with a diverse array of unsaturated substrates.

From the Republic of China, Professor Rai-Shung Liu of National Tsinghua University examines the "Synthetic Application of Cyclopentadienyl Molybdenum(II)- and Tungsten(II)-Allyl and Diene Compounds in Organic Synthesis." In addition to a brief overview of other work in the field, Professor Liu shows us some of the many intriguing transformations he has uncovered in his studies of these easily prepared molybdenum and tungsten species. Finally, in Chapter 5 Professor Norio Miyaura of Hokkaido University reviews the "Synthesis of Biaryls via the Cross-Coupling Reaction of Arylboronic Acids." The Suzuki–Miyaura reaction is one of the most important carbon–carbon bond-forming process developed in the last half of this century and the current offering should be well-received by practitioners of organic synthesis.

Lanny S. Liebeskind
Editor

OXA- AND AZAMETALLACYCLES OF NICKEL:

FUNDAMENTAL ASPECTS AND SYNTHETIC APPLICATIONS

Antonio M. Echavarren and Ana M. Castaño

Advances in Metal-Organic Chemistry
Volume 6, pages 1–47.
Copyright © 1998 by JAI Press Inc.
All rights of reproduction in any form reserved.
ISBN: 0-7623-0206-2

I. INTRODUCTION

Metallacycles are carbocyclic systems with one or more carbon atoms replaced by a metal. The growing interest in the synthesis and properties of these organometallic compounds are due to their involvement as intermediates in catalytic reactions, such as metathesis, dimerization, cyclotrimerization, and other processes.[1] Many metallacycles can be prepared by cycloaddition reactions of alkynes and alkenes.[2] Additionally, their enhanced stability allows the study of reactions that often lead to decomposition with their acyclic analogs. In particular, the β-hydride elimination is usually a slow process in the rigid 5- and 6-membered ring metallacycles because the required M-C-C-H angle around 0° cannot be attained.[1]

1a : X = O
1b : X = NR

In this article we review the chemistry of oxa- and azanickelacycles **1** and their six-membered analogs and related complexes. The first part of this account describes the synthesis of nickelacycles by oxidative additions to cyclic anhydrides. Alternative methods based on the intramolecular insertion of the OH or NH bonds into the double bond of α,β-unsaturated carboxylic acids or amides and oxidative coupling reactions are also included. Related syntheses of palladacycles are also described. The second part reviews the reactivity of these metallacycles. Our efforts directed toward the development of an amino acid synthon based on the chemistry of nickelacycles are described in the third part of this chapter.

II. SYNTHESIS OF OXA- AND AZANICKELACYCLES

A. Oxidative Additions of Cyclic Anhydrides to Ni(0) Complexes

The reaction of carboxylic acid chlorides with low-valent metal complexes leads to the formation of acyl complexes,[3] which can undergo decarbonylation to give σ-bonded metal alkyls, alkenyls, or aryl complexes.[4] This decarbonylation has been used to synthesize arylsilanes from aryl acid chlorides in the presence of nickel(0)[5] or palladium(0)[6] complexes as catalysts. Similarly, acyl cyanides undergo decarbonylation with palladium(0) complexes to afford nitriles or alkenes, depending on the substrate.[7]

Some methods based on decarbonylation reactions promoted by nickel complexes have also been developed.[8] Thus, coupling reactions proceeding through acylnickel(II) complexes lead to carbonyl compounds resulting from partial decarbonylation of the starting acid chloride (Eq. 1)[9–12] or related derivatives.[13] On the other hand,

$$2\ CH_3(CH_2)_6COCl \xrightarrow{\ Ni(CO)_2(PPh_3)_2\ } CH_3(CH_2)_6CO(CH_2)_6CH_3 \qquad (1)$$

decarboxylation of thioesters coupled to a β-hydride elimination in the presence of catalytic amounts of *in situ* generated nickel(0) complexes has been applied as a method for synthesizing alkenes from aliphatic carboxylic acids.[14] Aryl thioesters also undergo decarbonylation with catalytic amounts of palladium complexes or stoichiometric amounts of Wilkinson complex $Rh(PPh_3)_3Cl$.[15]

Some activated esters[16] also undergo oxidative addition to $Ni(COD)(PPh_3)_2$ (COD = *cis,cis*-1,5-cyclooctadiene) followed by decarbonylation to yield alkylnickel(II) complexes, which undergo β-hydride elimination and reductive elimination (Scheme 1).

The oxidative addition of simple carboxylic acid anhydrides to nucleophilic organometallic complexes has also been studied. For example, nickel(0) complex Ni(COD)(bpy) (bpy = 2,2′-bipyridine) reacts with

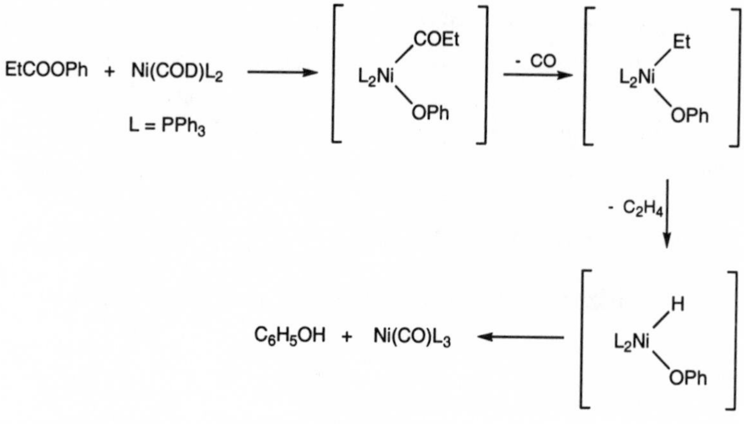

Scheme 1.

benzoic anhydride to form an oxidative addition intermediate that disproportions to form a nickel(II) dibenzoate and an unstable diacyl complex, which decomposes to afford biphenyl and a nickel dicarbonyl complex (Scheme 2).[17]

Most of the interest has been focused on the reactions of cyclic anhydrides with nickel(0) complexes. Thus, the reaction of five-membered ring anhydrides with $Ni(CO)_2(PPh_3)_2$ leads to alkenes (Scheme 3).[18] The reaction takes place by oxidative addition of the anhydride to

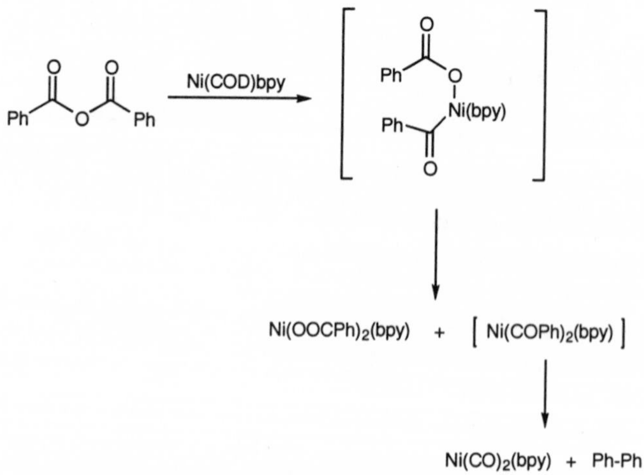

Scheme 2.

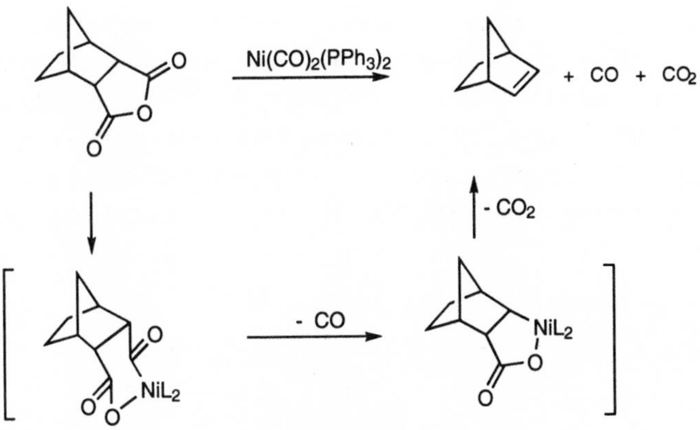

Scheme 3.

the nickel(0) complex, followed by decarbonylation to form a five-membered ring oxanickelacycle, which undergoes β-elimination of carbon dioxide to form the corresponding alkene. This method has been used to synthesize some bridgehead substituted norbornenes[19] and other bicyclic alkenes.[20]

The intermediate nickelacycles decompose by β-hydride elimination to form mixtures of alkenes.[18] For example, *trans*-dimethylsuccinic anhydride reacts with $Ni(CO)_2(PPh_3)_2$ to form a 10:10:1 mixture of *trans*- and *cis*-2-butene and 1-butene (Scheme 4). The formation of these alkenes indicates that an intermediate nickel hydride, such as the one shown in Scheme 4, is involved because control experiments demonstrate that no isomerization of the alkenes takes place under the reaction conditions.[18]

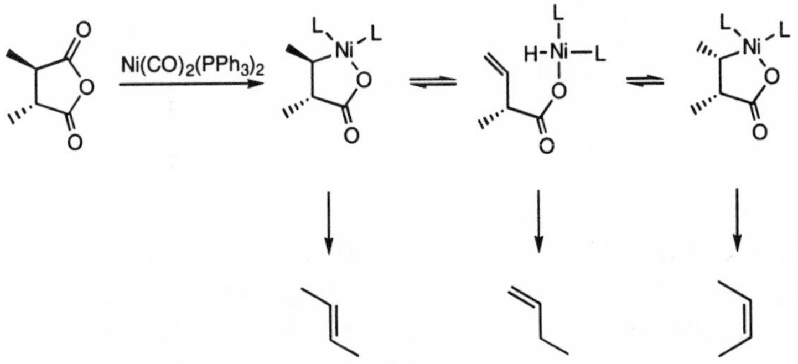

Scheme 4.

Nickelacycles analogous to those involved as intermediates in the previous reactions are isolated by using nickel(0) complexes with bidentate N-donor ligands.[21] Thus, succinic anhydride reacts readily at 15 °C with Ni(COD)(bpy) in THF solution to form a red-brick six-membered nickelacycle 2, which undergoes decarbonylation to afford a five-membered ring metallacycle 3 (Eq. 2). The expelled CO is trapped by the

$$\text{(2)}$$

       2   **3**

starting Ni(0) complex leading to the formation of complex Ni(bpy)(CO)$_2$.

Phthalic anhydride undergoes oxidative addition and decarbonylation only under refluxing conditions in THF to yield nickelacycle 4.[21] Similarly, the anhydride of cis-4-cyclohexene-1,2-dicarboxylic acid reacts with Ni(COD)bpy or a complex prepared from Ni(COD)$_2$ and TMEDA (N,N,N′,N″-tetramethylethylendiamine) to yield the corresponding five-membered ring nickelacycles.[22] However, the conjugated isomer 2-cyclohexene-1,2-dicarboxylic anhydride gives only (η^2-alkene)nickel(0) complexes under the same reaction conditions.[22] Diphenic anhydride furnishes the seven-membered ring derivative 5. The structure of a similar complex with the Ni coordinated to TMEDA was confirmed by X-ray.[23]

       4   **5**

The reaction also takes place with six-membered ring derivatives, such as glutaric anhydride.[24] In this case the primary nickelacycle 6 undergoes a ring-contracting reaction promoted by ligand exchange with 1,2-bis(diphenylphosphinoethane) (dppe) by β-elimination followed by insertion of the resulting nickel hydride with the opposite regiochemistry to yield five-membered ring nickelacycle 7 (Eq. 3).

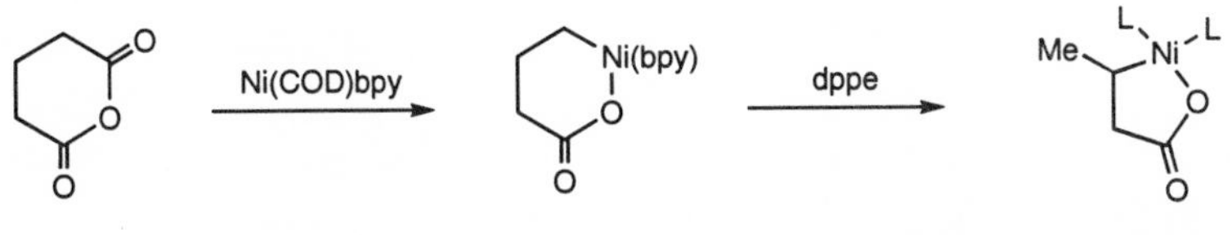

$$(3)$$

This isomerization reaction leads to the buildup of a new stereogenic center.[24] Therefore, addition of a chiral diphosphine may lead to the formation of an optically active product. Addition of the chiral bidentate diphosphine (S,S)-2,3-bis(diphenylphosphino)butane (S,S)-chiraphos) to a CH_2Cl_2 solution of the six-membered ring nickelacycle derived from glutaric anhydride leads to a mixture of diastereomers with a 16% diastereomeric excess favoring the R configuration (Scheme 5). However, on standing at 24 °C, the mixture of diastereomers equilibrates favoring the thermodynamically more stable S nickelacycle isomer with 52% diastereomeric excess.

Maleic anhydride acts as a η^2-ligand with nickel(0) complexes and does not yield any nickelacycle.[21] On the other hand, itaconic anhydride furnishes nickelacycle **8** in low yield, along with the η^2-complex **9**, isolated as the major product.[23] Complex **8** is also prepared by an alternative route based on the oxidative cycloaddition of carbon dioxide and allene.[25]

Platinum(0) complexes prepared from $Pt(COD)_2$ and tricyclohexyl-phosphine also react with succinic anhydride to afford the platinacycle

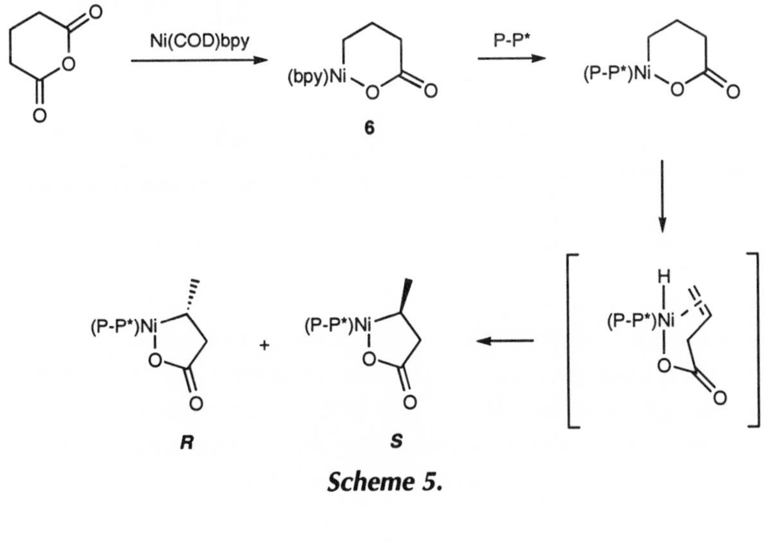

Scheme 5.

8 : L$_2$ = TMEDA   **9** : L$_2$ = TMEDA

resulting from oxidative addition.[24(b)] Similar results are obtained in the reaction between Pt(C$_2$H$_4$)(PPh$_3$)$_2$ and perfluorosuccinic anhydride.[26] On the other hand, palladium(0) complexes such as Pd(PCy$_3$)$_2$ and Pd(dba)$_2$[27] (dba = dibenzylideneacetone) fail to react with cyclic anhydrides.[24(b)]

B. Oxidative Cycloaddition of Alkenes and Carbon Dioxide or Isocyanates Promoted by Ni(0) Complexes

The formation of metallacycles by reaction of an organometallic complex with two coordinated unsaturated organic ligands is a common step in many catalytic processes.[1] Among these processes, the cyclotrimerization of alkynes catalyzed by cobalt complexes is probably the best known and most applied in organic synthesis.[28] A key step in this reaction is oxidative cycloaddition (also called oxidative coupling)[29] of two alkynes coordinated to cobalt(I) to give a cobalt(III) metallacyclopentadiene.[30] A similar reaction can be used to synthesize nickelacycles. Thus, five-membered ring oxanickelacycles can be prepared by the formal oxidative cycloaddition of acyclic and cyclic alkenes with CO$_2$ and nickel(0) complexes, such as Ni(COD)bpy (Eq. 4).[31-34] Related

$$CH_2=CH_2 + CO_2 \xrightarrow{NiL_n} \quad \textbf{1a} \tag{4}$$

complexes of titanium[35] and iron[36] have been prepared from alkenes and carbon dioxide. Allenes also react with nickel(0) complexes to afford the corresponding metallacycles.[25]

Because alkenes bind to Ni(0) phosphine complexes more strongly than CO$_2$, it is likely that this cycloaddition proceeds by reaction of Ni(alkene)L with CO$_2$.[37] An *ab initio* calculation indicates that this reaction is exothermic and proceeds with a moderate activation barrier.[38]

The reaction between 1,3-dienes and carbon dioxide leads to the formation of carboxylato (η^3-allyl)nickel(II) complexes, which may equilibrate with seven-membered ring nickelacycles in which the metal coordinates the allyl in a η^1-fashion (Eq. 5).[39-41]

$$(5)$$

The cycloaddition of alkenes is quite general and has been extended to the formation of azanickelacycles **12** by using isocyanates instead of CO_2 in the reaction with alkenes[42] or 1,3-butadienes[43] (Eq. 6). These

$$(6)$$

reactions have been carried out under catalytic conditions from isocyanates and alkenes[42(b),(d),(f),(g),(h)] or 1,3-dienes[43(b)] leading to the formation of α,β-unsaturated amides in moderate to good yields by using 10–20 mol% of nickel(0) catalysts. These α,β-unsaturated amides arise by decomposition of the intermediate azanickelacycles of type **12** by a β-hydride elimination reaction which regenerates the catalytically active nickel(0) complex.

Oxanickelacycles of type **1a** have also been synthesized by an alternative procedure based on inserting of carbon dioxide into the Ni–C bond of certain reactive nickelacycles. Thus, reaction of complex **13** with carbon dioxide leads to the formation of metallacycle **14** resulting from an insertion into the Ni-C(aryl) bond of **13** (Eq. 7).[44] Complex **13** is

$$(7)$$

prepared by ring opening of cyclopropabenzene with $Ni(COD)(PBu_3)_2$.[44] Five-membered ring nickelacycle **15** reacts similarly, yielding oxametallacycle **16** by selective insertion into the Ni-C(aryl) bond (Eq. 8).[45]

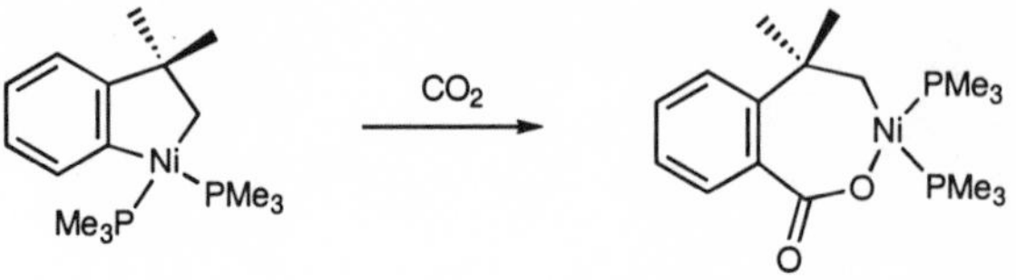

$$(8)$$

        15   **16**

Analogously, alkynes react with carbon dioxide or isocyanates to form the corresponding α,β-unsaturated oxanickelacycles **17** (Eq. 9)[33(c),46,47] or azanickelacycles **18** (Eq. 10), respectively.[48] Related palladacycles

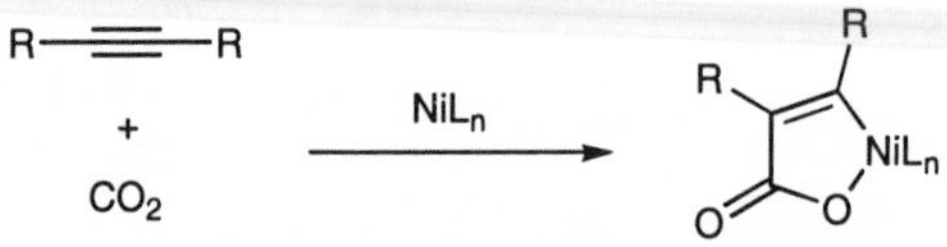

$$(9)$$

17

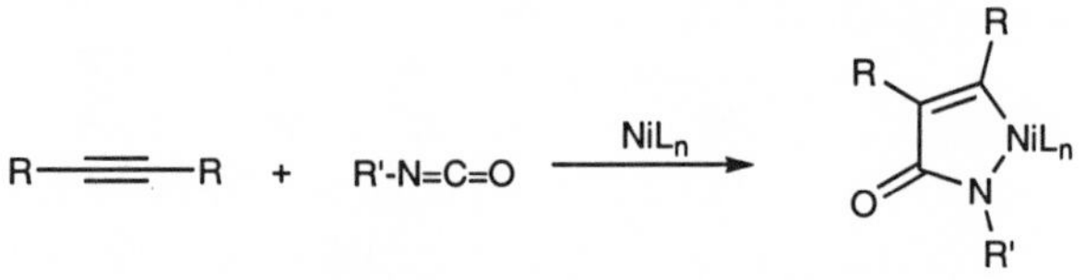

$$(10)$$

18

have been proposed as intermediates in the palladium-catalyzed dicarbonylation of alkynes.[49]

Cyclohexyne nickel(0) complex **19**, prepared by the reductive elimination of 1,2-dibromocyclohexene with sodium amalgam in the presence of $Ni(C_2H_4)L_2$ ($L_2 = Cy_2PCH_2CH_2PCy_2$),[50-52] also reacts with carbon dioxide or carbon disulfide to give complexes **20** and **21**, respectively (Eq. 11).[50]

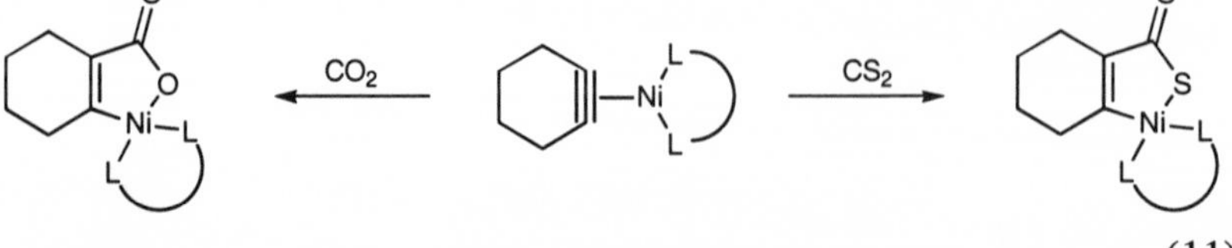

20   **19**   **21** (11)

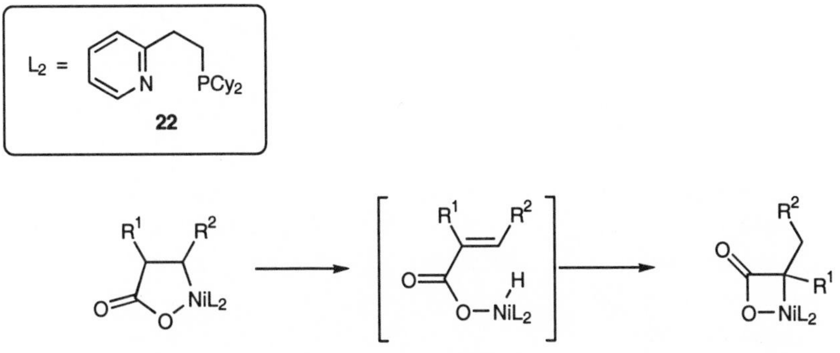

Scheme 6.

Although all these methods furnish five- to seven-membered ring metallacycles, four-membered ring derivatives can also be prepared by ring contraction with the appropriate ligands. Thus, by using pyridinephosphine **22** as the bidentate ligand, nickelacycles **23** undergo a facile β-hydride elimination to form intermediate complexes **24**, which undergo insertion to afford metallacycles **25** (Scheme 6).[33(f),53] Interestingly, this β-hydride elimination could also be promoted by adding of $BeCl_2$.

C. Synthesis of Nickel- and Palladacycles from α,β-Unsaturated Carboxylic Acids and Amides

The reaction of α,β- or β,γ-unsaturated carboxylic acids or amides with nickel(0) or palladium(0) complexes also leads to metallacycles. Thus, reactions of acrylic or methacrylic acid with $Ni(COD)_2$ and PCy_3 lead to the formation of the corresponding five-membered ring derivatives **26** (Eq. 12).[54] Similarly, reaction of α,β-unsaturated carboxylic acids with a nickel(0) complex with bidentate ligand **22** leads to complexes **27** (Eq. 13).[33(f)]

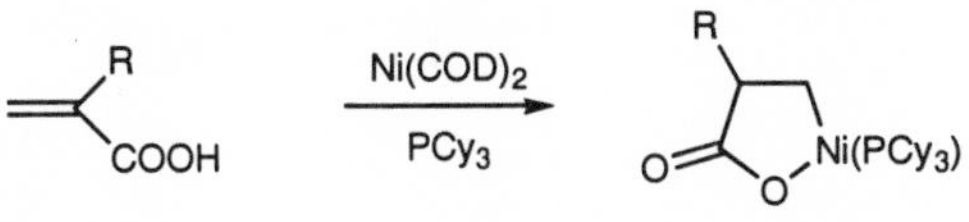

$$\text{(12)}$$

R = H , Me   **26**

$$R = Me, Ph \qquad\qquad 27 : L_2 = 22 \tag{13}$$

The reaction also proceeds with β,γ-unsaturated carboxylic acids and nickel(0) complexes to yield six-membered ring nickelacycles **28**, which after coordination with a bidentate ligand afford the ring-contracted metallacycles **29** (Eq. 14).[54(a),55] 2-Cyclopentencarboxylic acid reacts by a similar pathway with Ni(COD)$_2$ in the presence of pyridine as a ligand to afford bicyclic metallacycle **30** in 85% yield (Eq. 15).[33(f)]

$$\tag{14}$$

$$\tag{15}$$

Palladacycles are also prepared from β,γ-unsaturated carboxylic acids and coordinatively unsaturated palladium(0) complexes, such as Pd(PCy$_3$)$_2$ or *in situ* generated Pd(PMe$_3$)$_2$ to yield palladacycles **31** and **32** (Scheme 7).[54,56] Again, ring contraction affords the more stable five-membered ring derivatives, such as **33**.[55]

This method also allows preparing azanickelacycles starting from unsaturated carboxamides. Thus, α,β-unsaturated primary and secondary amides lead to the formation of azametallacycles **34** (Eq. 16).[54(c),57]

$$\tag{16}$$

$R^1 = H, Me$
$R^2 = H, Me, Ph$
$PR_3 = PCy_3, P(t\text{-Bu}_2)Et$

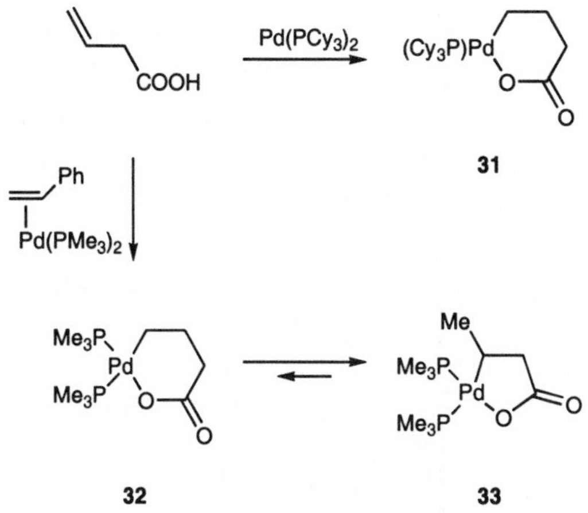

Scheme 7.

The reaction of a N-dideuterated primary amide leads to the selective formation of dideuterated azanickelacycle **35** (Eq. 17).[54(c)]

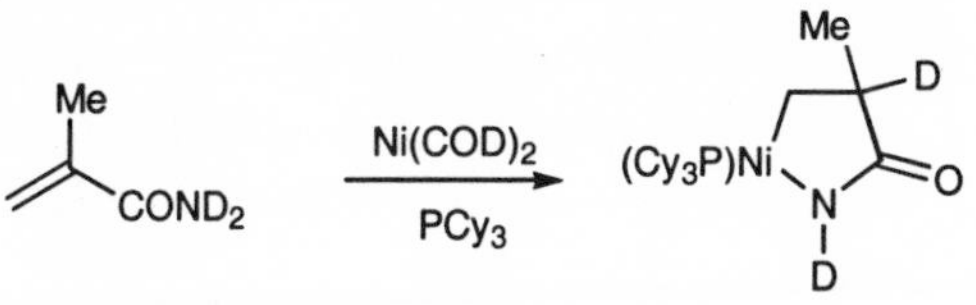

$$(17)$$

Analogously, reaction of a β,γ-unsaturated primary carboxamide with a nickel(0) complex affords nickelacycle **36** (Eq. 18).[54(a),(b)] As before,

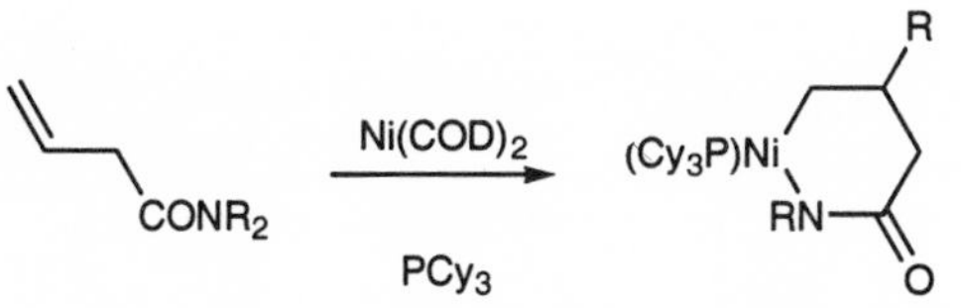

$$(18)$$

R = H , D

36 : R = H
37 : R = D

deuterated nickelacycle **37** is obtained regioselectively from the corresponding dideuterated primary amide.[54(a),(b)]

Although reactions in the presence of PCy$_3$ as the ligand lead to the formation of nickelacycles, treatment of acrylic acid or acrylamide with

Ni(COD)$_2$ and PPh$_3$ leads instead to the formation of η^2-olefin complexes.[58] The use of PEt$_3$, with a basicity similar to that of PCy$_3$,[59,60] also fails to yield the desired nickelacycles due to coordination of two phosphine ligands to the metal.[59] These results indicate that the cyclometallation reaction requires a strongly donating ligand that is also very bulky. On the other hand, increased steric hindrance on the alkene leads only to the isolation of η^2-olefin complexes or diminished yields of nickelacycles.[54]

The formation of nickelacycles from α,β-unsaturated carboxylic acids or amides can be rationalized as shown in Scheme 8. Thus, formation of nickelacycles[61] from carboxylic acids or amides proceed by oxidative addition of the polar Z-H[62–64] followed by insertion of the alkene into the Ni-H bond.[54]

Alternatively, coordination of nickel(0) or palladium(0) to the alkene could be followed by protonation of the nucleophilic C-2 of the alkene and ring formation (Scheme 9).[54] The protonation of C-2 agrees with the resonance stabilization shown in Scheme 9 for the intermediate Ni(0) complexes, as suggested by the ^{1}H and ^{13}C NMR spectra and X-ray data of nickel(0) complexes of methyl methacrylate.[58(c),(e)] Accordingly, coordination of nickel(0) with strongly donor phosphines, such as PCy$_3$, should increase the polarization of the alkene double bond, favoring protonation and formation of the metallacycle.

Although both mechanisms are consistent with the deuteration experiments, (Eqs. 17 and 18), the mechanism proposed in Scheme 9 agrees with the isolation of (η^2-alkene)nickel(0) complexes in many of these

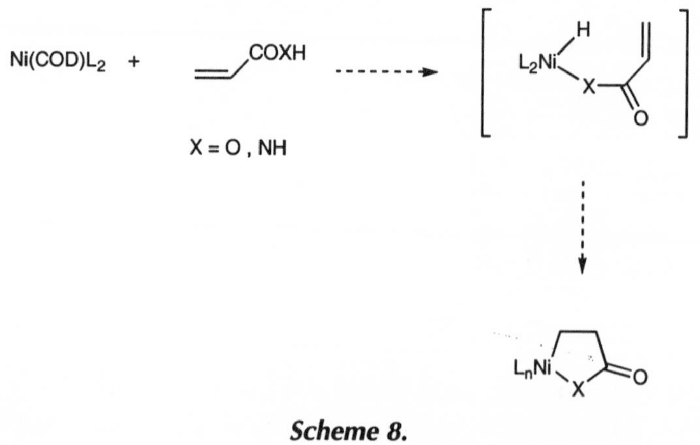

Ni(COD)L$_2$ +   ⟍COXH   ‑‑‑‑‑‑→   [L$_2$Ni structure]

X = O , NH

L$_n$Ni structure

Scheme 8.

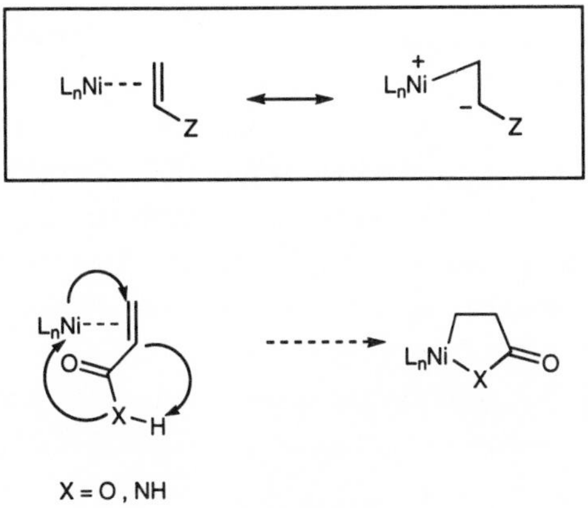

X = O , NH

Scheme 9.

reactions.[58(c),(e)] Additionally, the nickel hydride formed as an intermediate in the mechanism outlined in Scheme 8 could insert into one of the double bonds of 1,5-cyclooctadiene leading to its isomerization.[65] However, isomerization of this dialkene was not detected in these experiments.[54(a)] Mechanisms similar to that shown in Scheme 9 could be proposed for the formation of six-membered ring metallacycles from β,γ-unsaturated carboxylic acids or amides and nickel(0) or palladium(0) complexes.[54]

D. Other Methods for Synthesizing Oxa- and Azanickelacycles

Alternative methods for synthesizing heteronickelacycles have been recently developed on the basis of the formally inserting oxygen or nitrogen into Ni-C bonds. Thus, an interesting selective oxidation of nickelacyclopentane **38**[66] with nitrous oxide leads to oxanickelacycle **39** (Scheme 10).[67] The formal insertion of nitrogen into the Ni-C bond of

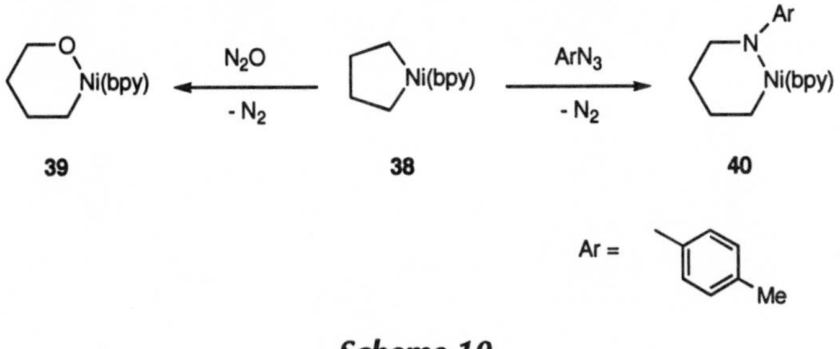

Scheme 10.

complex **38** is achieved by using *p*-tolylazide leading to the formation of azanickelacycle **40**.[68]

The analogous sulfur nickelacycles are prepared by an entirely different method. Thus, the oxidative addition of ethylene sulfide to a nickel(0) complex, formed *in situ* from **41** by reductive elimination of butane, gives thianickelacycle **42** (Scheme 11).[69] Similarly, the oxidative addition of the four-membered heterocycle thietane with Ni(COD)bpy leads to five-membered ring complex **43**.[69]

Very recently, an oxanickelacyclopropane, synthesized by a Wittig reaction on coordinated CO_2,[37] is the smallest nickelacycle prepared.

III. REACTIVITY OF OXA- AND AZANICKELACYCLES

A. Insertion Reactions

The carbonylation of nickelacycles proceeds readily at room temperature to give the corresponding cyclic anhydride by reductive elimination. For example, nickelacycle **44**, prepared by oxidative addition of the corresponding cyclic anhydride and decarbonylation, regenerates the starting material after treatment with carbon monoxide (Scheme 12).[24(c)] Acid hydrolysis of the reaction mixtures allows the formation of dicarboxylic acids, as illustrated for nickelacycles **30**, **45**,[33(f)] and **46**[70] (Scheme 12).

Although the carbonylation reactions of nickelacycles prepared from cyclic anhydrides is of no synthetic interest, the corresponding reactions of metallacycles synthesized from unsaturated carboxylic acids or alke-

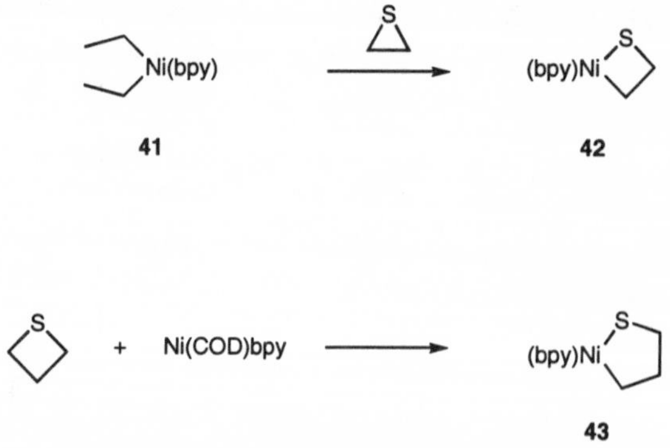

Scheme 11.

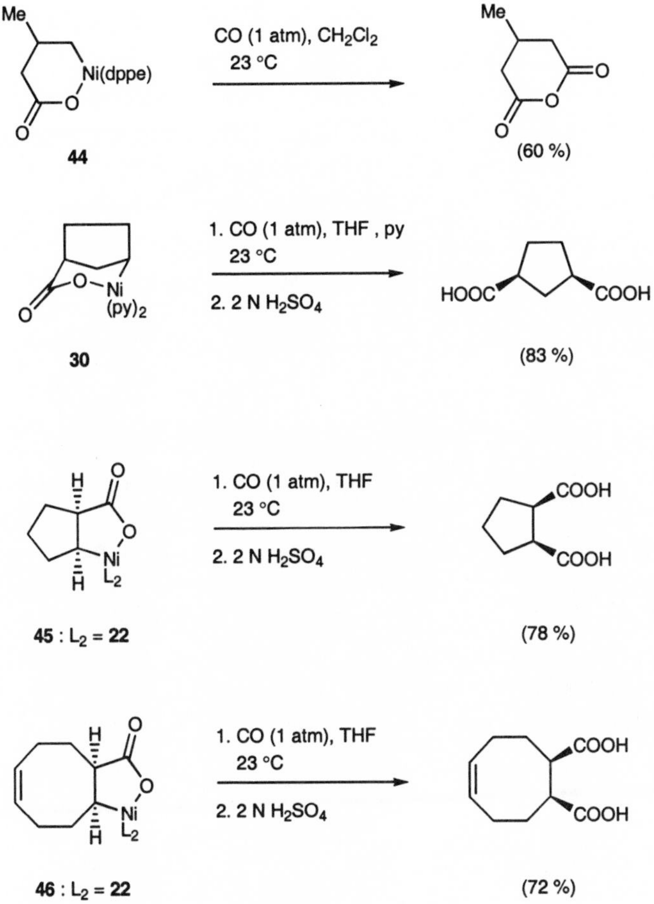

Scheme 12.

nes allow synthesizing succinic acid derivatives by formation of one or two carbon–carbon bonds (Scheme 13).[33(c),36,54]

Similarly, carbonylation of azanickelacycles, such as **47**, leads to the formation of cyclic imides (Eq. 19).[42(d),(e),(f),(g),(h),48(a)] The carbonylation

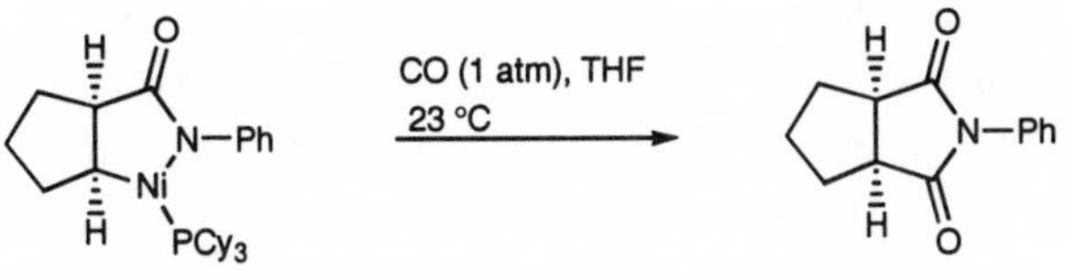

(19)

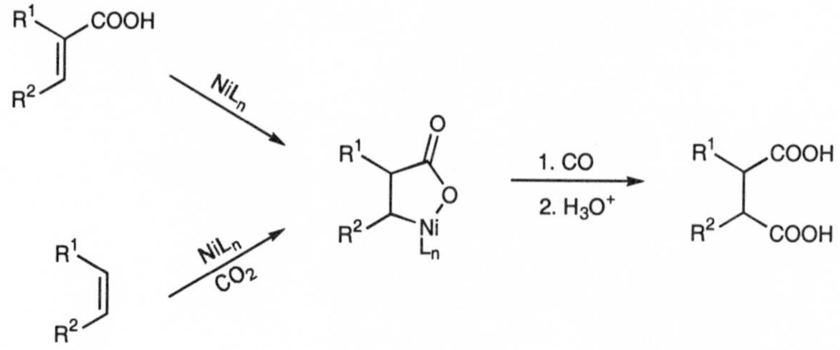

Scheme 13.

of palladacycles, such as **31** and **33** (see Scheme 7), also proceeds readily and has been used for the catalytic formation of the corresponding anhydrides.[55,67(a)–69,71]

Insertion of carbon dioxide is also possible. For example, seven-membered ring nickelacycle **11**, derived from 1,3-butadiene, undergoes further carboxylation in the presence of pyridine to form a nickel(II) dicarboxylate complex **48**, which yields a diester after treatment with methanol under acidic conditions (Eq. 20).[40(c)] Similar results are obtained in the carboxylation of (η^4-diene)iron(0) complexes.[72]

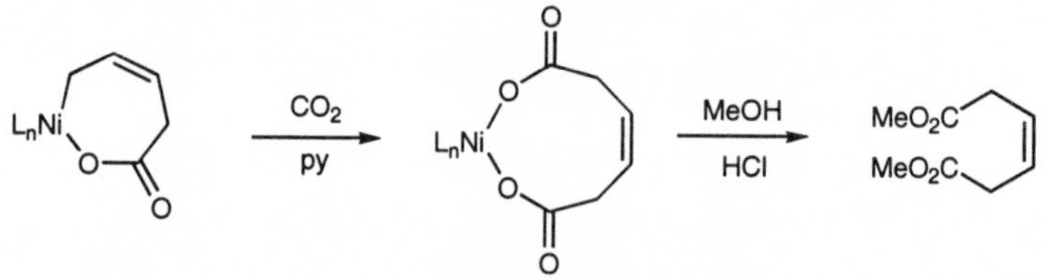

$$(20)$$

Carboxylation of five-membered ring nickelacycles **45** and **46** is less straightforward (Scheme 14). Reaction of the nickelacycles with carbon dioxide proceeds in the presence of pyridine to form 1,1-dicarboxyxlic acid, instead of the expected 1,2-derivatives.[33(f),70(a)]

Although the role of additive $BeCl_2$ in the carboxylation of **45** is not clear, $BeCl_2$ apparently promotes ring contraction to form an intermediate four-membered ring metallacycle. The same behavior has been mentioned before (see Scheme 6) in the contraction of nickelacyles **23**. In fact, carboxylation of **23** in the presence of $BeCl_2$ leads to the formation of malonic acids after acid hydrolysis of the six-membered ring nickel(II) dicarboxylate **49** (Scheme 15).[70(b)]

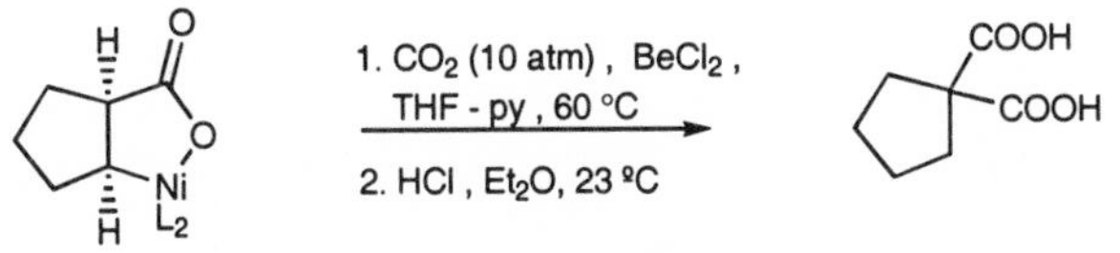

45 : L$_2$ = **22**   (71 %)

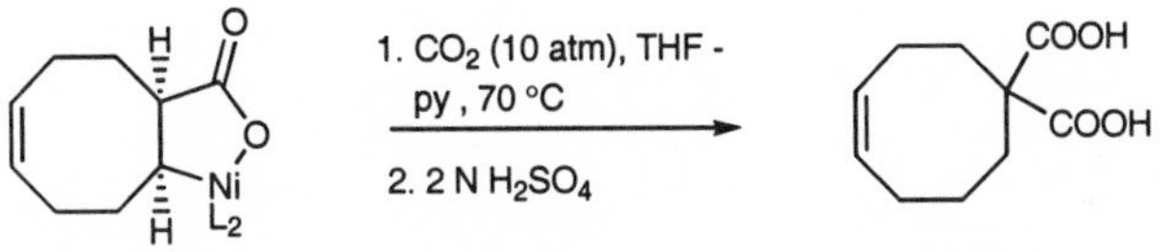

46 : L$_2$ = **22**   (60 %)

Scheme 14.

The insertion reactions of ethene into the nickel-carbon bond of four-membered oxanickelacycles **25** leads to the expected derivatives **50**, which depending on the workup conditions, yield protonolytic or β-hydride elimination products (Scheme 16).[70(b)] On the other hand, azanickelacycle **51**, formed in the nickel(0)-promoted reaction of ethene and phenylisocyanate, reacts with a second molecule of ethene to yield an insertion product **52**, which afford a secondary carboxamide by β-hydride elimination (Eq. 21). The process is carried out catalytically with moderate efficiency.[42(i),73]

The insertion of alkynes has been carried out only with unsaturated nickelacyles.[74] Thus nickelacycles **17** (see Eq. 9) react with some alkynes

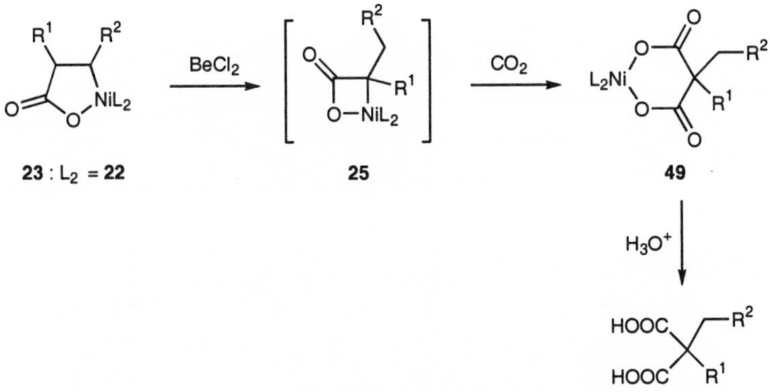

23 : L$_2$ = **22**   **25**   **49**

Scheme 15.

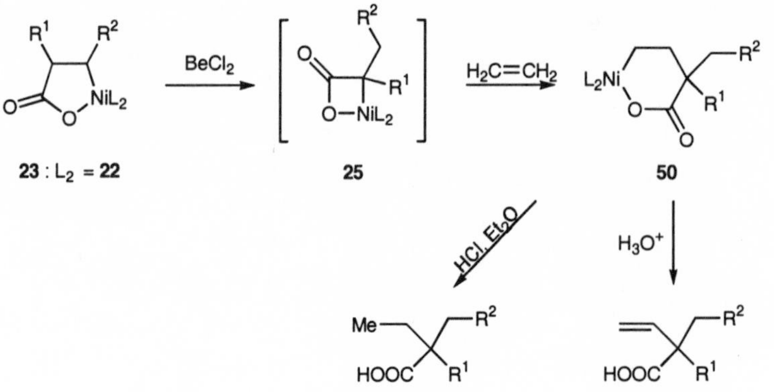

Scheme 16.

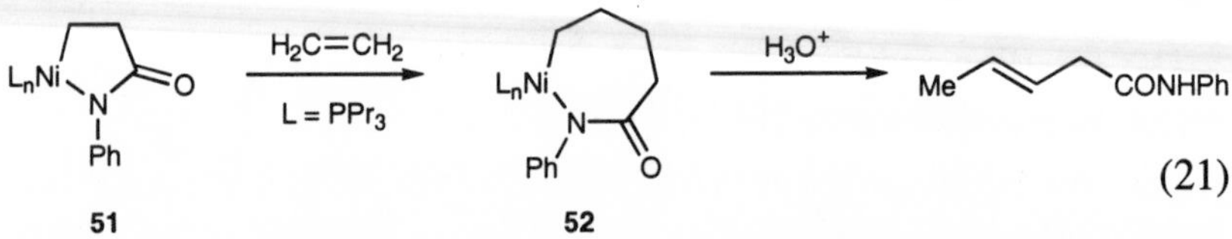

$$(21)$$

to form intermediate seven-membered ring metallacycles **53**, which lead to the formation of pyrones by reductive elimination (Eq. 22). [33(c),40(c),75]

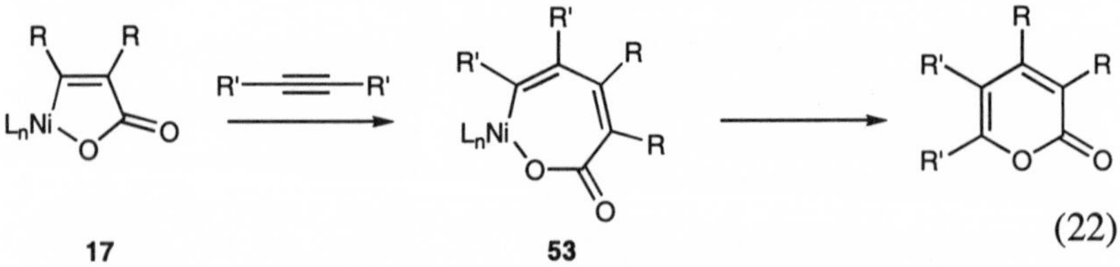

$$(22)$$

Insertion of alkynes also takes place with unsaturated azanickelacycles. As shown in Scheme 17, nickelacycle **54**, prepared at 45% yield, reacts with diphenylacetylene to form **55**, which reductively eliminates to form the heterocycle (Scheme 17). The process allows synthesizing substituted 2-pyridones, although the yields are rather low.[46,48(a),76,77] Better results are obtained by using CpCo(CO)$_2$ as the promoter for the oxidative coupling of alkynes with isocyanates to form pyridones.[78]

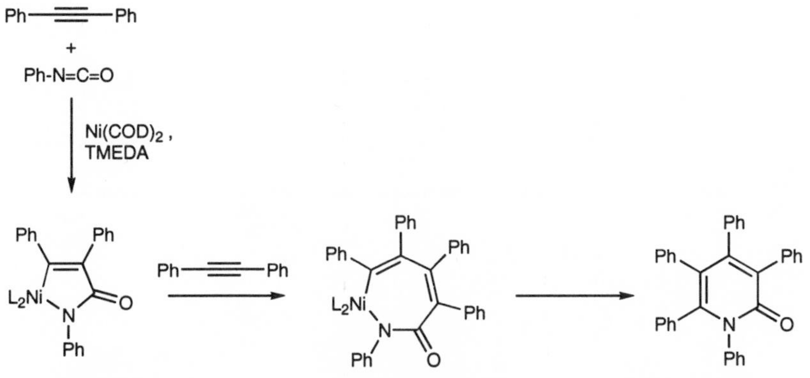

Ph——≡——Ph

+

Ph-N=C=O

| Ni(COD)₂ ,
| TMEDA

54 : L₂ = TMEDA   **55** : L₂ = TMEDA   (12 %)

Scheme 17.

B. Alkylation, Electrophilic Cleavage, and Oxidation Reactions

The reactions of the five-membered ring nickelacycle derived from succinic anhydride **3** with certain alkyl halides in DMF in the presence of anhydrous MnI₂ have been reported.[79,32] This reaction tolerates the presence of reactive functional groups on the electrophile and is accelerated by ultrasonic irradiation. Satisfactory yields are obtained with primary and some secondary alkyl iodides, whereas tosylates and aryl iodides are unreactive. These reactions proceed more sluggishly in the absence of MnI₂, although the role of this additive is not clear. Based on some IR studies that indicate coordination of MnI₂ with the nickelacycle carbonyl it has been suggested that bimetallic intermediates are formed.[79] Scheme 18 summarizes some of the reactions carried out by using nickelacycle **3** as a synthetic equivalent of propionic acid (see Eq. 2).[79]

Methylation of nickelacycle **30**, obtained in the reaction between Ni(COD)(py)₂ and 2-cyclopentencarboxylic acid, unexpectedly leads to the formation of *cis*-2-methylcyclopentanecarboxylic acid (Scheme 19).[33(f)] This product probably arises by methylation of isomeric nickelacycle **56**, formed by β-hydride elimination and insertion with the opposite regiochemistry. When the same reaction is carried out with a large excess of iodomethane, *cis*-3-methylcyclopentanecarboxylic acid is obtained. Methylation reactions of related azanickelacycles have also been reported.[42(a),46,48(a),54(a)]

Simple electrophilic cleavage of the nickel–carbon bond takes place readily with proton acids.[33(f),70(a)] Cleavage with different electrophiles

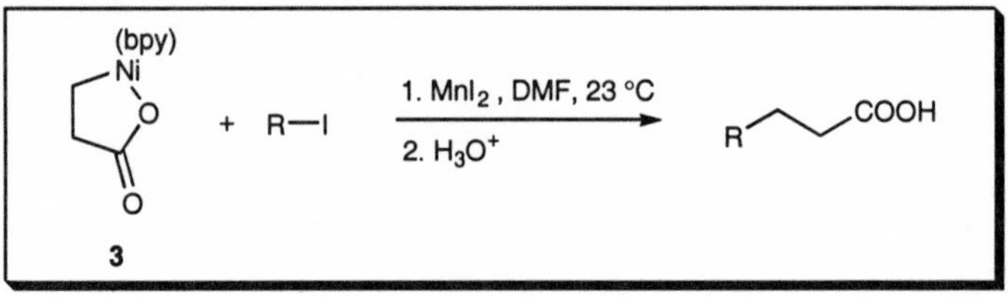

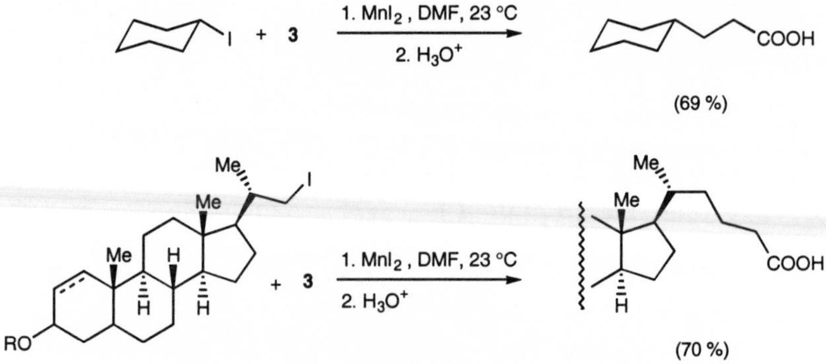

Scheme 18.

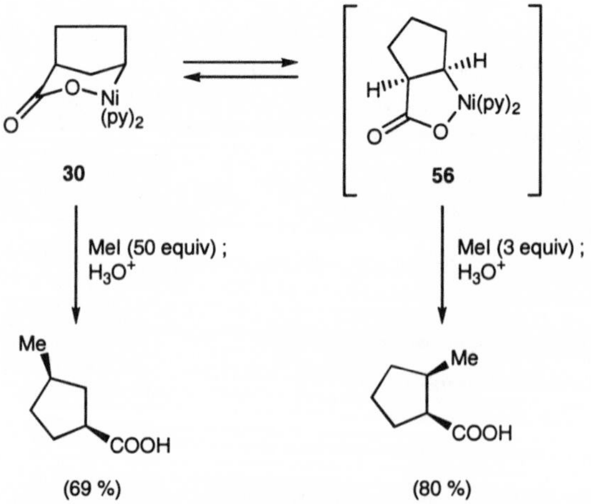

Scheme 19.

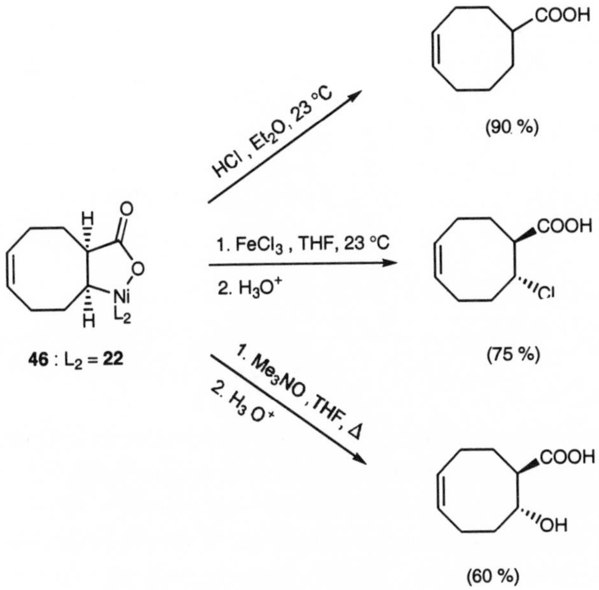

Scheme 20.

has also been studied. Thus, reaction of the nickelacycle **46** with FeCl$_3$ gives the chloride resulting from electrophilic cleavage with inversion of the configuration (Scheme 20).[70(a)] Similarly, oxidation with trimethyl-

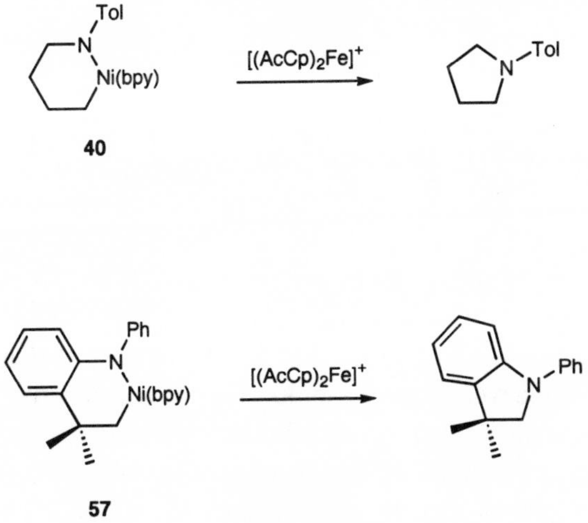

Scheme 21.

amine *N*-oxide yields the β-hydroxyacid, again with an inversion of the configuration.[70(a)] Oxidation of nickelacycle **30** with trimethylamine *N*-oxide affords a single stereoisomer, although the stereochemistry was not determined in this case.[33(f)]

An interesting oxidatively induced reductive elimination of azanickelacycles, such as **40** and **57**, proceeds with 1,1′-diacetylferrocenium tetrafluoroborate, which suggests that the reaction proceeds through a Ni(III) intermediate (Scheme 21).[80]

IV. DEVELOPMENT OF CHIRAL NICKELACYCLES AS AMINO ACID SYNTHONS

The development of strategies for preparing enantiomerically pure amino acids has attracted considerable interest in recent years.[81] Much of the work has been focused on the preparation of glycine synthons, although alanine synthons suitable for synthesizing a variety of amino acids are of considerable synthetic interest. We decided to synthesize chiral nickelacycles **58** as potential alanine synthetic equivalents.[82] These nickelacycles could be prepared from readily available L-aspartic acid (Asp-OH) anhydrides by oxidative addition to nickel(0) complexes followed by decarbonylation. Additionally, the use of L-glutamic acid (Glu-OH) anhydrides would yield the chiral six-membered ring nickelacycles **59**. These metallacycles could be considered synthetic equivalents for α-aminobutyric acids. The preparation of the enantiomers of **58** and **59** could be also carried out from commercially available D-aspartic and D-glutamic acids, respectively.

58   **59**

Utilizing aspartic and glutamic anhydrides as starting materials for the preparing nickelacycles raises the question of regiochemical control in the oxidative addition step. An additional problem with this approach for synthesizing aminoacids is the compatibility of the highly reactive nickel(0) complexes and the intermediate metallacycles with the NH functionality and the N-protective groups.

The required N-protected aspartic acids are usually prepared by acylation with acid chlorides or chloroformates under Schotten–Baumann conditions.[83] However, this procedure fails when the less reactive acylating agents are allowed to react with the starting aminoacids because of their extensive hydrolysis under the reaction conditions. For example, attempted acylation of aspartic acid with the acid chlorides derived from (R)- or (S)-2-methoxy-2-phenyl-3,3,3-trifluoropropanoic acids (Mosher acids, MTPA)[84] under Schotten–Baumann conditions leads only to hydrolysis products. Other methods for acylating unreactive amines were similarly fruitless.[84(a)] On the other hand, the acylation of di-*n*-butyl aspartate leads to a mixture of diastereomers with 80% diastereomeric excess (de). Furthermore, saponification of the diester with anhydrous KOH[85] leads to the desired Mosher amides with only 33% de. Similar results are obtained using the diallyl ester of aspartic acid. These results led us to synthesize L-aspartic acid bis(trimethylsilyl) ester **60**.[86] Although silylation of α-amino acids has been carried out with a variety of reagents,[87-90] L-aspartic acid either fails to react or undergoes partial racemization under the reaction conditions. However, **60** could be synthesized by heating a suspension of L-aspartic acid in neat bis(trimethylsilyl)amine[91] under refluxing conditions without any additives. Diester **60** was isolated in quantitative yield from the resulting solution by evaporating the solvent. The reaction of **60**, which is stable at −15 °C for several months, with a variety of acylating agents (acid chlorides, chloroformates, aminoacid chlorides or fluorides and even di-*t*-butyl dicarbonate) proceeds in good to excellent yields to give the corresponding N-protected aspartic acid derivative as a single isomer (Eq. 23). In this

60

(23)

way, a series of enantiomerically pure N-protected aspartic acids was prepared to synthesize the anhydrides required for preparing nickelacycles **58**.

A. Reactions of Aspartic and Glutamic Anhydrides with Ni(0) Complexes

The strategy for developing chiral metallacycles of Ni or Pd from aspartic and glutamic acids anhydrides is shown in Scheme 22. Func-

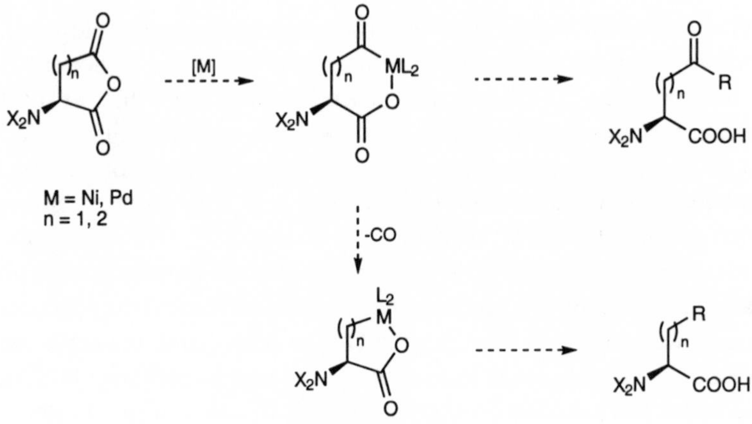

Scheme 22.

tionalization of these metallacycles by reaction either with electrophiles or the common nucleophilic organometallic reagents (B, Sn, Mg, etc.) or by insertion reactions would result in a simple procedure for synthesizing nonnatural aminoacids with different structural elements from inexpensive, readily available starting materials.

Initially we focused our attention on preparing palladium metallacycles. The Pd(0)-catalyzed Stille cross-coupling reaction of acid chlorides with organostannanes is a well documented reaction of considerable synthetic interest for synthesizing ketones.[92–94] However, the related reaction with carboxylic acid anhydrides as the starting materials is unprecedented.[95] Unfortunately, all of our attempts to prepare palladacycles derived from aspartic anhydrides using the most common Pd(0) complexes [Pd(PPh₃)₄, Pd(dba)₂ /phoshine][96] failed. Thus, for example, no decarbonylation or coupling reaction were observed when Pd(PCy₃)₂(dba),[97] a highly reactive Pd(0) complex, was employed, although the reaction with phthaloyl aspartic anhydride showed two new intermediates when the reaction was monitored by NMR. However, after acidic hydrolysis of the crude mixture, phthaloyl aspartic acid was obtained as the only isolated product. All attempts to transform carboxylic acid anhydrides with palladium(0) complexes under catalytic conditions were similarly negative.[95]

Because of the lack of reactivity of anhydrides with Pd(0) complexes, we turned to the more nucleophilic Ni(0) complexes. A number of N-protected anhydrides were chosen to study the regioselectivity and functional group compatibility in metallacycle formation. The effect of different ligands on the Ni complex was also tested. The regiochemical outcome of the process was determined after acid hydrolysis of the crude mixture of nickelacyles to give the corresponding alanines or aminobutyric acids (Scheme 23).[82]

The reaction of *N*-benzyloxycarbonyl aspartic acid anhydride **61** and Ni(COD)bpy under the reaction conditions used for succinic anhydride (THF, 23 °C, 6 h) failed to give the desired nickelacycle, although the oxidative addition took place as indicated by the dark red suspension formed upon mixing. Acid hydrolysis yields *N*-bencyloxycarbonyl-L-aspartic acid exclusively. When the reaction is carried out in refluxing THF, an equimolecular mixture of α- and β-alanines (Z-L-α-Ala-OH and Z-β-Ala-OH) is obtained, as a result of a nonselective oxidative addition (Scheme 24 and Table 1, Entry 1).

A series of nitrogen and phosphorous ligands and several protecting groups were used to control the regiochemistry of the oxidative addition. The reactions were carried out by adding the anhydride to a mixture of Ni(COD)$_2$ (1.5 equiv) and ligand (1.5 equiv) under Ar. Most of the reactions were run in refluxing THF. An excess of Ni(COD)L$_2$ complex (0.5 equiv) was used to trap the CO released in the decarbonylation reaction, thus forming Ni(CO)$_2$L$_2$ complexes. Selected results are shown in Table 1.

The best results were obtained by using the phthaloyl derivative (±)**63** because this protective group allows the regioselective preparation of both α- and β-phthaloyl alanines by simply changing the ligand on nickel (Entries 9 and 11). Furthermore, with this protective group the decar-

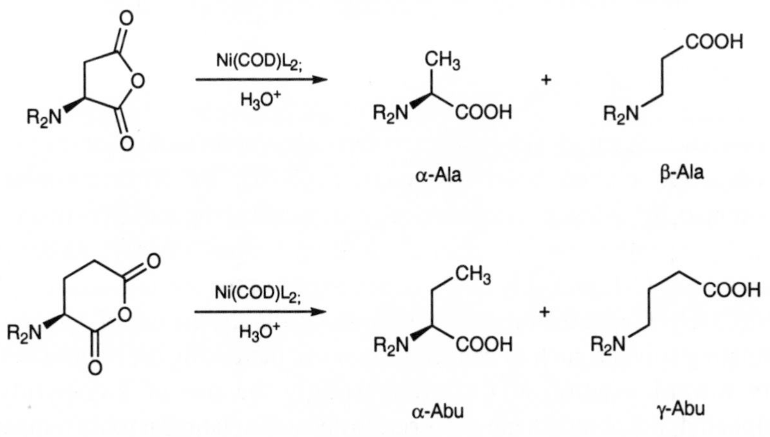

Scheme 23.

Table 1.

| | | | α-Ala | β-Ala | |
Entry	R,R'	Ligand, $L_2{}^a$	α-Alab	β-Ala	Yield (%)^b
1	PhCH$_2$OCO, H (61)	bpy	1	1	88
2		Me$_2$Phen	2	1	61
3		TMEDA	8.8	1	60
4		dppf	1	3	54
5		PCy$_3$	1.2	1	78
6	CF$_3$CO, H (62)	TMEDA	2.3	1	86
7	Pht^f (63)	bpy	1	3	91
8		TMEDA	1	1.1	82
9		Me$_2$Phen	11.3	1	91
10		Me$_2$Phen^c	8	1	76
11		PCy$_3$	1	9.2	89
12		PPh$_3{}^{c,d}$	4	1	80
13		PCy$_3{}^c$	1	11.5	81
14		PCy$_3{}^{c,e}$	1	28	91
15		^{c}	1	34	88
16		PMe$_3{}^{c,d}$	1	4	80

Notes: ^{b}Yields were determined by ^{1}H NMR. The only other constituent of the crude reaction mixture was the N-protected aspartic acid.
^{a}Abbreviations: 2,2'-bypyridine (bpy), 2,9-dimethyl-1,10-phenanthroline (Me$_2$Phen), N,N,N',N'-tetramethylethylendiamine (TMEDA), 1,1'-bis(diphenylphosphino)ferrocene (dppf).
^{c}Reaction run at 23 °C.
^{d} 3 equiv of phosphine were used.
^{e}Reaction run in benzene.
^{f}Pht = Phthaloyl.

bonylation reaction takes place at room temperature with all of the nickel complexes studied. Selective attack at C-4 by the Ni(0) complex is obtained by using a more sterically demanding ligand, 2,9-dimethyl-1,10-phenanthroline (Me$_2$Phen).[98] Adding 1 equiv of PCy$_3$ directs the reaction to C-1, probably due to coordination of the unsaturated complex Ni(COD)PCy$_3$ to the imide function. Accordingly, the use of a noncoordinating solvent, such as benzene, results in increasing the regioselectivity toward reaction at C-1. Interestingly the use of 2-(2-pyridyl)-diphenylphosphinoethane gives rise to decarbonylation at room temperature. This is in contrast to other bidentate phosphines (dppe, dppf) which

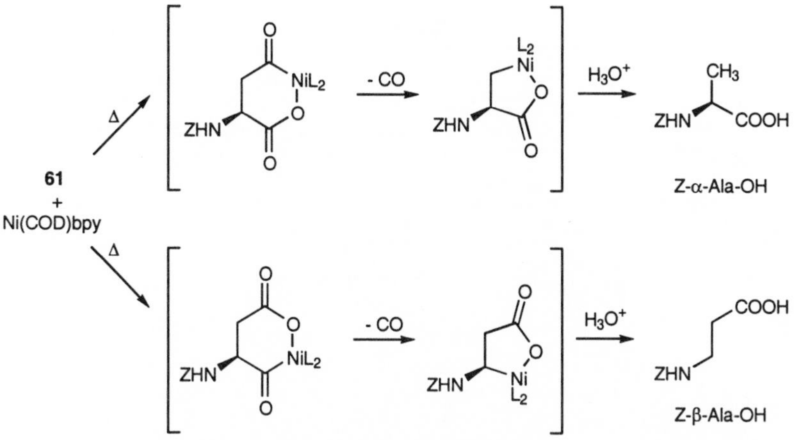

Scheme 24.

do not promote decarbonylation at room temperature and with the previous report that bidentate phosphines do not induce decarbonylation of succinic and glutamic anhydrides because of formation of inert $Ni(dppe)_2$.[24(b)] No reaction occurs when COD, 2,2′-biquinoline or 4,4′-dimethyl-2,2′-bipyridine is employed as the ligand for nickel. Complex reaction mixtures and poor regioselectivity were observed with $Ni(PPh_3)_4$ as the reagent.[99,100]

Similar results were achieved with glutamic acid anhydrides **64** and **65** (Table 2). However, although regioselective formation of α-aminobutyric acid was possible (Entries 1, 2, and 4), the formation γ-aminobutyric acids proceeded with only moderate regioselectivity (Entry 6).

Having established regiochemical control in the oxidative addition step, preliminary studies on the functionalization of intermediate nickelacycles were undertaken. Treatment of the nickelacycles prepared *in situ* from the corresponding aspartic anhydrides with DCl and *N*-bromosuccinimide (NBS) give deutero- and bromoalanines, respectively, with good to excellent yields when the nickelacycles are generated at room temperature (Eq. 24). The yields of deutero and bromo derivatives

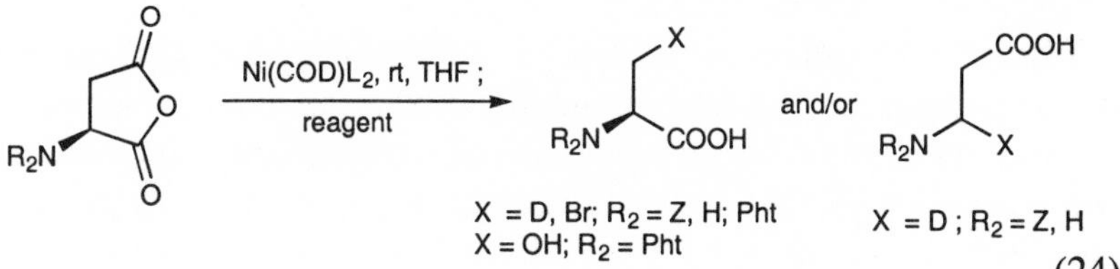

$$X = D, Br; R_2 = Z, H; Pht$$
$$X = OH; R_2 = Pht$$

$$X = D; R_2 = Z, H$$

(24)

Table 2.

Entry	R,R'	Ligand, L_2	α-Abu	γ-Abu	Yield (%)[a]
1	PhCH$_2$OCO, H (**64**)	bpy	>5	1	75
2		PCy$_3$	>50	<1	40
3		Me$_2$Phen	4	1	6b
4		TMEDA	>50	<1	96[b]
5	Pht (**65**)	bpy	3	1	46
6		PCy$_3$	1	3	91

Notes: [a]Yields were determined by ^{1}H NMR. The only other constituent of the crude reaction mixture was the N-protected glutamic acid. Additionally small amounts (ca. 4–20%) of Z-pyroglutamic acid were also obtained in several runs.
         [b]Isolated yield 82%.

decrease substantially at higher temperatures. In these experiments unfunctionalyzed α- or β-alanines were the major products resulting from hydrogen extraction.

Reaction with oxidizing reagents (O$_2$, N-methylmorpholine N-oxide) yields serine together with the corresponding aspartic acid.[101] Previous syntheses of serine from aspartic acid in the optically active series were carried out by Baeyer–Villiger oxidation of a methyl ketone derived from protected aspartic acids.[102] The Ni-mediated procedure allows synthesizing (±)-N-phthaloyl serine from aspartic acid in just two steps. No oxidation was observed by using ammonium cerium(IV) nitrate or t-butylhydroperoxide as the oxidants. On the other hand, treatment with benzoyl peroxide,[103] followed by hydrolysis, gives the starting carboxylic acid. Control experiments showed that decarbonylation of the initially formed oxidative addition product took place because acid hydrolysis of the reaction mixture afforded of α- and β-N-phthaloylalanines. Thus, addition of benzoyl peroxide promotes the carbonylation of the intermediate nickelacycle, probably by oxidation of the dicarbonyl Ni(0) complex Ni(CO)$_2$Me$_2$Phen.

When 1,4-benzoquinone is used, a mixture of 4-hydroxyphenyloxy α- and β-aspartates (hydroquinone monoesters) is obtained. Complex mixtures are obtained by using 1,4-naphthoquinone. No functionalization was achieved in any case with nickelacycles derived from glutamic anhydrides, and only products from β-elimination were observed.

When enantiomerically pure aspartic and glutamic anhydrides were used, scalemic alanines and aminobutyric acids were obtained. Thus, the reaction of optically active L-*N*-benzyloxycarbonyl glutamic anhydride (L-**64**) with Ni(COD)(TMEDA) gives scalemic *N*-bencyloxycarbonyl α-aminobutyric acid (Z-L-α-Abu-OH, L/D 75/25) as determined by the optical rotation of the methyl ester. The reaction of L-N-phthaloyl aspartic anhydride (L-**61**) with Ni(COD)(Me$_2$Phen) gives Pht-L-α-Ala-OH with 61% ee when the reaction is carried out at room temperature for 8 h. Increasing the ratio of NiL$_2$/anhydride from 3:2 to 2:1 decreases the enantiomeric excess (50% ee). The best result (73% ee) was obtained by using shorter reaction times with a 3:2 ratio of NiL$_2$ anhydride. The loss of enantiomerical integrity can take place by a base-catalyzed process on the starting anhydride or the intermediate nickelacycles[104] and/or by a β-hydride elimination-insertion process (Eq. 25). In fact, L-*N*-phthaloyl

$$ \tag{25} $$

aspartic anhydride suffers partial racemization upon treatment with the basic ligand Me$_2$Phen in THF at room temperature (80% ee after 48 h).[105]

To gain more information about this undesired process, (*S*)- and (*R*)-MTPA aspartic anhydrides were allowed to react with several Ni(0) complexes. Although phosphine-containing ligands yielded alanines with moderate to good diastereomeric excesses (de's) (70–89%) depending on the ligand and reaction conditions, nitrogen ligands gave low de's (30–50% for TMEDA) or complete racemization (Me$_2$Phen). In the case of the starting anhydrides, only free TMEDA led to partial racemization in the control experiments. These results suggest that there is a large difference in the reactivity of the intermediate nickelacycles toward β-hydride elimination and subsequent insertion, resulting in racemization at C-2.

B. Isolation of Chiral Nickelacycles

The reaction of anhydride **63** with Ni(COD)Me2Phen in THF at room temperature leads regioselectively to nickelacycle **66** as an orange solid. Complex Ni(CO)$_2$Me$_2$Phen, resulting from the reaction of excess Ni(0) with the expelled CO, was isolated from the filtrate. Because of its insolubility, complex **66** was spectroscopically characterized after ligand exchange with the diphosphine ligand dppe in CH$_2$Cl$_2$ to afford **67**. On the other hand, nickelacycle **68** underwent extensive decomposition during the reaction and could not be characterized spectroscopically.

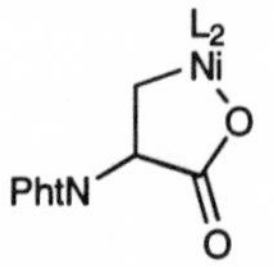

66, L$_2$ = Me$_2$Phen
67, L$_2$ = dppe

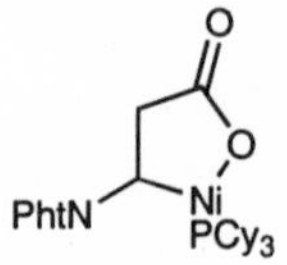

68

Attempts to isolate primary decarbonylated nickelacycles, such as **69**, from the N-protected glutamic anhydrides were similarly fruitless. However, after ligand exchange with dppe, the five-membered ring nickelacyles **70** and **71** derived from **69** were isolated as a ca. 1:1 mixture.[106] This agrees with the ring contraction observed in nickelacycles derived from glutaric anhydride upon treatment with bidentate phosphines (see Eq. 3).[24(c)]

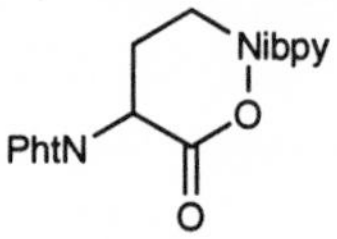

69

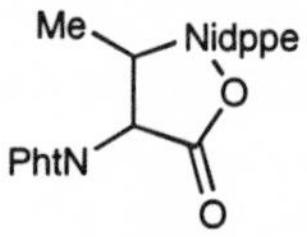

70, *erythro*
71, *threo*

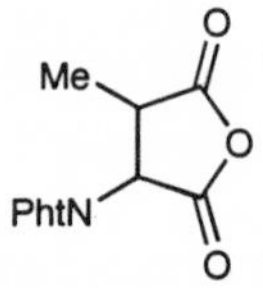

72, *erythro*
73, *threo*

C. Insertion and Alkylation Reactions

The reaction of the mixture of racemic **70** and **71** with CO (1 atm) in CDCl$_3$ yields a mixture of *erythro* and *threo* (±)-N-phthaloyl 3-methylaspartic anhydrides **72** and **73** in a 1:1 ratio. Upon standing at room temperature for several days, however, this ratio changes to 1:2, which demonstrates that the carbonylation is reversible under these conditions.

The conversion of (2S)-glutamic acid into *threo*-(2S,3S)-3-methylaspartic acid is a process produced in nature by the cobalamin enzyme glutamate mutase of the bacterium *Clostridium tetanomorphum*.[107] This rearrangement is the first step in using glutamic acid as a source of energy and proceeds by migration of a glycine portion. In our case, using nickelacycles as intermediates allows for synthesizing 3-methylaspartic acids in a single-pot operation starting from glutamic acid anhydride.

When isocyanides as a surrogate of CO are employed, differentiation of the carboxyl functions is achieved, yielding a ca. 3:1 mixture of *threo* and *erythro* methylasparagines (Eq. 26).

$$R = t\text{-Bu, Cy} \qquad (26)$$

The isomeric ratio obtained in the previous reaction is the result of an equilibrium process because diastereomerically pure *threo* (±)-*N*-phtha-loyl-3-methylaspartic anhydride yields a similar mixture, together with elimination products (Scheme 25). Several β-hydride and/or CO_2 elimi-nation-insertion processes account for isomerization at C-3 and also at C-2. As a result, when enantiomerically pure anhydride **65** is used, a racemic mixture of 3-methylaspartic acid derivatives is obtained.

The reaction of nickelacycle **66** with CO and *t*-BuCN proceeds in the same way after addition of $(PhCOO)_2$ and gives phthaloyl aspartic acid and phthaloyl *t*-butylasparagine after acid hydrolysis, respectively. Be-sides insertion of CO and RNC, the nickelacycles also react with some alkynes. Reaction of *in situ* generated **66** with phenylacetylene at room temperature, followed by acid hydrolysis, leads to a 1.5:1 mixture of **74** and **75** with 83% yield (Eq. 27). When this reaction is performed at 66

$$74, R = Ph \qquad 75, R = Ph$$
$$76, R = (CH_2)_5CH_3 \qquad (27)$$

°C, a 1:1.5 ratio of the same two products is obtained. The formation of alkyne **74** may be explained by an elimination reaction of the alkenyl Ni(II) complex. Alternatively, this alkyne may be the result of the reaction of nickelacycle **66** with the alkyne to form an alkynyl Ni(II) complex, which may undergo reductive elimination to form **74**. Alkenyl derivative **75** probably arises from insertion of the alkyne into the Ni-C

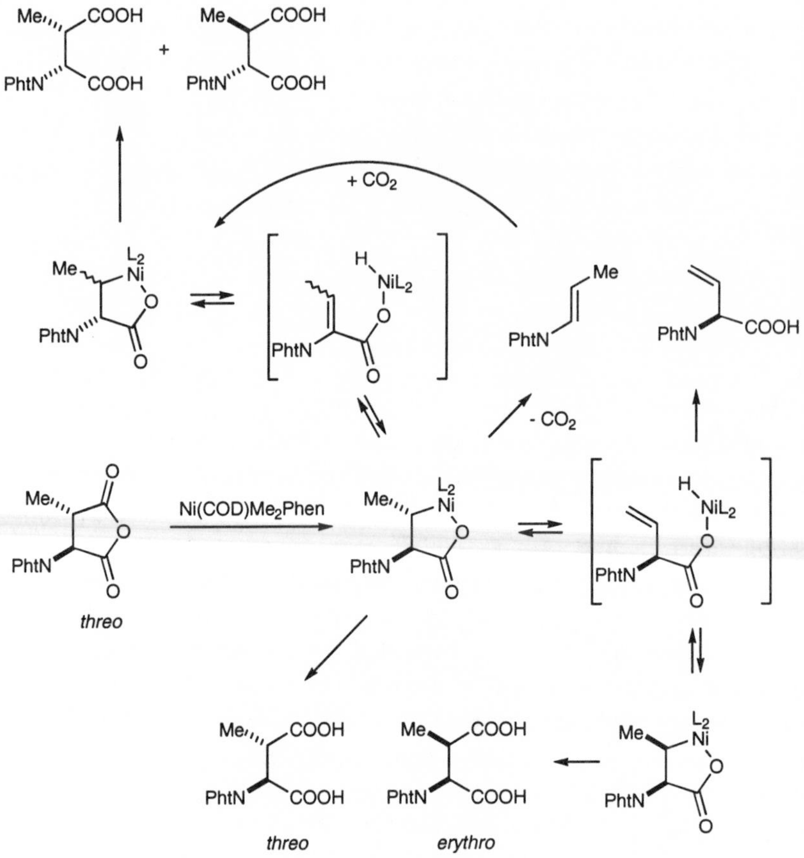

Scheme 25.

bond, followed by hydrolysis of the alkenyl Ni(II) complex. The formation of **67** is facilitated by a weak acid. Thus, reaction between **66** and phenylacetylene in the presence of pyrazole as a weakly acidic proton donor leads exclusively to alkene **75**, which was isolated as the methyl ester with 63% yield. On the other hand, reaction of anhydride **63** Ni(COD)Me$_2$Phen followed by addition of 1-octyne yields only alkynyl derivative **76**, which is isolated with only a 36% yield (Eq. 27). Internal alkynes, such as 4-octyne or dimethyl acetylendicarboxylate do not react.

Nickelacycle **66** also reacts with aliphatic halides. Usually, alkylation reactions were carried without isolating any intermediate by treating anhydride **63** with 1.5 equiv each of Ni(COD)$_2$ and ligand in THF at 23 °C for 5 h, followed by addition of excess alkyl halides (10–20 equiv) and stirring at 23° or 40 °C. Best results were obtained using Me$_2$Phen (1.5 equiv) as the ligand for nickel (Table 3). Triphenylphosphine (3 equiv) also led to alkylation, although in lower yields.

The alkylation reactions proceed with moderate to good yields with primary iodides (Entries 1, 3, 6, and 7) or bromides (Entries 2, 4, and 5) in the absence of additives or polar solvents. The only byproducts detected were small amounts of (±)-α- and β-N-phthaloyl alanines and (±)-N-phthaloyl aspartic acid. In a few instances, small amounts of N-vinylphthalimide, resulting from double decarboxylation, were also detected. Shorter reaction times were required when the reaction was performed at 40 °C. Secondary iodides also gave alkylated derivatives (Entries 8 and 9). However, no alkylation was observed when more hindered substrates, such as α-cholestanyl iodide and menthyl iodide or bromide were used. These halides were recovered unchanged or suffered elimination to afford mixtures of alkenes after several days at room temperature. On the other hand, no alkylation took place with methyl p-toluenesulfonate, ethyl triflate, and propylene oxide under the same reaction conditions. Similarly, alkenyl and aryl halides were unreactive with nickelacycles.

When the reaction of Entry 3 (Table 3) is performed with enantiomerically pure **63**, the alkylated derivative, L-N-phthaloyl norvaline, is obtained with only 50% ee. The isolation of N-vinylphthalimide as a minor byproduct, also points to the possible involvement of a reversible β-decarboxylation in the racemization process.

In contrast with the results obtained with simple alkyl halides, benzyl bromide leads to the formation of **77** and the ketone **78** in variable ratios (Scheme 26). A similar result has been reported in the reactions between the oxidative addition product of Ni(COD)bpy or Ni(COD)TMEDA with cis-4-cyclohexen-1,2-dicarboxylic anhydride and alkyl iodides.[22] With allyl bromide as the electrophile, ketone **79** is the only product isolated. However, when the reaction is performed with isolated nickelacycle **66** in the absence of Ni(CO)$_2$Me$_2$Phen, allylated alanine **80** is formed exclusively (60% yield) (Scheme 26). These results show that the carbonyl nickel complex is not inert because with certain reagents it transfers CO to the nickelalactone **66**. Alternatively, the formation of ketones in these reactions could be explained by alkylation of the primary oxidative addition product or by carbonylation of allyl or benzyl bromide to give acyl bromides which react with **66** to give the observed products.[108] However, this last reaction pathway seems unlikely because acetyl or benzoyl chloride do not react with *in situ* generated nickelacycle **66**.

Table 3.[a]

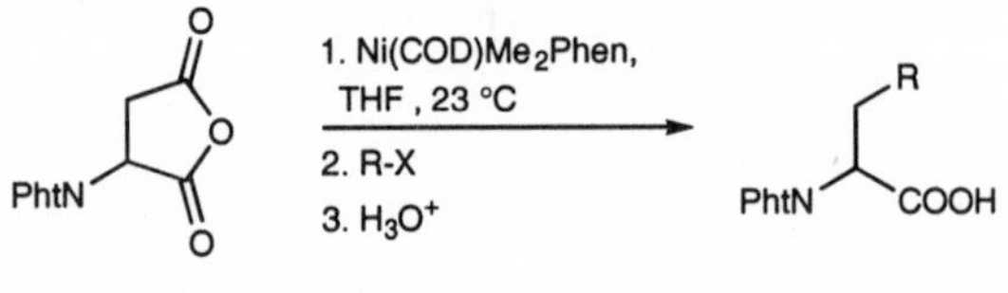

Entry	R-X	Equiv of RX	Reaction Time	R	Yield (%)[b]
1	MeI	20	13	Me	69 (80)
2	EtBr	20	13	Et	81 (90)
3	EtI	20	15	Et	77 (79)
4	n-BuBr	18	144	n-Bu	51 (73)
5	n-BuBr	18	14[c]	n-Bu	49 (70)
6	n-BuI	10	40	n-Bu	49 (78)
7	n-BuI	10	15[c]	n-Bu	47 (71)
8	i-PrI	20	48	i-Pr	52 (86)
9	c-C_6H_{11}I	9	42	c-C_6H_{11}	42 (58)

Notes: [a]The reactions were performed at 23 °C.
[b]Yields determined by ^{1}H NMR (corrected for conversion).
[c]The reaction was carried out at 40 °C.

It has been proposed that a number of processes catalyzed or mediated by Ni of interest for organic synthesis proceed by radical intermediates.[109–111] In particular, η^3-allyl Ni(II) complexes react with alkyl halides in polar solvents by a process for which a radical chain mechanism has

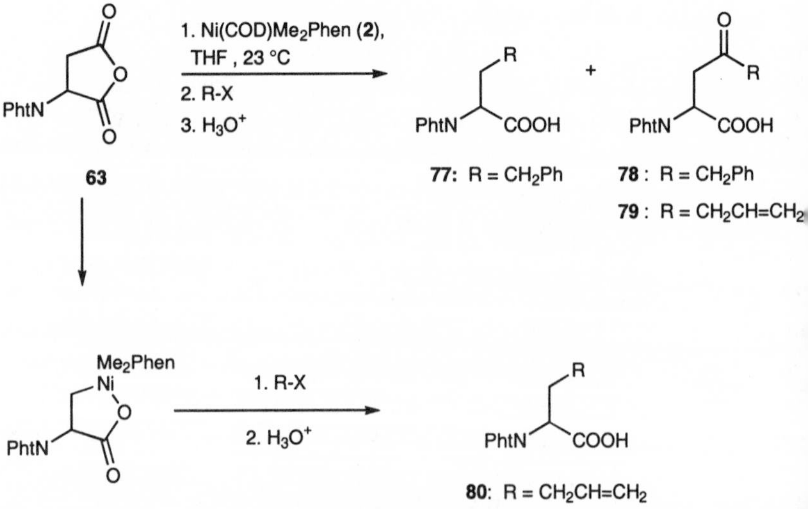

Scheme 26.

been proposed.[110] The alkylation of nickelacycle **66** might also proceed by a radical pathway because the lack of reaction of **66** with alkyl sulfonates apparently excludes a simple S_N2 mechanism. By using alkyl halides **81–85** in the reaction with nickelacycle **66**, prepared *in situ* from **63**, the results summarized in Scheme 27 were obtained. These alkyl halides are known to generate alkyl radicals, which undergo rearrangement reactions with known rates, thus behaving like "radical clocks."[112] Bromo- (**81**) and (iodomethyl)cyclopropane (**82**) give only rearranged linear product **86**, albeit in low yields (26 and 45%, respectively). On the other hand, 5-hexenyl iodide (**83**) gives only unrearranged product **87** with 30% yield. The only by-products of these reactions were (±)-α- and β-*N*-phthaloyl alanines and (±)-*N*-phthaloyl aspartic acid. With 2-allyloxyethyl bromide (**84**) and the iodide (**85**) a 4.5:1 ratio of rearranged cyclic **88** and unrearranged acyclic derivative **89** was obtained (35 and 85% yield, respectively) (Scheme 27). Although the excess halides were recovered unchanged in the reactions with **83** and **84**, in the reaction with iodide **85** the recovered halide was contaminated with approximately 15% of 3-(iodomethyl)-tetrahydrofurane. Control experiments showed

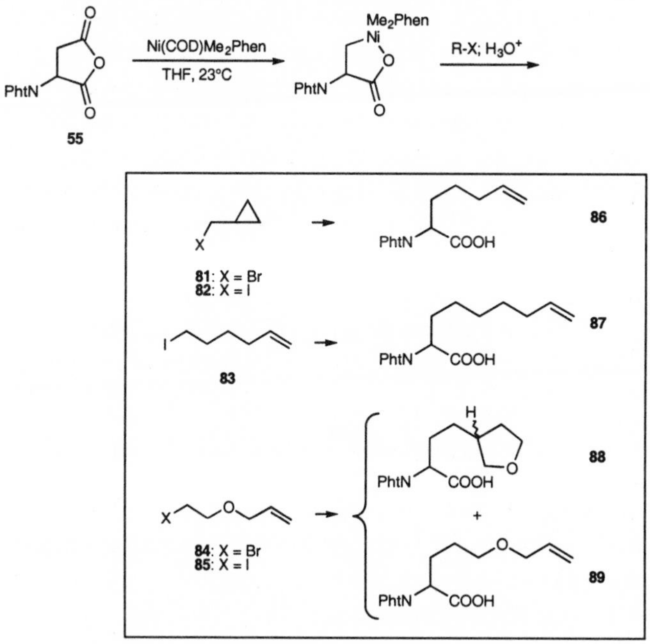

Scheme 27.

that complexes $Ni(COD)Me_2Phen$ and $Ni(CO)_2Me_2Phen$ promote, although rather inefficiently (3–5%), the cyclization of **85**, which suggests that the cyclized iodide recovered in this experiment is produced by a different process. No reaction was observed with 5-hexenyl or 2-allyloxyethyl mesylates.

If the reaction with alkyl halides proceeds via formation of free radicals, an apparent rate for radical trapping of $2 \times 10^6 \, M^{-1}s^{-1}$ can be determined from the results of Scheme 27.[110(a)] This rate is an order of magnitude slower than the rate determined in the alkylation of η^3-allyl Ni(II) halide dimers in polar solvents.[113,114] These results suggest that a radical intermediate is involved in the carbon–carbon bond-forming step. Unlike η^3-allyl Ni(II) complexes, however, the alkylation of nickelacycle **66** is not accelerated by the irradiation with light or by the addition of common radical initiators. Additionally, the alkylations are not inhibited by the addition of substantial amounts (0.2–1.0 equiv) of m-dinitrobenzene as a radical inhibitor.[110,111] Furthermore, although η^3-allyl Ni(II) halide complexes react smoothly with alkenyl and aryl halides,[110] no reaction was observed between **66**, or related nickelacycles, with Csp^2 electrophiles, such as alkenyl, or with aryl halides. The results obtained with the radical clocks and the lack of reactivity observed with alkyl sulfonates can be explained by assuming that the reaction of nickelacycle **66** with the alkyl halides proceeds by an electron transfer process,[115] yielding a radical pair that collapses with an approximate rate of 2×10^6 $M^{-1}s^{-1}$ to afford the alkylated derivatives. This radical pair can also undergo rearrangement to yield the corresponding rearranged alkylated products. Probably, steric hindrance in the transition state excludes a classical S_N2 reaction.[116] However, the reactions performed in the absence of the Ni(0) complex $Ni(CO)_2L_2$ are slower, giving rise to the alkylated compounds in lower yields. Although the lower solubility of pure **66** in THF can account for this effect, a more important role for complex $Ni(CO)_2L_2$ in the alkylation reaction cannot be excluded.

V. SUMMARY AND OUTLOOK

As shown throughout this article, nickelacycles of general structure **1** can be prepared in a number of ways, mainly by oxidative addition to cyclic anhydrides followed by decarbonylation and by oxidative coupling of $CO_2/RNCO$ with alkenes and alkynes. These oxa- and azanickelacycles react by β-elimination, oxidation, insertion, and alkylation reactions. Although most of the chemistry uncovered so far requires stoichiometric amounts of nickel complexes, catalytic processes have also been developed, pointing to an open field in this area. Nickelacycles of type **1** are regarded as synthetic equivalents for propionic acid, and their scope in this sense has been explored. The use of more functionalized nickelacycles such as **58** and **59**, as potential synthetic equivalents for α-alanine and α-aminobutyric acid respectively, has also been studied, showing chemistry analogous to that of **1**. A limitation of these nickelacycles as amino acids equivalents is the racemization observed at C-2 during the process. However, this problem could be probably overridden by the proper choice of ligands on nickel and protective groups for the amino function. The reactivity of these nickelacycles with alkyl, benzyl, and allyl halides makes them interesting from a synthetic viewpoint. The development of new chemistry of these and related nickelacycles could be of interest to exploit their ready preparation from easily available starting materials.[117a,b] Additionally, the development of catalytic transformations of chiral, nonracemic cyclic anhydrides based on the chemistry described by using highly reactive Ni(0) or Pd(0) complexes could allow for the development of one of the most straightforward syntheses of nonproteinogenic amino acids by starting from inexpensive chiral amino acids which are available in both enantiomeric series.

VI. SELECTED EXPERIMENTAL PROCEDURES

A. Preparation of Nickelacycles:
($\pm$)-4-(2,9-Dimethyl-1,10-phenanthrolinenickela)-2-phthalimidobutyrolactone (66) and
($\pm$)-4-[(1,2-bis(diphenylphosphino)ethane)-nickela]2-phthalimidobutyrolactone (67)

A solution of anhydride **63** (143 mg, 0.58 mmol) and Me_2Phen (182 mg, 0.87 mmol) in THF (10 mL) was added to $Ni(COD)_2$ (240 mg, 0.87 mmol). The resulting mixture was stirred at 23 °C for 1.5 h to give a dark red suspension. The suspension was filtered off and washed with Et_2O (3 $\times$ 3 mL) to give an air-sensitive orange solid in ca. 80% yield (determined by acid hydrolysis and isolation of *N*-phthaloyl-α-alanine): IR (Nujol): 1768, 1708, 1662, 1385, 1325, 725 cm^{-1}. The purple filtrate was concentrated and cooled to -15 °C to give a purple solid, characterized as dicarbonyl (2,9-dimethyl-1,10-phenanthroline) nickel: IR (Nujol) 1983, 1890 cm^{-1}; ^1H NMR (300 MHz, $CDCl_3$) δ 8.24 (d, $J = 8.2$ Hz, 2H), 7.76 (s, 2H), 7.69 (d, $J = 8.1$ Hz, 2H), 3.26 (s, 6H); $^{13}C\{^1H\}$ NMR (75 MHz, $CDCl_3$) δ 196.76, 159.65, 144.21, 134.65, 127.54, 125.32, 124.24, 27.66.

Complex **66** was transformed into **67** as follows: complex **66** (obtained from 0.29 mmol of **63** and 0.44 mmol of Ni(COD)Me$_2$Phen) was treated with a solution of 1,2-bis(diphenylphosphino)ethane (dppe) (175 mg, 0.44 mmol) in CH$_2$Cl$_2$ (3 mL) and stirred at 23 °C for 16 h. The yellow suspension was filtered, and the solid was recrystallized from CH$_2$Cl$_2$ (−15 °C for 3 days) to give **59** in ca. 30% yield, after washing with Et$_2$O (2 × 1 mL) as a yellow air-sensitive solid: IR (Nujol): 1710, 1650, 1380, 1325 cm^{-1}; ^{1}H NMR (300 MHz, CDCl$_3$) δ 8.02–7.30 (m, 24H), 4.96 (ddd, J = 11.9, 7.2, 1.1 Hz, 1H, CH a), 2.4–2.0 (m, 2H, PCH$_2$), 1.9–1.8 (m, 2 H, PCH$_2$), 1.52–1.42 (m, 1H, CH β), 1.04–0.98 (m, 1H, CH β′); ^{1}H{^{31}P} NMR (300 MHz, CDCl$_3$) δ 8.01–7.30 (m, 24H), 4.96 (dd, J = 11.9, 7.2 Hz, 1H), 2.3-2.1 (m, 2H), 1.9–1.7 (m, 2H), 1.47 (dd, J = 11.8, 9.8 Hz, 1H), 1.02 (dd, J = 9.7, 7.2 Hz, 1 H); ^{13}C{^{1}H} NMR (125 MHz, CDCl$_3$) δ 181.65 (d, J = 10 Hz), 167.74, 133.59 128.65 (m, 10 C, P-Ph), 133.16, 131.71, 129.95, 122.79, 53.89, 29.67 (dd, J = 30.4, 20.4 Hz, P-CH$_2$), 22.21 (dd, J = 29.4, 11.1 Hz, P-CH$_2$), 21.86 (dd, J = 57.7, 23.8 Hz, CH$_2$ β); ^{31}P NMR (121 MHz, CDCl$_3$) δ 59.33 (d, J = 7.8 Hz, 1P), 35.20 (d, J = 7.8 Hz, 1P). Hydrolysis of these nickelacycles with 1.2 M aqueous HCl at 23 °C gave (±)-N-phthaloyl-α-alanine as the major product in ca. 80% yield: ^{1}H NMR (200 MHz, DMSO-d_6) δ 7.95–7.81 (m, 4H), 4.86 (q, J = 7.3 Hz, 1H), 1.55 (d, J = 7.3 Hz, 3H); ^{13}C{^{1}H} NMR (50 MHz, DMSO-d_6) δ 171.00, 167.12, 134.71, 131.29, 123.25, 46.96, 14.78. This compound was identical with a sample prepared from (±)-alanine.

B. Alkylation of Nickelacycle 66 (Table 3). General Procedure

Anhydride **63** (186 mg, 0.76 mmol) was added to a solution of Ni(COD)$_2$ (275 mg, 1.0 mmol) and Me$_2$Phen (208 mg, 1.0 mmol) in THF (15 mL). The reaction mixture was stirred at 23 °C for 5 h. Excess alkyl halide (10–20 equiv) was added and stirring was continued at 23° or 40 °C. The mixture was treated with 1.2 M aqueous HCl and extracted with EtOAc (3 ×). The combined EtOAc solution was extracted with 5% aqueous NaHCO3, acidified with 1.2 M HCl, and extracted with EtOAc (3 ×). The solution was dried (Na$_2$SO$_4$) and evaporated to give the crude N-protected amino acids.

ACKNOWLEDGMENTS

Financial support of this work by the Dirección General de Investigación Científica y Técnica (Projects PB87-0201-C03-02, PB91-0612-C03-02, and PB94-0163) is gratefully acknowledged. Ana M. Castaño acknowledges the receipt of a fellowship by the Comunidad Autónoma de Madrid and a contract by the Ministerio de Educación y Ciencia. We also thank Enrique Gómez-

Bengoa for preliminary work on the synthesis of nickelacycles from α,β-unsaturated acids.

REFERENCES

1. (a) Collman, J. P.; Hegedus, L. S.; Norton, J. R.; Finke, R. G. *Principles and Applications of Organotransition Metal Chemistry*; University Science Books, Mill Valley: CA, 1987; Chapter 9; (b) Hegedus, L. S. *Transition Metals in the Synthesis of Complex Organic Molecules*; University Science Books, Mill Valley: CA, 1994; pp 113–123.

2. For recent reviews, see Broene, R. D.; Buchwald, S. L. *Science* **1993**, *261*, 1696; Doxsee, K. M.; Mouser, J. K. M.; Farahi, J. B. *Synlett* **1992**, 13.

3. Stille, J. K. In *The Chemistry of Metal-Carbon Bond*; Hartley, F. R.; Patai, S., Eds.; Wiley: Chichester, 1985; Vol. 2, Chap. 9.

4. For a review of the decarbonylation reactions with the Wilkinson catalyst, see Baird, M. C. In *The Chemistry of Acid Derivatives*; Patai, S., Ed.; Wiley: New York, 1979; Supplement B, Part 2, p. 825.

5. Obora, Y.; Tsuji, Y.; Kawamura, T. *J. Am. Chem. Soc.* **1993**, *115*, 10414.

6. Rich, J. D. *J. Am. Chem. Soc.* **1989**, *111*, 5886. Rich, J.D. *Organometallics* **1989**, *8*, 2609.

7. Murahashi, S.-I.; Naota, T.; Nakajima, N. *J. Org. Chem.* **1986**, *51*, 898.

8. See Jolly, P. W. In *Comprehensive Organometallic Chemistry*; Wilkinson, G.; Stone, F. G. A.; Abel, E. W., Eds.; Pergamon: Oxford, 1982; Vol. 8, Chap. 56.6.

9. Chiusoli, G. P.; Costa, M.; Pecchini, G.; Cometti, G. *Transition Met. Chem.* **1977**, *2*, 270.

10. Ryang, M.; Rhee, I.; Murai, S.; Sonoda, N. *Organotransition Met. Chem.* **1975**, 243.

11. Flood, T. C.; Sarhangi, A. *Tetrahedron Lett.* **1977**, *18*, 3861.

12. Tamao, K.; Kumada, M. In *The Chemistry of the Metal-Carbon Bond*; Hartley, F.R., Ed.; Wiley: Chichester, 1978; Vol. 4, Chap. 9.

13. Goto, T.; Onaka, M.; Mukaiyama, T. *Chem. Lett.* **1980**, 51.

14. Goto, T.; Onaka, M.; Mukaiyama, T. *Chem. Lett.* **1980**, 709.

15. Osakada, K.; Yamamoto, T.; Yamamoto, A. *Tetrahedron Lett.* **1987**, *28*, 6321.

16. Yamamoto, T.; Ishizu, J.; Kohara, T.; Komiya, S.; Yamamoto, A. *J. Am. Chem. Soc.* **1980**, *102*, 3758.

17. Uhlig, E.; Nestler, B. *Z. Chem.* **1981**, *21*, 451.

18. Trost, B. M.; Chen, F. *Tetrahedron Lett.* **1971**, *12*, 2603.

19. Grunewald, G. L.; Davis, D. P. *J. Org. Chem.* **1978**, *43*, 3074.

20. (a) Wiesner, K.; Ho, P.-T.; Jain, R. C.; Lee, S. F.; Oida, S.; Phillipp, A. *Can. J. Chem.* **1973**, *51*, 1448; (b) Dauben, W. G.; Rivers, G. T.; Twieg, R. J.; Zimmerman, W. T. *J. Org. Chem.* **1976**, *41*, 887.

21. Uhlig, E.; Fehske, G.; Nestler, B. *Z. Anorg. Allg. Chem.* **1980**, *465*, 141.

22. Fischer, R.; Walther, D.; Kempe, D.; Sieler, J.; Schönecker, B. *J. Organomet. Chem.* **1993**, *447*, 131.

23. Döring, M.; Kosemund, D.; Uhlig, E.; Görls, H. *Z. Anorg. Allg. Chem.* **1993**, *619*, 1512.

24. (a) Sano, K.; Yamamoto, T.; Yamamoto, A. *Chem. Lett.* **1983**, 115; (b) Sano, K.; Yamamoto, T.; Yamamoto, A. *Bull. Chem. Soc. Jpn.* **1984**, *57*, 2741; (c) Yamamoto, T.; Sano, K.; Yamamoto, A. *J. Am. Chem. Soc.* **1987**, *109*, 1092.

25. Hoberg, H.; Oster, B.W. *J. Organomet. Chem.* **1984**, *266*, 321.

26. Blake, D. M.; Shields, S.; Wyman, L. *Inorg. Chem.* **1974**, *13*, 1595.

27. Pd(dba)$_2$ actually refers to the dinuclear complex Pd$_2$(dba)$_3$.dba: (a) Takahashi, Y.; Ito, Ts.; Sakai, S.; Ishii, Y. *J. Chem. Soc., Chem. Commun.* **1970**, 1065; (b) Ukai, T.; Kawazura, H.; Ishii, Y.; Bonnet, J. J.; Ibers, J. A. *J. Organomet. Chem.* **1974**, *65*, 253.

28. (a) Vollhardt, K. P. C. *Acc. Chem. Res.* **1977**, *10*, 1; (b) Vollhardt, K. P. C. *Angew. Chem., Int. Ed. Engl.* **1984**, *23*, 539.

29. Crabtree, R. H. *The Organometallic Chemistry of the Transition Metals*; 2nd Ed; Wiley: New York, 1994; pp. 155–159.

30. (a) Wakatsuki, Y.; Kuramitsu, T.; Yamazaki, H. *Tetrahedron Lett.* **1974**, 4549; (b) McAlister, D. R.; Bercaw, J. E.; Bergman, R. G. *J. Am. Chem. Soc.* **1977**, *99*, 1666.

31. Behr, A. *Carbon Dioxide Activation By Metal Complexes*; Chemie: Weinheim, 1988; Chap. 3.

32. Walther, D.; Braünlich, G.; Ritter, U.; Fischer, R.; Schönecker, B. In *Organic Synthesis via Organometallics*; Dötz, K.H.; Hoffmann, R.W., Eds.; Vieweg: Braunschweig, 1991; pp. 77–93.

33. (a) Hoberg, H.; Schaefer, D. *J. Organomet. Chem.* **1983**, *251*, C51; (b) Hoberg, H.; Schaefer, D. *J. Organomet. Chem.* **1982**, *236*, C28; (c) Hoberg, H.; Schaefer, D.; Burkhart, G.; Krüger, C.; Romao, M. J. *J. Organomet. Chem.* **1984**, *266*, 203; (d) Hoberg, H.; Peres, Y.; Milchereit, A. *J. Organomet. Chem.* **1986**, *307*, C41; (e) Hoberg, H.; Peres, Y.; Krüger, C.; Tsay, Y.-C. *Angew. Chem., Int. Ed. Engl.* **1987**, *26*, 771; (f) Hoberg, H.; Ballesteros, A.; Sigan, A.; Jegat, C.; Milchereit, A. *Synthesis* **1991**, 395; (g) Hoberg, H.; Bärhausen, D. *J. Organomet. Chem.* **1989**, *379*, C7.

34. (a) Dinjus, E.; Walther, D.; Schütz, H. *Z. Chem.* **1983**, *23*, 408; (b) Walther, D.; Dinjus, E.; Sieler, J.; Andersen, L.; Lindqvist, O. *J. Organomet. Chem.* **1984**, *276*, 99.

35. Cohen, S. A.; Bercaw, J. E. *Organometallics* **1985**, *4*, 1006.

36. Hoberg, H.; Jenni, K.; Angermund, K.; Krüger, C. *Angew. Chem., Int. Ed. Engl.* **1987**, *26*, 153.

37. For stable Ni(0) CO$_2$ complexes, see (a) Aresta, M.; Nobile, C. F.; Albano, V. G.; Forni, E.; Manassero, M. *J. Chem. Soc., Chem. Commun.* **1975**, 636; (b) Wright, C. A.; Thorn, M.; McGill, J. W.; Sutterer, A.; Hinze, S. M.; Prince, R. B.; Gong, J. K. *J. Am. Chem. Soc.* **1996**, *118*, 10305 and references cited therein.

38. Sakaki, S.; Mine, K.; Taguchi, D.; Arai, T. *Bull. Chem. Soc. Jpn.* **1993**, *66*, 3289.

39. (a) Walther, D.; Dinjus, E. *Z. Chem.* **1982**, *22*, 228; (b) Dinjus, E.; Walther, D.; Schütz, H.; Schade, W. *Z. Chem.* **1983**, *23*, 303; (c) Walther, D.; Dinjus, E.; Sieler, J.; Thanh, N. N.; Schade, W.; Leban, I. *Z. Naturforsch.* **1983**, *38B*, 835; (d) Walther, D.; Dinjus, E. *Z. Chem.* **1984**, *24*, 63; (e) Walther, D.; Dinjus, E.; Görls, H.; Sieler, J.; Lindqvist, O.; Andersen, L. *J. Organomet. Chem.* **1985**, *286*, 103.

40. (a) Hoberg, H.; Schaefer, D. *J. Organomet. Chem.* **1983**, *255*, C15; (b) Hoberg, H.; Schaefer, D.; Oster, B. W. *J. Organomet. Chem.* **1984**, *266*, 313; (c) Hoberg, H.; Apotecher, B. *J. Organomet. Chem.* **1984**, *270*, C15.

41. Behr, A.; Kanne, U. *J. Organomet. Chem.* **1986**, *317*, C41.

42. (a) Hoberg, H.; Sümmermann, K.; Milchereit, A. *J. Organomet. Chem.* **1985**, *288*, 237; (b) Hernandez, E.; Hoberg, H. *J. Organomet. Chem.* **1986**, *315*, 245; (c) Hernandez, E.; Hoberg, H. *J. Organomet. Chem.* **1987**, *328*, 403; (d) Hoberg, H.; Hernandez, E.; Guhl, D. *J. Organomet. Chem.* **1988**, *339*, 213; (e) Hoberg, H. *J. Organomet. Chem.* **1988**, *358*, 507; (f) Hoberg, H.; Sümmermann, K.; Hernandez, E.; Ruppin, C.; Guhl, D. *J. Organomet. Chem.* **1988**, *344*, C35; (g) Hoberg, H.; Guhl, D. *J. Organomet. Chem.* **1989**, *375*, 245; (h) Hoberg, H.; Nohlen, M. *J. Organomet. Chem.* **1990**, *382*, C6; (i) Hoberg, H.; Hernandez, E. *J. Chem. Soc., Chem. Commun.* **1986**, 544.

43. (a) Hoberg, H.; Bärhausen, D. *J. Organomet. Chem.* **1990**, *397*, C20. (b) Hoberg, H.; Bärhausen, D.; Mynott, R.; Schroth, G. *J. Organomet. Chem.* **1991**, *410*, 117.

44. Neidlein, R.; Rufinska, A.; Schwager, H.; Wilke, G. *Angew. Chem., Int. Ed. Engl.* **1986**, *25*, 640.

45. (a) Carmona, E.; Palma, P.; Paneque, M.; Poveda, M. L.; Gutiérrez-Puebla, E.; Monge, A. *J. Am. Chem. Soc.* **1986**, *108*, 6424; (b) Carmona, E.; Gutiérrez-Puebla, E.; Marín, J. M.; Monge, A.; Paneque, M.; Poveda, M. L.; Ruiz, C. *J. Am. Chem. Soc.* **1989**, *111*, 2883; (c) Carmona, E.; Paneque, M.; Poveda, M. L.; Gutiérrez-Puebla, E.; Monge, A. *Polyhedron* **1989**, *8*, 1069. (d) Cámpora, J.; Gutiérrez, E.; Monge, A.; Palma, P.; Poveda, M. L.; Ruíz, C.; Carmona, E. *Organometallics* **1994**, *13*, 1728. (e) Cámpora, J.; Paneque, M.; Poveda, M. L.; Carmona, E. *Synlett* **1994**, 465.

46. Hoberg, H.; Schaefer, D.; Burkhardt, G. *J. Organomet. Chem.* **1982**, *228*, C21.

47. Dérien, S.; Duñach, E.; Périchon, J. *J. Am. Chem. Soc.* **1991**, *113*, 8447.

48. (a) Hoberg, H.; Oster, B. W. *J. Organomet. Chem.* **1982**, *234*, C35; (b) Hoberg, H.; Oster, B. W. *Synthesis* **1982**, 324; (c) Hoberg, H.; Oster, B. W. *J. Organomet. Chem.* **1983**, *252*, 359.

49. Zargarian, D.; Alper, H. *Organometallics* **1991**, *10*, 2914.

50. Bennett, M. A.; Johnson, J. A.; Willis, A. C. *Organometallics* **1996**, *15*, 68.

51. For the synthesis of analogous platinum and palladium complexes, see (a) Bennett, M. A.; Fick, H.-G.; Warnock, G. F. *Aust. J. Chem.* **1992**, *45*, 135; (b) Bennett, M. A.; Rokicki, A. *Aust. J. Chem.* **1985**, *38*, 1307; (c) Bennett, M. A.; Yoshida, T. *J. Am. Chem. Soc.* **1978**, *100*, 1750; (d) Robertson, G. B.; Whimp, P. O. *J. Am. Chem. Soc.* **1975**, *97*, 1051; (e) Bennett, M. A.; Robertson, G. B.; Whimp, P. O.; Yoshida, T. *J. Am. Chem. Soc.* **1971**, *93*, 3797.

52. For a review on benzyne complexes: Bennett, M.; Wenger, E. *Chem. Ber.-Recueil* **1997**, *130*, 1029.

53. For another example of ligand control on the β-hydride elimination, see Hoberg, H.; Guhl, D. *Angew. Chem., Int. Ed. Engl.* **1989**, *28*, 1035.

54. (a) Yamamoto, T.; Sano, K.; Osakada, K.; Komiya, S.; Yamamoto, A.; Kushi, Y.; Tada, T. *Organometallics* **1990**, *9*, 2390; (b) Sano, K.; Yamamoto, T.; Yamamoto, A. *Chem. Lett.* **1982**, 695; (c) Yamamoto, T.; Igarashi, K.; Komiya, S.; Yamamoto, A. *J. Am. Chem. Soc.* **1980**, *102*, 7448.

55. Osakada, K.; Doh, M.-K.; Ozawa, F.; Yamamoto, A. *Organometallics* **1990**, *9*, 2197.

56. Doh, M. K.; Jung, M. J.; Lee, D. J.; Osakada, K.; Yamamoto, A. *J. Korean Chem. Soc.* **1993**, *37*, 423; *Chem. Abstr.* **1994**, *119*, 271.

57. Yamamoto, T.; Igarashi, K.; Ishizu, J.; Yamamoto, A. *J. Chem. Soc., Chem. Commun.* **1979**, 554.

58. (a) Yamamoto, T.; Yamamoto, A.; Ikeda, S. *J. Am. Chem. Soc.* **1971**, *93*, 3350; (b) Yamamoto, T.; Yamamoto, A.; Ikeda, S. *J. Am. Chem. Soc.* **1971**, *93*, 3360; (c) Ishizu, J.; Yamamoto, T.; Yamamoto, A. *Bull. Chem. Soc. Jpn.* **1978**, *51*, 2646; (d) Yamamoto, T.; Ishizu, J.; Komiya, S.; Nakamura, Y.; Yamamoto, A. *J. Organomet. Chem.* **1979**, *171*, 103; (e) Komiya, S.; Ishizu, J.; Yamamoto, A.; Yamamoto, T.; Takenaka, A.; Sasada, Y. *Bull. Chem. Soc. Jpn.* **1980**, *53*, 1283.

59. Rahman, M. M.; Liu, H. Y.; Eriks, K.; Prock, A.; Giering, W. P. *Organometallics* **1989**, *8*, 1.

60. Tolman, C. A. *Chem. Rev.* **1977**, *77*, 313.

61. Similar metallacycles have also been prepared from unrelated substrates: (a) Reaction of γ,δ-unsaturated sulfinilic acids with platinum(0) complexes: Hallock, J. S.; Galiano-Roth, A. S.; Collum, D. B. *Organometallics* **1988**, *7*, 2486; (b) Reaction of γ,δ-unsaturated carboxylic acids with rhodium(I) complexes: Marder, T. B.; Chan, D. M.-T.; Fultz, W. C.; Milstein, D. *J. Chem. Soc., Chem. Commun.* **1988**, 996; (c) Activation of α, β, γ, δ, or ε C-H bonds with platinum(II) complexes: Kao, L.-C.; Sen, A. *J. Chem. Soc., Chem. Commun.* **1991**, 1242.

62. Oxidative addtions of Z-H to nickel(0) complexes: (a) Jonas, K.; Wilke, G. *Angew. Chem., Int. Ed. Engl.* **1969**, *8*, 519; (b) Yamamoto, T.; Sano, K.; Yamamoto, A. *Chem. Lett.* **1982**, 907.

63. Lead references of oxidative addition of N-H and O-H bonds to other transition metal complexes: (a) Nelson, J. H.; Schmitt, D. L.; Henry, R. A.; Moore, D. W.; Jonassen, H. B. *Inorg. Chem.* **1970**, *9*, 2678; (b) Roundhill, D.M. *J. Chem. Soc., Chem. Commun.* **1969**, 567; (c) Roundhill, D. M. *Inorg. Chem.* **1970**, *9*, 254; (d) Fornies, J.; Green, M.; Spencer, J. L.; Stone, F. G. A. *J. Chem. Soc., Dalton Trans.* **1977**, 1006; (e) Hedden, D.; Roundhill, D. M.; Fulz, W. C.; Rheingold, A. L. *J. Am. Chem. Soc.* **1984**, *106*, 5014; (f) Casalnuovo, A. L.; Calabrese, J. C.; Milstein, D. *J. Am. Chem. Soc.* **1988**, *110*, 6738; (g) Müller, U.; Keim, W.; Krüger, C.; Betz, P. *Angew. Chem., Int. Ed. Engl.* **1989**, *28*, 1011; (h) Ladipo, F. T.; Merola, J. S. *Inorg. Chem.* **1990**, *29*, 4172; (i) Schaad, D. R.; Landis, C. R. *J. Am. Chem. Soc.* **1990**, *112*, 1628; (j) Hursthouse, M. B.; Mazid, M. A.; Robinson, S. D.; Sahajpal, A. *J. Chem. Soc., Chem. Commun.* **1991**, 1146; (k) Schaad, D. R.; Landis, C. R. *Organometallics* **1992**, *11*, 2024.

64. (a) For a review on the hydroamination of alkenes: Roundhill, M. D. *Chem. Rev.* **1992**, *92*, 1; (b) See also Gagné, M. R.; Stern, C. L.; Marks, T. J. *J. Am. Chem. Soc.* **1992**, *114*, 275 and references cited therein.

65. (a) Miura, Y.; Kiji, J.; Furukawa, J. *J. Mol. Catal.* **1976**, *1*, 447; (b) Yamamoto, T.; Ishizu, J.; Kohara, T.; Komiya, S.; Yamamoto, A. *J. Am. Chem. Soc.* **1980**, *102*, 3758.

66. Binger, P.; Doyle, M. J.; Krüger, C.; Tsay, Y.-H. *Z. Naturforsch.* **1979**, *34B*, 1289.

67. (a) Matsunaga, P. T.; Hillhouse, G. L.; Rheingold, A. L. *J. Am. Chem. Soc.* **1993**, *115*, 2075; (b) Koo, K.; Hillhouse, G. L.; Rheingold, A. L. *Organometallics* **1995**, *14*, 456; (c) Oxidation with O_2 and subsequent reductive elimination of the resulting oxametallacycles: Han, R.; Hillhouse, G. L. *J. Am. Chem. Soc.* **1997**, *119*, 8135.

68. Matsunaga, P. T.; Hess, C. R.; Hillhouse, G. L. *J. Am. Chem. Soc.* **1994**, *116*, 3665.

69. Matsunaga, P. T.; Hillhouse, G. L. *Angew. Chem., Int. Ed. Engl.* **1994**, *33*, 1748.

70. (a) Hoberg, H.; Ballesteros, A. *J. Organomet. Chem.* **1991**, *411*, C11; (b) Hoberg, H.; Ballesteros, A.; Sigan, A.; Jégat, C.; Bärhausen, D.; Milchereit, A. *J. Organomet. Chem.* **1991**, *407*, C23

71. For a somewhat related insertion of carbon monoxide into the C-Ni bond of a thianickelacycle, see Sellmann, D.; Häussinger, D.; Knoch, F.; Moll, M. *J. Am. Chem. Soc.* **1996**, *118*, 5368.

72. Hoberg, H.; Jenni, K.; Krüger, C.; Raabe, E. *Angew. Chem., Int. Ed. Engl.* **1986**, *25*, 810.

73. For an asymmetric version of this reaction which leads to low enantiometric excesses, Tillack, A.; Selke, R.; Fischer, C.; Bilda, D.; Kortus, K. *J. Organomet. Chem.* **1996**, *518*, 79.

74. For the insertion of alkynes into Ni-C bond of phosphine or imine-coordinated Ni(II) complexes, see (a) Martínez, M.; Muller, G.; Panyella, D.; Rocamora, M.; Solans, X.; Font-Bardía, M. *Organometallics* **1995**, *14*, 5552; (b) Ceder, R. M.; Granell, J.; Muller, G.; Font-Bardía, M.; Solans, X. *Organometallics* **1996**, *15*, 4618.

75. Walther, D.; Braünlich, G.; Kempe, R.; Sieler, J. *J. Organomet. Chem.* **1992**, *436*, 109.

76. Inoue, Y.; Itoh, Y.; Kazama, H.; Hashimoto, H. *Bull. Chem. Soc. Jpn.* **1980**, *53*, 3329.

77. (a) Tsuda, T.; Morikawa, S.; Sumiya, R.; Saegusa, T. *J. Org. Chem.* **1988**, *53*, 3140; (b) Tsuda, T.; Morikawa, S.; Hasegawa, N.; Saegusa, T. *J. Org. Chem.* **1990**, *55*, 2978.

78. (a) Earl, R. A.; Volhardt, K. P. C. *J. Am. Chem. Soc.* **1983**, *105*, 6991. (b) Earl, R. A.; Volhardt, K. P. C. *J. Org. Chem.* **1984**, *49*, 4786.

79. (a) Schönecker, B.; Walther, D.; Fischer, R.; Nestler, B.; Braünlich, G.; Eibisch, H.; Droescher, P. *Tetrahedron Lett.* **1990**, *31*, 1257; (b) Fischer, R.; Walther, D.; Braünlich, G.; Undeutsch, B.; Ludwig, W.; Bandmann, H. *J. Organomet. Chem.* **1992**, *427*, 395; (c) Braünlich, G.; Walther, D.; Eibisch, H.; Schönecker, B. *J. Organomet. Chem.* **1993**, *453*, 295.

80. (a) Koo, K.; Hillhouse, G. L. *Organometallics* **1995**, *14*, 4421; (b) Koo, K.; Hillhouse, G. L. *Organometallics* **1996**, *15*, 2669; (c) Koo, K.; Hillhouse, G. L. *Organometallics* **1998**, *17*, 2924; (d) See ref. 67(c) for the oxidation to a ox-anickelacycle and its reductive elimination.

81. Reviews: (a) Williams, R. M. *Synthesis of Optically Active α-Amino Acids*; Pergamon: Oxford, 1989; (b) Williams, R. M.; Hendrix, J. A. *Chem. Rev.* **1992**, *92*, 889; (c) Duthaler, R. O. *Tetrahedron* **1994**, *50*, 1539.

82. (a) Castaño, A. M.; Echavarren. A. M. *Tetrahedron Lett.* **1990**, *31*, 4783; (b) Castaño, A. M.; Echavarren. A. M. *Organometallics* **1994**, *13*, 2262.

83. Green, T. W.; Wuts, P. G. M. *Protective Groups in Organic Synthesis*; Wiley: New York, 1991. Wünsch, E. In *Methoden der Organischen Chemie (Houben-Weyl)*; Thieme: Stuttgart, 1974; Vols. 15/1, 15/2, and references therein.

84. (a) Dale, J. A.; Dull, D. L.; Mosher, H. S. *J. Org. Chem.* **1969**, *34*, 2543; (b) Ward, D. E.; Rhee, C. K. *Tetrahedron Lett.* **1991**, *32*, 7165; (c) König, W. A.; Nippe, K.-S.; Mischnick, P. *Tetrahedron Lett.* **1990**, *31*, 6867; (d) Jeanneret-Gris, G.; Pousaz, P. *Tetrahedron Lett.* **1990**, *31*, 75.

85. Gassman, P. G.; Hodgson, P. K. G.; Balchunis, R. J. *J. Am. Chem. Soc.* **1976**, *98*, 1275.

86. Castaño, A. M.; Echavarren. A. M. *Tetrahedron* **1992**, *48*, 3377.

87. (a) Kricheldorf, H. R. *Liebigs Ann. Chem.* **1972**, *763*, 17; (b) Kricheldorf, H. R. *Chem. Ber.* **1970**, *103*, 3353; (c) Hils, J.; Rühlman, K. *Chem. Ber.* **1967**, *100*, 1638; (d) Barlos, K.; Papaioannou, D.; Theodoropoulos, D. *J. Org. Chem.* **1982**, *47*, 1324.

88. (a) Rühlman, K. *Chem. Ber.* **1961**, *94*, 1876; (b) Hils, J.; Hagen, V.; Ludwig, H.; Rühlman, K. *Chem. Ber.* **1966**, *99*, 776; (c) Mason, P. S.; Smith, E. D. *J. Gas Chromatgr.* **1966**, *4*, 398.

89. Rogozhin, S.V.; Davidovich, Y. A.; Yurtanov, A. I. *Synthesis* **1975**, 113.

90. (a) Becu, C.; Reyniers, M.-F.; Anteunis, M. J. O.; Callens, R. *Bull. Soc. Chim. Belg.* **1990**, *99*, 779; (b) Anteunis, M. J. O.; Becu, C.; Becu, F.; Callens, R. *Bull. Soc. Chim. Belg.* **1990**, *99*, 361; (c) Findeisen, K.; Fauss, R. (Baeyer, A.-G.) German Patent 3 505 746 (1985); *Chem. Abstr.* **1986**, *105*, 133013q.

91. Bruynes, C. A.; Jurriens, T. K. *J. Org. Chem.* **1982**, *47*, 3966.

92. (a) Stille, J. K. *Angew. Chem., Int. Ed. Engl.* **1986**, *26*, 508; (b) Mitchell, T. N. *Synthesis* **1992**, 803; (c) Ritter, K. *Synthesis* **1993**, 735.

93. Farina, V. In *Comprehensive Organometallic Chemistry II*; Abel, E. W., Stone, F. G. A., Wilkinson, G., Eds.; Pergamon: Oxford, 1995; Vol. 12; Chap. 3.4; pp. 228–229.

94. For recent examples, see (a) Pérez, M.; Castaño, A. M.; Echavarren, A. M. *J. Org. Chem.* **1992**, *57*, 5047; (b) Echavarren, A. M.; Pérez, M.; Castaño, A. M.; Cuerva, J. M. *J. Org. Chem.* **1994**, *59*, 4179.

95. Yamamoto has described the oxidative addition of acyclic carboxylic acids anhydrides to Pd(0) complexes: Nagayama, K.; Kawataka, F.; Sakamoto, M.; Shimizu, Y.; Yamamoto, A. *Chem. Lett.* **1995**, 367.

96. Amatore, C.; Broeker, G.; Jutand, A.; Khalil, F. *J. Am. Chem. Soc.* **1997**, *119*, 5176.

97. (a) Huser, M.; Youinou, M.-T.; Osborn, J. A. *Angew. Chem., Int. Ed. Engl.* **1989**, *28*, 1386; (b) Huser, M. Doctoral Thesis, Université Louis Pasteur, 1988.

98. This is the regioselectivity observed in the addition of amines to phthaloyl aspartic anhydride, sterically controlled: King, F. E.; Kidd, D. A. A. *J. Chem. Soc.* **1951**, 243.

99. Sugimoto, T.; Misaki, Y.; Kajita, T.; Yoshida, Z.-I.; Kai, Y.; Kasai, N. *J. Am. Chem. Soc.* **1987**, *109*, 4106.

100. Henningsen, M. C.; Jeropoulos, S.; Smith, E. H. *J. Org. Chem.* **1989**, *54*, 3015.

101. For related oxidations of nickelacycles, see (a) Oxidation with O_2: Herrera, A.; Hoberg, H. *Synthesis* **1981**, 831. Julia, M.; Lauron, H.; Verpeaux, J.-N. *J. Organomet. Chem.* **1990**, *387*, 365; see also ref. 67(c); (b) Oxidation with trimethylamine *N*-oxide: see ref. 33(f) and 70(a); (c) Oxidation with N_2O: see ref. 67(a).

102. Gani; D.; Young, D. W.; *J. Chem. Soc., Perkin Trans. 1* **1983**, 2393.

103. For the use of benzoyl peroxide in oxidations catalyzed by Ni, see Doyle, M. P.; Patrie, W. J.; Williams, S. B. *J. Org. Chem.* **1979**, *44*, 2955.

104. Simple complexation with metal cations is known to enhance the CH acidity of amino acids: Buckingham, D. A.; Stewart, I.; Sutton, P. A. *J. Am. Chem. Soc.* **1990**, *112*, 845. Angus, P. M.; Golding, B. T.; Sargeson, A. M. *J. Chem. Soc., Chem. Commun.* **1993**, 979.

105. For the increased tendency towards racemization of phthaloyl amino acids, see Anderson, G. W.; Callahan, F. M.; Zimmerman, J. E. *Acta Chim. Hung.* **1965**, *44*, 51; Liberek, B. *Tetrahedron Lett.* **1963**, 1103.

106. (a) Castaño, A. M., Echavarren, A. M. *Tetrahedron Lett.* **1993**, *34*, 4361; (b) Echavarren, A. M., Castaño, A. M. *Tetrahedron* **1995**, *51*, 2369.

107. Dowd, P.; Choi, S.-C.; Duah, F.; Kaufman, C. *Tetrahedron* **1988**, *44*, 2137. Dowd, P.; Wilk, B.; Wilk, B. K. *J. Am. Chem. Soc.* **1992**, *114*, 7949 and references cited therein.

108. Chiusoli, G. P. *Acc. Chem. Res.* **1973**, *6*, 422.

109. (a) Kochi, J. K. *Organometallic Mechanisms and Catalysis*; Academic: New York; 1978; (b) Jolly, J. P. In *Comprehensive Organometallic Chemistry*; Wilkinson, G.; Stone, F. G. A.; Abel, E. W., Eds.; Pergamon: Oxford, 1982; Vol. 4, Chaps. 4.1 and 4.2.

110. (a) Hegedus, L. S.; Thompson, D. H. P. *J. Am. Chem. Soc.* **1985**, *107*, 5663; (b) Hegedus, L. S.; Miller, L. L. *J. Am. Chem. Soc.* **1975**, *97*, 459.

111. Jolly, J. P. In *Comprehensive Organometallic Chemistry*; Wilkinson, G.; Stone, F. G. A.; Abel, E. W., Eds.; Pergamon: Oxford, 1982; Vol. 6, Chap. 37.6. Billington, D. C. In *Comprehensive Organic Synthesis*; Trost, B. M.; Fleming, I., Eds.; Pergamon: Oxford, 1991; Vol. 3, Chap. 2.1.

112. (a) Newcomb, M.; Curran, D. P. *Acc. Chem. Res.* **1988**, *21*, 206; (b) Newcomb, M. *Tetrahedron* **1993**, *49*, 1151.

113. On the basis of the 57 : 43 ratio of linear to cyclic products obtained in ref. 15(a) with 2-allyloxyethyl bromide and the more recently reported value for the rate of this radical rearrangement [9×10^6, see ref. 112(b)] a 1.9×10^7 M^{-1}s^{-1} can be determined [see ref. 110 (a)].

114. For the rates of colligation of alkyl radicals with Ni(cyclam)$^{2+}$, see Kelley, D. G.; Marchaj, A.; Bakac, A.; Espenson, J. H. *J. Am. Chem. Soc.* **1991**, *113*, 7583

115. For a discussion on the relevance of electron transfer to the mechanism of the alkylation of Ni(I) complexes related to the coenzyme F430, see Stolzenberg, A. M.; Stershic, M. T. *J. Am. Chem. Soc.* **1988**, *110*, 5397; Lahiri, G. K.; Stolzenberg, A. M. *Inorg. Chem.* **1993**, *32*, 4409; Helvenston, M. C.; Castro, C. E. *J. Am. Chem. Soc.* **1992**, *114*, 8490.

116. For a study about electron transfer and S$_N$ mechanism in alifatic nucleophilic substitution, see Lund, H.; Daasbjerg, K.; Lund, T.; Pedersen, S.U. *Acc. Chem. Res.* **1995**, *28*, 313.

117. (a) Deming, T. J. *Nature* **1997**, *390*, 386; (b) Deming, T. J. *J. Am. Chem. Soc.* **1998**, *120*, 4240.

TRANSITION-METAL-CATALYZED CYCLOADDITION REACTIONS OF BICYCLO[2.2.1]HEPTA-2,5-DIENES (NORBORNADIENES)

Mark Lautens and William Tam

Advances in Metal-Organic Chemistry
Volume 6, pages 49–101.
Copyright © 1998 by JAI Press Inc.
All rights of reproduction in any form reserved.
ISBN: 0-7623-0206-2

I. INTRODUCTION

Bicyclo[2.2.1]hepta-2,5-diene, norbornadiene (NBD) was first reported in the patent literature in 1951.[1] Unlike other molecules containing two isolated double bonds, these olefins are homoconjugated[2] due to a 'through-space' interaction,[2(a)] as indicated by its photoelectron spectrum (Fig. 1).[2(b)] The significant strain energy (25.6 kcal/mol vs. norbornene which has a strain energy of 17.6 kcal/mol)[3] is responsible for many of the reactions it undergoes, including cycloadditions under thermal, photochemical, Lewis acidic, or metal-catalyzed conditions. The cycloadditions with NBD can be categorized into two main types: (A) one of the two olefins in NBD is involved in the cycloaddition; (B) both of the olefins are involved in the cycloaddition.

A. Cycloadditions Involving One of the Two Olefins in NBD

Reactions of NBD follow the pattern typically observed in bridged bicyclic molecules, namely, that reactions occur preferentially on the more accessible *exo* face, although there are a few exceptions which yield exclusively the *endo* products.[7(c)] The different modes of cycloaddition which are designated as [2 + n], are summarized following, and some representative examples are shown in Scheme 1.

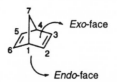

Figure 1. Structure of norbornadiene.

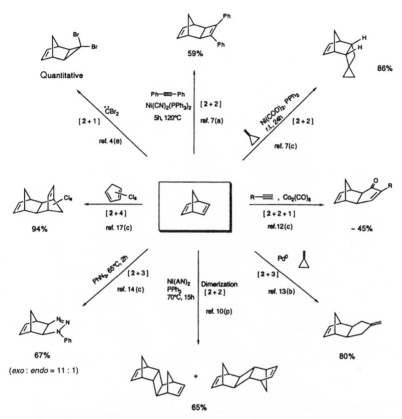

Scheme 1. Cycloaddition involving one of the olefins in NBD.

- Isomerization of NBD-[**2 + 2**] (ref. 18)
- [**2 + 1**] (ref. 4)
- [**2 + 2**] (refs. 5, 6, 7, and 8; see also ref. 9)
- [**2 + 2**] dimerization and trimerization of NBD (ref. 10 and 11)
- [**2 + 2 + 1**] (ref. 12)
- [**2 + 3**] (refs. 13, 14, 15, and 16)
- [**2 + 4**] Diels–Alder reaction (ref. 17)

B. Cycloadditions Involving Both Olefins in NBD

The homoconjugated[2] olefins in NBD also undergo cycloadditions of the [**2 + 2** + n] type which are summarized following with some representative examples shown in Scheme 2.

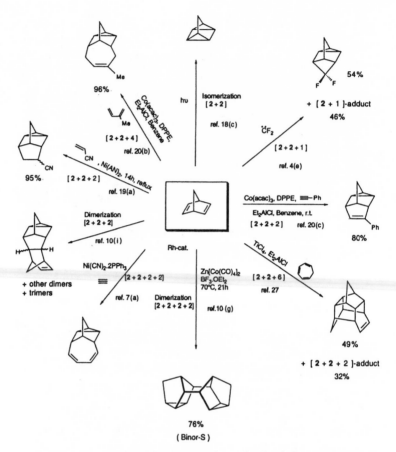

Scheme 2. Cycloadditions involving both olefins in NBD.

- [2 + 2 + 1] [ref. 4(b–e)]
- [2 + 2 + 2] homo Diels–Alder reaction (refs. 5, 6, 19–23; see also refs. 24, 25)
- [2 + 2 + 2] and [2 + 2 + 2 + 2] dimerization and trimerization of NBD (ref. 10)
- [2 + 2 + 2 + 2] [ref. 7(a)]
- [2 + 2 + 4] (ref. 26)
- [2 + 2 + 6] (ref. 27)

The catalysts and the conditions used govern which type of cycloaddition occurs. For example, when NBD is reacted with methylenecyclopropane, either a [2 + 2] or [2 + 3] cycloaddition occurs depending on the reaction conditions. In the presence of a Pd^0 catalyst in refluxing benzene, a mixture of norbornadiene **1** and methylenecyclopropane **2** forms the [2 + 3] cycloadduct **3** with 80% yield.[13(b)] On the other hand, when a Ni^0 catalyst is used a [2 + 2] cycloadduct **4** is obtained with 86% yield and with no detection of any [2 + 3] cycloadduct (Eq. 1).[7(c)]

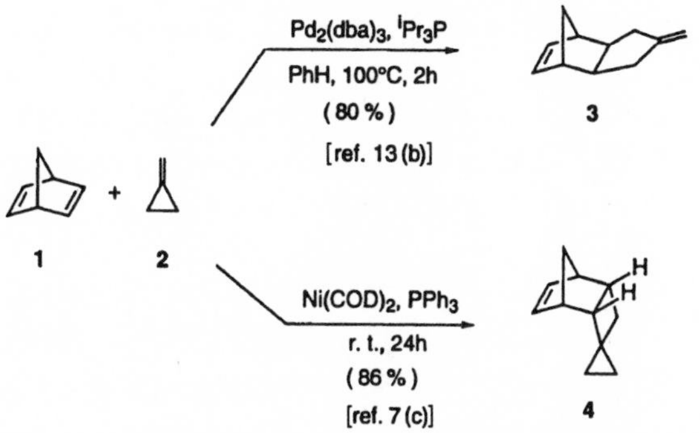

(1)

The nature of the coupling partner used in the cycloaddition is also very important in determining which reaction pathway occurs. For example, in the presence of $Ni(CN)_2(PPh_3)_2$ as the catalyst, NBD reacts with dimethyl acetylenedicarboxylate, an electron deficient alkyne, to give a [2 + 2 + 2] homo Diels–Alder adduct **5**. Under the same catalytic conditions, NBD reacts with diphenylacetylene to give [2 + 2] cycloadduct **6** (Eq. 2).[7(a)]

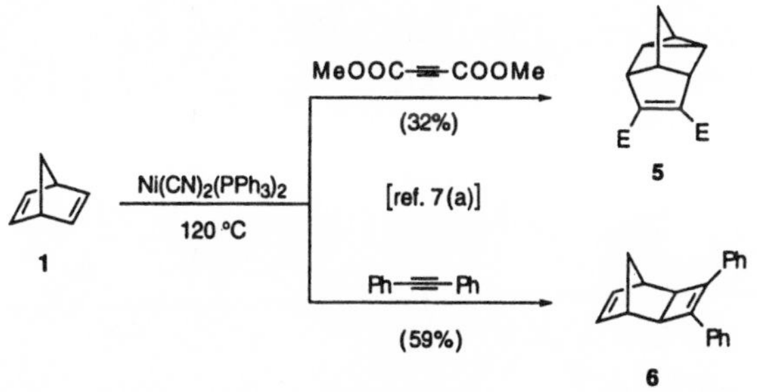

(2)

In this article, we focus the discussion on our investigations on [2π + 2π + 2π] (homo Diels–Alder, HDA) and [2π + 2π] cycloaddition reactions of norbornadiene that were carried out during the past seven years.

Our overall objective is to explore the utility of the [2π + 2π + 2π] (homo Diels–Alder) cycloaddition reactions of norbornadiene as a route to synthesizing linearly and angularly fused polycyclic natural products through a sequence of cycloaddition and fragmentation. As shown in Scheme 3, selective fragmentation of the homo Diels–Alder cycloadduct could lead to a variety of interesting polycyclic compounds. Before we embarked on the synthesis of a specific target compound, however, it was clear that many questions on the cycloaddition reaction needed to be answered. This was the case in spite of the fact that the HDA reaction had been known for many years, as outlined in the following background information.

The first homo Diels–Alder (HDA) reaction was reported in 1958 by Ullman.[22(a)] At 205 °C, a "small amount" of a 1:1 adduct **7** was obtained from norbornadiene and maleic anhydride (Eq. 3). The reaction was

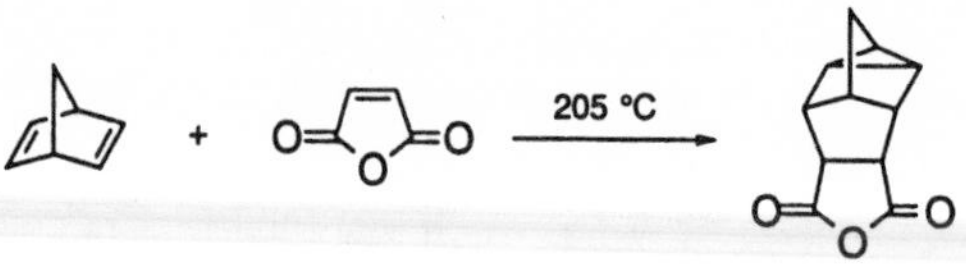

$$
\text{(3)}
$$

7

called the *homo* Diels–Alder reaction because of the sp^3 center separating the olefins of the bicyclic diene. Because the diene was not conjugated, a cyclopropane was created in lieu of a double bond. A pentacyclic adduct was created in one step from a cyclic and a bicyclic precursor.

Following Ullman's observation, Blomquist and Meinwald reported a similar reaction between NBD and tetracyanoethylene (TCNE), leading to deltacyclane **8** in quantitative yield (Eq. 4).[22(b)] Other Diels–Alder dienophiles were less reactive than TCNE and consequently gave much lower yields. The reaction conditions required for reactive dienophiles (refluxing benzene, or stirring at room temperature for 3 to 4 days)

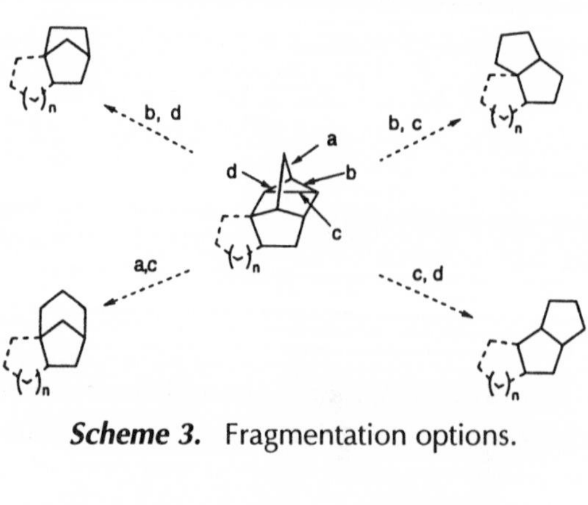

Scheme 3. Fragmentation options.

suggested that this reaction would be of limited applicability with less reactive partners. In fact, on prolonged heating of a mixture of NBD and acrylonitrile at 200 °C, only 12% of the corresponding deltacyclane **9** was isolated (Eq. 5).[22(c)]

(4)

(5)

The limitations of the thermal homo Diels–Alder reaction led several groups to consider the possibility of using transition metals to promote the cycloaddition.[22,23] During the past thirty-five years, transition-metal-catalyzed cycloaddition of NBD with various dienophiles has been extensively studied.[19–21] Many different cycloaddition pathways have been observed, including those previously illustrated in Schemes 1 and 2. One confusing aspect of these studies is that the same catalytic system catalyzes many different pathways, giving rise to a complex mixture of products. Our goal was to find catalysts or to modify known catalysts to generate the desired products selectively.

II. COBALT- AND NICKEL-CATALYZED [2π + 2π + 2π] CYCLOADDITION REACTIONS WITH NORBORNADIENES (THE HOMO DIELS–ALDER REACTION)

A. Cobalt-Catalyzed [2π + 2π + 2π] Homo Diels–Alder Reaction with NBD

Unactivated acetylenes are poor dienophiles and react only with difficulty with dienes to give [4 + 2] cycloadducts. This limits their use as dienophiles in the Diels–Alder reaction.[28] To overcome this problem, several imaginative alternatives employing alkyne equivalents have been developed.[29] Utilizing these synthetic equivalents in the Diels–Alder reaction represents an indirect solution to this reactivity dilemma be-

cause modifications of the substrates before and/or after the cycloaddition are usually required. Thus, recent developments by Wender[30(a)] and
Livinghouse,[30(b)] which showed that certain low-valent metal complexes
promote intramolecular Diels–Alder reaction with unactivated alkynes,
are important and provide a practical, direct, and mild dienyne cycloaddition approach to 1,4-cyclohexadiene-containing polycycles.

Once we became interested in determining the utility of the homo
Diels–Alder (HDA) in synthesis, the search for a highly active catalyst
was undertaken. In the late 1970s, Lyons showed that unactivated acetylenes react with norbornadiene in the presence of low-valent cobalt
complexes to yield a mixture of cycloadducts in which the major product
arises via a [2 + 2 + 2] coupling (Eq. 6).[20(a,b)]

$$
\text{(structures)} \quad + \quad \equiv\!\!-\text{Ph} \quad \xrightarrow[\text{Et}_2\text{AlCl, PhH, r.t.}]{\text{Co(acac)}_3,\ \text{DPPE}} \quad \text{(product)} \quad (6)
$$

$$\underset{\mathbf{10}}{\text{Ph}}$$

A severe limitation in the reported reaction was that substitution of
the aryl group in the acetylene for a simple alkyl group resulted in low
yields of the desired coupling products. The major adduct arises from a
homocoupling of two norbornadienes[20(a),(b)] (forming NBD dimers[10]).
The lack of reactivity and the resulting competitive side processes had
to be overcome before the reaction would be sufficiently reliable for use
in synthesis.

In our initial experiments[20(c)] we isolated deltacyclene **10** with 75–
80% yield when a benzene solution of norbornadiene and phenylacetylene is treated with a catalytic quantity (1–5 mol%) of commercially
available Co(acac)$_3$ with added 1,2-bis(diphenylphosphino)ethane
(DPPE) (1:1 wrt Co) is reduced with Et$_2$AlCl (large excess). Any change
in solvent, ligand, or the use of other acetylenes leads to unacceptable
yields and very slow reactions. For example, 1-hexyne gives less than
10% of the desired homo Diels–Alder adduct even after prolonged
heating. We reasoned that adventitious moisture might interfere with the
reduction step. The reported preparation of Ni(COD)$_2$[31] indicates that
drying commercially available "anhydrous" Ni(acac)$_2$ before the reduction step is required. Analogously, Co(acac)$_3$ was first subjected to
azeotropic drying before use. Our studies on the cobalt-catalyzed homo
Diels–Alder reaction with NBD and unactivated acetylenes using this
improvement are shown in Scheme 4 and Table 1.[20(c)]

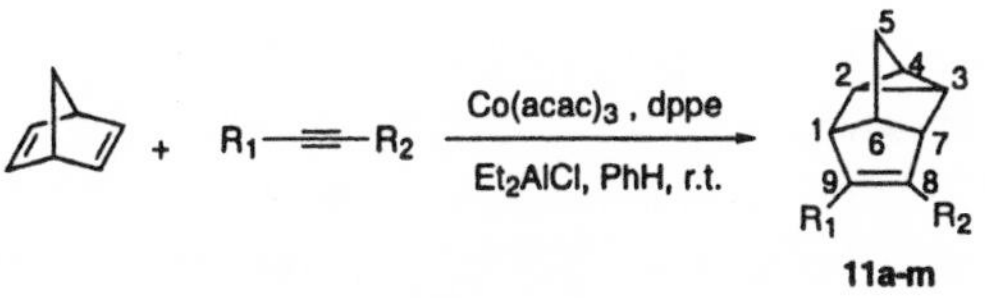

Scheme 4.

Table 1. Cobalt-Catalyzed HDA Reaction of NBD with Unactivated
Alkynes

Deltacyclene **11**	R_1	R_2	Yield (%)
a	H	Ph	Quantitative
b	H	^nBu	91
c	H	^iPr	58
d	H	^tBu	50
e	H	$(CH_2)_4OPMB$	74
f	H	$(CH_2)_3OTBS$	90
g	H	$CH_2CH(OAc)CH_2CH_3$	82
h	H	CH_2OR	N.R.
i	H	$SiMe_3$	50
j	H	CH_2SiMe_3	75
k	Et	Et	55
l	Ph	Ph	58^a
m	$SiMe_3$	$SiMe_3$	N.R.

Note: ^{a}Reaction run at 60 °C instead of r.t. No HDA adduct was obtained when the reaction was carried out
at r.t.

Typical Procedure: To a flame-dried flask containing norbornadiene,
benzene and, the acetylene, azeotropically dried Co(acac)₃ (1–5 mol%)
and DPPE (1 equiv wrt Co) were added under a N_2 atmosphere. After
stirring for 5 min., the Co(acac)₃ and dppe dissolved yielding a dark green
solution. Then Et_2AlCl (4–6 equiv wrt Co) was added over 2–10 min. at
a rate such that the temperature did not exceed 45 °C. (Caution: the
reaction is very exothermic and there is an induction period.) The reaction
mixture usually turned dark brown. After stirring at room temperature for
2–24 h, the reaction mixture was filtered through a plug of silica gel. (For
a large scale reaction, quenching the excess diethylaluminum chloride
with isopropanol is necessary.) Removal of the solvent and purification
by bulb-to-bulb distillation or flash chromatography provided the product.

We also examined the use of Co(acac)₂ instead of Co(acac)₃ in the
cycloaddition. Although a longer 'induction period' is usually observed,

the rate and yield of deltacyclenes are similar. These cycloadditions are very clean processes and yield a single [2 + 2 + 2] product. Internal acetylenes are less reactive than terminal acetylenes with our improved catalytic system. Reaction with 3-hexyne at room temperature affords only 55% of the corresponding deltacyclene. Diphenylacetylene reacts only at 60 °C to afford the HDA adduct (at room temperature, only NBD dimers were obtained), whereas bis(trimethylsilyl)acetylene does not react even under refluxing benzene conditions.

Cheng and co-workers[20(e)] recently reported a cobalt(II) iodide complex which upon reduction also catalyzes the cycloaddition of unactivated internal alkynes (Eq. 7).

$$
\begin{array}{c}
\text{[Co}_2\text{(PPh}_3\text{)}_2\text{] (2.5\%)} \\
\xrightarrow{\hspace{2cm}} \\
\text{Zn (25\%)} \\
\text{Cl}\diagdown\diagup\text{Cl , 80 °C}
\end{array}
\tag{7}
$$

R$_1$ ─≡─ R$_2$

R$_1$ = R$_2$ = Et : 87%
R$_1$ = Me, R$_2$ = ^{n}Pr : 84%

Dimerization of NBD[10(g),(m)] and trimerization of alkynes[32] catalyzed by cobalt complexes have been reported in the literature. In the absence of alkyne, we found that >60% of NBD dimers are obtained under our cobalt catalytic conditions. Slow trimerization of alkynes to the corresponding benzene derivatives was also observed under the same conditions in the absence of NBD. Although the exact nature of the catalytic intermediate in the HDA reaction is not known, these results and the observed high yields of deltacyclenes strongly indicate that the coordination of both NBD and alkyne are greatly favored over the coordination of two NBDs or two alkyne units to the same intermediate.

Low-valent cobalt [Co^{-1}, Co^0 and Co^{+1}] are known to complex with unconjugated dienes (e.g., 1,5-COD, NBD, etc.) and acetylenes.[33] The Co^{2+} or Co^{3+} precatalyst is presumably reduced by Et$_2$AlCl to either Co^0 or Co^{1+}. Co^{1+}, which has the electronic configuration [Ar]3d^8 and commonly forms five-coordinate complexes in a trigonal bipyramidal form. Thus, although very little is known about the structure of the catalytic intermediates, an intermediate with structures I$_1$ and/or I$_2$ seems reasonable (Fig. 2).

Previous reports on some low-valent cobalt complexes with NBD showed that the two olefins in NBD occupy an axial and equatorial position (with bite angle ~90°) in the trigonal bipyramid.[33,34] For chelat-

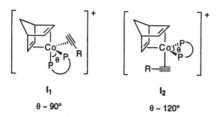

I_1

$\theta \sim 90°$

I_2

$\theta \sim 120°$

Figure 2. Possible intermediates.

ing bidentate phosphine ligands, $R_2P(CH_2)_nPR_2$, the two phosphorus atoms can occupy either two equatorial positions (giving intermediate I_2) or one equatorial and one axial position (giving intermediate I_1), depending on the bite angle θ and the chain length between the two phosphorus atoms.[33–38] For DPPE ($n = 2$), the normal bite angle θ is approximately 90°.[38(a)] Thus, intermediate I_1 (with $\theta = 90°$) is more likely to form. Further discussion of the mechanism is presented following and in Section IIC.

Proposed Mechanism of Cobalt-Catalyzed HDA Reaction with Unactivated Terminal Acetylenes

The individual steps in the cycloaddition are not known but several proposals have been put forward. Our studies build on these proposals. Following the complexation of NBD and the acetylene to the low-valent cobalt, formation of metallacycle M_1 is likely to occur (Scheme 5). Isolation of 5–10% of side product N (when R = TMS) provides further support for the formation of such a metallacycle, but insertion into the

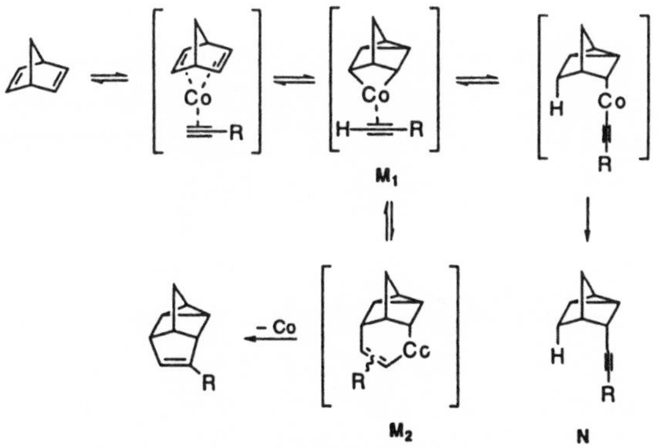

M_1

M_2

N

Scheme 5. Possible mechanism for cobal-catalyzed HDA reaction.

terminal C-H bond, rather than carbametallation, must occur. Insertion of the acetylene into the metal-carbon bond of M_1 would give metallacycle M_2. Reductive elimination of this metal complex would form the deltacyclene. Structures similar to these proposed intermediate metallacycles (M_1 and M_2) were observed previously with other metals.[39]

B. Nickel-Catalyzed [$2\pi + 2\pi + 2\pi$] Homo Diels–Alder Reaction with NBD

Electron-deficient olefins are known to participate in the HDA reaction and lead to deltacyclanes with an additional, newly created stereocenter (Eq. 8).

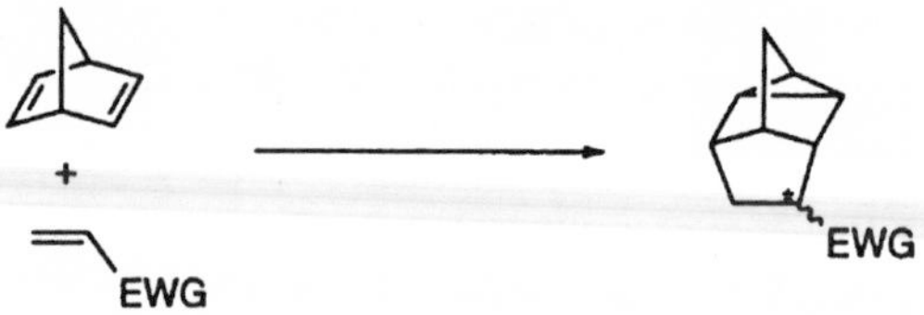

$$(8)$$

Previous workers reported that norbornadiene reacts with acrylonitrile, methacrylonitrile, or methyl methacrylate after prolonged heating at 180 °C to give a [2 + 2 + 2] adduct with low to moderate yields.[22(c),(f),(g),(h)] In the presence of $Ni(CO)_4$, $Ni(CO)_2(PPh_3)_2$, or $Ni(AN)_2/2PPh_3$, (AN = acrylonitrile), the reactions proceed at 80–100 °C with improved yields.[19(a–c)] A preference for [2 + 2 + 2] over [2 + 2] cycloaddition is also observed. However, the reported stereoselectivities are poor (*exo/endo* ratios of 1.5 : 1 to 4 : 1 are common).[19(a–c)] The temperature, the specific electron-withdrawing group, and the steric bulk of the phosphine influence the stereoselectivity, although no clear pattern has emerged.[7(b),19(b),(c),22(g)]

We investigated the cycloaddition between methyl vinyl ketone (MVK) and norbornadiene (Scheme 6), which had previously been accomplished with only low yield. Other reports indicate that MVK

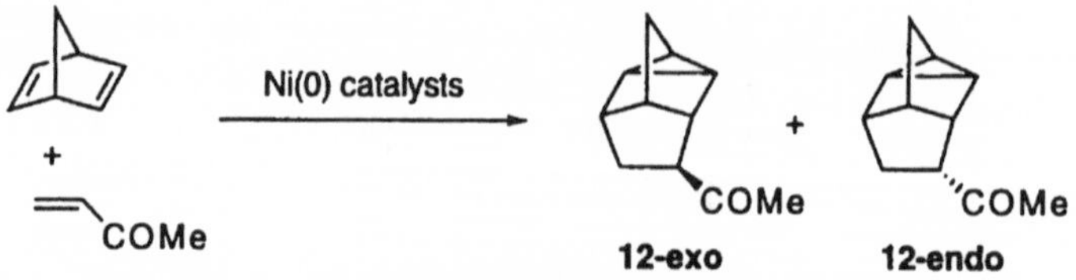

Scheme 6.

polymerization, rather than cycloaddition, occurs.[19(a)] In the presence of 5 mol% $Ni(CO)_2(PPh_3)_2$ at 80 °C, deltacyclane **12** was isolated with 90% yield as a 2 : 1 mixture of *exo/endo* isomers (Table 2, Entry 1).[19(d)] Improvement in the *exo* stereoselectivity occurs when the reaction temperature is lowered to 60 °C although the yield of **12** decreases (Entry 2). Any further decrease in temperature leads to long reaction times and low (<10%) yields. Attempts to activate the catalyst using Me_3N-O (to oxidize a CO ligand to a loosely coordinating CO_2 ligand) or CuI (to complex a phosphine and free a coordination site) yields the cycloadducts but the temperature could not be lowered and no improvement in stereoselectivity occurs (Entries 3 and 4). A more reactive complex is required. Yoshikawa demonstrated that low-valent nickel complexes generated by reduction of $Ni(acac)_2$ with sodium borohydride are catalysts but no yields were given.[19(b)] Sodium borohydride is incompatible with an enone, and therefore we required an alternative reducing agent. Triethylaluminum[31] proved very convenient, and reaction of 5–10 mol% $Ni(acac)_2$ with 2 equiv. of the reducing agent in the presence of a phosphine ligand (2 equiv.), under the usual conditions gave **12** with 62% yield as a 19 : 1 mixture of stereoisomers (Entry 5). Importantly, the reaction takes place at room temperature. $Ni(COD)_2$[31] and 2 equiv. of triphenylphosphine is even more effective in promoting both high yields and good stereoselectivity in the cycloaddition (Entry 6). These findings may be contrasted within the low selectivity observed by Noyori in the reaction between NBD and a variety of acrylates under similar conditions.[19(c)] We found that the *exo/endo* ratios are highly dependent on the dienophile and the phosphine ligands (see later).

Table 2. Nickel-catalyzed HDA Reaction of NBD with MVK

Entry	Catalyst[a]	Temperature	Exo:Endo	Yield (%)
1	$Ni(CO)_2(PPh_3)_2$[b]	80 °C	2 : 1	90
2	$Ni(CO)_2(PPh_3)_2$[b]	60 °C	9 : 1	62
3	$Ni(CO)_2(PPh_3)_2/Me_3NO$[b]	80 °C	5 : 1	44
4	$Ni(CO)_2(PPh_3)_2/CuI$[b]	80 °C	8 : 1	23
5	$Ni(acac)_2/Et_3Al/PPh_3$[c]	r.t.	19 : 1	62
6	$Ni(COD)_2/2PPh_3$[c]	r.t.	14 : 1	88

Notes: [a] 5 mol% of Ni(0).
[b] No solvent.
[c] The solvent was 1,2-dichloroethane.

To determine the effect on the selectivity and the compatibility of other functional groups with the reaction conditions, other electron deficient alkenes were examined. The yields and selectivities observed with various dienophiles under the previously established optimum conditions are shown in Scheme 7 and Table 3a.[19(d),(g),40]

Unlike thermal HDA reactions in which *endo* adducts predominate[22(g)] (this may be due to the more favorable 'endo transition state,' as in the usual Diels–Alder reaction[41]), nickel-catalyzed HDA reactions of acyclic electron deficient dienophiles give *exo* isomers as the major cycloadducts. The *exo/endo* selectivities are highly dependent on the nature of the dienophile, the phosphine ligands, and the reaction temperature.[40]

An interesting observation can be made by comparing some of our results with the previous study carried out by Noyori.[19(c)] Removal of one oxygen in the EWG on the dienophile results in a dramatic increase in the *exo/endo* selectivity [Table 3(b)]. Thus, changing the EWG on the dienophile from a methyl ester (COOMe) to a methyl ketone (COMe) or replacing a sulfone (SO_2Ph) with a sulfoxide (SOPh) results in a 10 to 20-fold improvement in the *exo/endo* selectivity.

Cyclic enones have not been previously examined as dienophiles in the homo Diels–Alder reaction. We found that this class of compounds is moderately reactive in the presence of $Ni(COD)_2$/PPh_3.[19(d)] One feature of the cycloaddition is that pentacyclic compounds **13** are created with high stereoselectivity in a single step (> 20 : 1 ratio of stereoisomers). However, in contrast to reactions with acyclic enones, the stereochemistry of the newly formed rings in **13a** and **13b** is *endo* (Eq. 9). The change

$$\text{(9)}$$

13a X = CH_2,   56%
  b X = $(CH_2)_2$, 23%
  c X = O,   58%

in stereoselectivity may be associated with the *s-cis* versus *s-trans* orientation of the dienophile described following.

Typical Procedure: $Ni(COD)_2$ (5–15%) was added to a flame-dried flask equipped with a magnetic stir bar and a rubber septum in the glove box. Triphenylphosphine (2 equiv. wrt Ni) was introduced with a positive flow

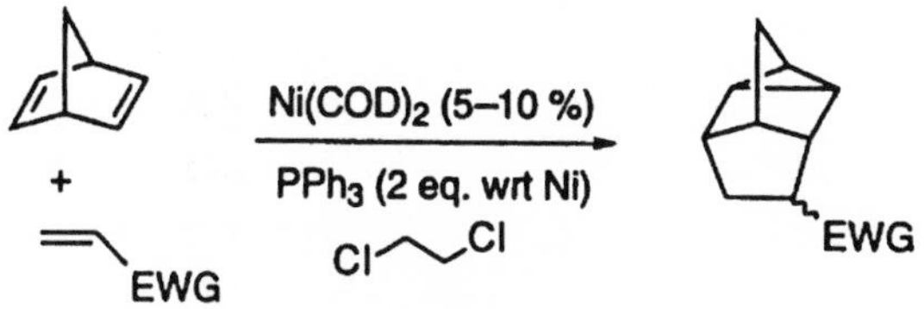

Scheme 7.

Table 3(a). Nickel-catalyzed HDA Reaction of NBD with
Electron-Deficient Olefins

Entry	EWG	Temperature	Yield (%)	Exo:Endo
1	COMe	r.t	88	14 : 1
2	CHO	r.t.	58	3 : 1
3	CO^{t}Bu	60 °C	69	1.5 : 1
4	CN	80 °C	82	4 : 1
5	SO$_2$Ph	r.t.	75	1 : 1
6	SOPh	r.t.	62	7 : 1
7	SOPh	r.t.	73	> 19 : 1[a]

Note: [a]P(OPh)$_3$ was used instead of PPh$_3$.

Table 3(b)

Entry	EWG	Exo:Endo
1[a]	COOMe	1.4 : 1
2	COMe	14 : 1
3	SO$_2$Ph	1 : 1
4	SOPh	19 : 1

Note: [a]Data taken from Noyori's study [ref. 19(c)].

of N_2(g), and a premixed solution of norbornadiene, 1,2-dichloroethane,
and the dienophile was added to the Ni(COD)$_2$/PPh$_3$ solid mixture via a
cannula. The reaction mixture was stirred at the appropriate temperature
under nitrogen for 5 to 48 h. Then the catalyst was oxidized by stirring the
mixture open to the air for 1–2 h. The reaction mixture was filtered through
a plug of silica gel which was washed repeatedly with CH$_2$Cl$_2$ to obtain a
solution free of metallic impurities. Removal of the solvent and purification
by bulb-to-bulb distillation or flash chromatography provided the product.

Proposed Mechanism of Nickel-Catalyzed HDA Reaction

Noyori's study of the nickel catalyzed reactions of quadricyclane **Q**
and norbornadiene **1** with electron-deficient olefins provided informa-

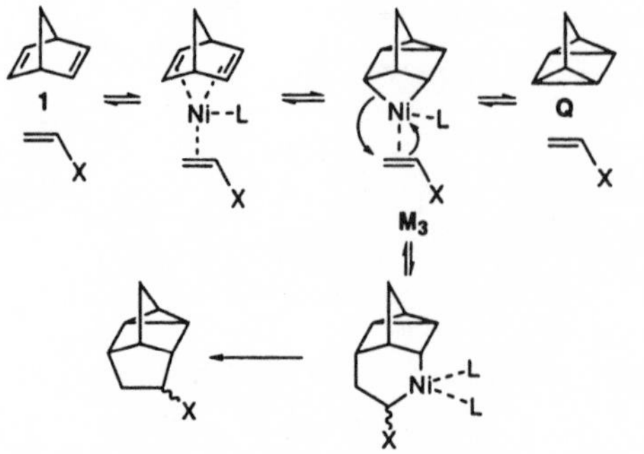

Scheme 8. Proposed mechanism of Ni-catalyzed HDA reaction.

tion about the mechanism of deltacyclane formation (Scheme 8).[19(c),25] A common intermediate M_3 was proposed to explain the formation of the same product(s) from different starting materials.

Our studies on the Ni-catalyzed HDA reaction of acyclic and cyclic enones provided some additional information. The reaction of MVK with NBD is highly *exo* selective (Scheme 7, Table 3(a)), whereas the cyclic enones and lactones give *endo* adducts (Eq. 9). Two differences are immediately evident for these substrates: the presence of the β-*cis*-substituent and the frozen *s-trans* geometry for cyclic dienophiles (Fig. 3).

Enones often coordinate initially to carbonyl group as in **A**, and the metal migrates to the olefin to give **B** or **B'** (Fig. 4).[36,42] The *s-cis* complex (**B**) of the metal to both the carbonyl and olefin of acyclic enones was also known for related complexes.[42] This suggested that the *s-cis* versus *s-trans* difference is important to the difference in the *exo/endo* selectivities observed.

Two possible metallacycle intermediates, M_4 and M_5, also have to be considered (Fig. 5). The β-*cis* methylene substituent in the chair conformation of M_4 (leading to the favored *endo* adduct) occupies the equatorial position with the ketone group in adjacent axial position. The *endo* axial substituent is tied back in the ring, decreasing 1,3-diaxial interactions. The interaction of the ligand on the nickel with the axial CH_2 for the *exo* precursor M_5 would be greater than the interaction between the ligand

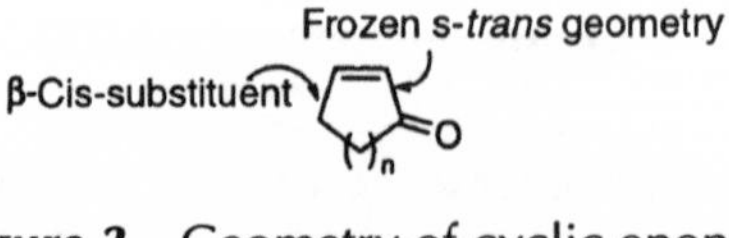

Figure 3. Geometry of cyclic enones.

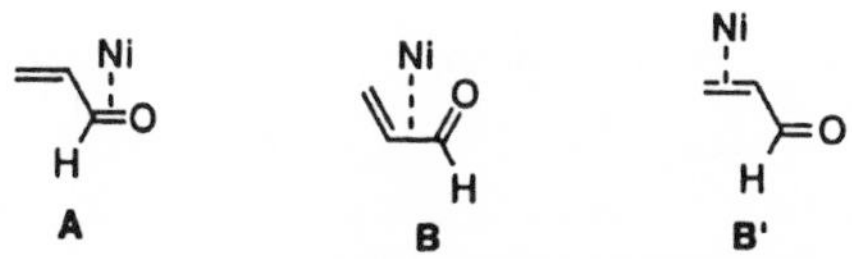

A   B   B'
(s-cis-geometry)   (s-trans-geometry)

Figure 4. Different types of coordinations between a metal complex and an enone.

and the axial H in M_4. This unfavorable interaction would result in forming the *endo* adduct for the cyclic enones.

A detailed study of the effect of the nature and the size of phosphine ligands on the *exo/endo* selectivities of Ni-catalyzed HDA reactions was carried out by Yoshikawa and co-workers.[19(b)] The authors attempted to correlate the 'cone angle' of the phosphines[37,42] and the observed *exo/endo* selectivities, but no clear trend emerged from these studies. Thus, many details of the mechanism in the metal-catalyzed HDA reaction still remain unsolved.

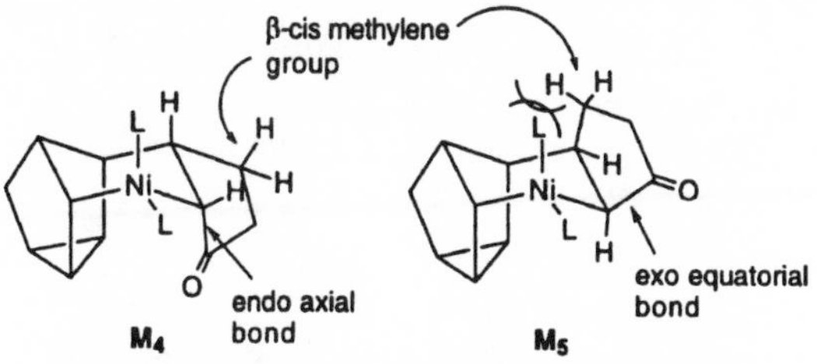

Figure 5. Two possible metallocycles in the Ni-catalyzed HDA reaction of NBD and a cyclic enone.

C. Asymmetric Induction Studies

Within the past fifteen years, efforts to prepare enantiomerically pure products has dominated new methodologies in organic chemistry. One approach is the use of chiral auxiliaries which is called *stoichiometric asymmetrical induction*.[43] *Catalytic asymmetrical* reactions represent an alternative which is highly desirable, and intense effort has recently led to several new important transformations.[44,45] Methods for achieving enantioselective cycloadditions are an interesting subset of these transformations.[46,47]

We addressed the question of enantioselectivity in the HDA cycloaddition by using chiral phosphines.[21(a),(e)] Two new rings (a cyclopropane ring and a five-membered ring) and a total of six new stereocenters are created in a single operation (Eq. 10). These studies are significant

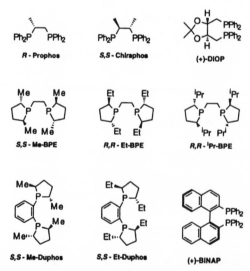

because they represent an emerging area of asymmetrical synthesis wherein substrates lacking strong polar groups (which provide complexation sites) give high enantioselectivities. Steric interactions in the diastereomeric transition states must be the major factor to consider.

We examined various chiral phosphines[38] (Fig. 6) in the reaction of NBD with several acetylenes (Table 4)[48] under the standard cycloaddition conditions as described previously. *S,S*-Chiraphos and *R*-Prophos usually give the best yields and optimize enantioselectivities for most of the tested acetylenes. BPE and Duphos ligands are less selective, and *R*-BINAP and (+)-DIOP gives no cycloadduct.[49]

Figure 6. Structures of various chiral phosphines.

Table 4. Cobalt-catalyzed Enantioselective HDA Reaction

Entry	R	L	Cobalt-Catalyst	Temp. (°C)	Yield (%)	ee (%)
1	Ph	DPPE	$Co(acac)_3$	20–35	80	–
2		S,S-Chiraphos		25–27	37	69 (R)
3		R-Prophos		25–32	93	48 (S)
4		S,S-Me-BPE	$Co(acac)_3$	20–27	91	65 (S)
5				–2–6	70	82 (S)
6			$Co(acac)_2$	20–30	93	70 (S)
7				–5–0	78	77 (S)
8		R,R-Et-BPE	$Co(acac)_3$	19–28	30	44 (R)
9			$Co(acac)_2$	20–26	86	60 (R)
10		R,R-^{i}Pr-BPE	$Co(acac)_3$	19–24	28	7 (R)
11			$Co(acac)_2$	18–24	42	20 (R)
12		S,S-Me-Duphos	$Co(acac)_3$	22–30	12	63 (S)
13			$Co(acac)_2$	18–25	28	47 (S)
14		S,S-Et-Duphos	$Co(acac)_3$	20–30	33	16 (S)
15			$Co(acac)_2$	22–30	18	14 (S)
16	^{n}Bu	DPPE	$Co(acac)_3$	20–30	91	–
17		S,S-Chiraphos		25–26	83	91 (R)
18		R-Prophos		25–27	87	78 (S)
19		S,S-Me–BPE		20–27	87	12 (S)
20		R,R-Et-BPE		4–29	10	11 (R)
21	^{i}Pr	S,S-Chiraphos	$Co(acac)_3$	25–33	75	36 (R)
22		R-Prophos		30–40	33	55 (S)
23	$^nC_4H_8OAc$	S,S-Chiraphos	$Co(acac)_3$	28–32	85	85 (R)
24	$^nC_4H_8OTBS$	S,S-Chiraphos	$Co(acac)_3$	28–32	67	18 (S)
25[a]	$^nC_4H_8OTBS$	S,S-Chiraphos	$Co(acac)_3$	28–30	60	80 (R)

Note: [a]Reaction run in THF/toluene (3:1) instead of benzene.

For phenylacetylene (Table 4, Entries 1–15), the $Co(acac)_3$/S,S-Me-BPE combination gives the highest ee (82%) and 70% isolated yield at –2 to 6 °C. At higher temperature with the same catalytic system, the yield increases but the ee decreases. At lower temperature (< –15 °C), no reaction is observed. For ligands in the BPE series (Entries 4–11), $Co(acac)_2$ instead of $Co(acac)_3$ at room temperature usually gives higher yields and higher ee's. But at a lower temperature (–5 to 6 °C), $Co(acac)_3$ gives a higher ee with a lower yield than $Co(acac)_2$ (Entries 5 and 7). R-Prophos also gives a high yield but the ee is only moderate. S,S-Chiraphos gives a higher ee than R-Prophos but the yield is rather low. Duphos ligands usually give unacceptable yields and ee's.

For 1-hexyne (Table 4, Entries 16–20), S,S-Chiraphos leads to th highest ee (91%) with good yield (83%). Although R-Prophos also give a high yield, the ee is lower than with S,S-Chiraphos. BPE ligands giv low enantioselectivity for alkyl-substituted acetylenes. No clear tren emerged from the studies of ligands, the initial oxidation state of th catalyst, or the temperature.

An acetylene bearing a remote oxygen also reacts with high enantiose lectivity (Entries 23–25). However, we found that the choice of protect ing group and solvent are critical variables. A side-chain oxygen, whic is capable of intramolecular coordination to the cobalt (Entry 24) disrupts the complexation of the phosphine, NBD, and acetylene, whic is required for high selectivity reactions. This effect was overcome b using an electron-withdrawing protecting group to inhibit complexatio of the oxygen (Entries 23) or by carrying out the reaction in THF, whic competes for a coordination site and displaces the OTBS group (Entrie 24 and 25).

We used a combination of spectroscopic and chemical techniques t determine the degree and sense of induction. A standard protocol wa developed: racemic and chiral deltacyclenes **14** were subjected to hy droboration-oxidation to give a single regio- and stereoisomeric alcoho **15**, and the resulting alcohols were converted to Mosher esters[50] **16** (Eq 11). The ee of deltacyclenes **14** were determined by measuring the de o the Mosher esters **16** from both ^{19}F and ^{1}H NMR. The absolute stereo chemistry of the cycloadducts can be assigned by using the method firs described by Mosher.[50,51] The assignment of the absolute stereochemistr was confirmed by X-ray crystallography (Fig. 7).[52]

$$(11$$

Mechanistic Aspects

Of the three modes of coordination between a bidentate phosphine ligand and a metal complex[37] (Fig. 8), the chelating mode (B) is mos common.[37,38] The active cobalt-catalyst generated by the reduction o Co^{2+} or Co^{3+} with Et_2AlCl is probably a five-coordinated cobalt complex Two olefins in NBD and the acetylene occupy three of the five coordi-

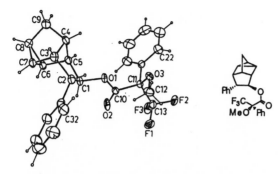

Figure 7. (X-ray crystal structure of Mosher ester **16** (R = Ph).

nation sites, leaving two vacant coordination sites. In our initial studies, we found that the best conditions for the cobalt-catalyzed HDA is with a Co : bidentate phosphine ratio of 1 : 1. With less than 0.7 equiv. or more than 1.5 equiv. of bidentate phosphine wrt Co, the yields and ee's decrease. These results suggest that the chelating mode (B) is the preferred intermediate for the HDA reaction.

Two possible modes of chelation in the active catalyst are possible, I_1 or I_2 (Fig. 2). One can rationalize that, if the natural bite angle $\theta^{38(a)}$ of the bidentate phosphine is closer to 90°, then intermediate I_1, with the two phosphorus atoms occupying one equatorial and one axial position, predominates. If the bite angle is closer to 120°, then intermediate I_2 forms preferentially.

Table 5 shows the bite angles of some bidentate phosphines measured by X-ray crystallography of some metal-phosphine complexes. Few of the examples have M = Co, but the change in metal does not significantly change the bite angle. We assume that the bite angles of the bidentate phosphine complexes are similar to the values in Table 5. The majority of ligands we used in the HDA reaction have θ = 90°, suggesting that intermediate I_1 (Fig. 9), which has two phosphine atoms occupying one

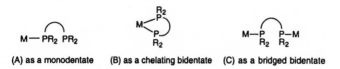

(A) as a monodentate   (B) as a chelating bidentate   (C) as a bridged bidentate

Figure 8. Three different modes of coordination between a bidentate phosphine ligand and a metal complex.

Table 5. Bite Angle of Some Bidentate Phosphines

Bidentate Phosphine	M	Bite Angle, θ	Reference
DPPE	Rh	84–90°	38(a)
	Co	81–82°	53(a)
DIOP	Rh	90–107°	38(a)
	Fe	99°	53(c)
Norphos	Fe	91°	53(d)
	Ni	92°	53(e)
S,S-Chiraphos	Rh	83°	53(f)
Me-BPE	Rh	83°	38(b)
(R)-(+)-BINAP	Rh	92°	38(h), 53(b)

equatorial and one axial position in the trigonal bipyramid, is more likely to form than I_2 (Fig. 2).

It is well known that saturated five-membered chelates adopt a puckered chiral conformation.[54(a)] In the absence of a substituent and when the donor atoms are symmetrically substituted, the ring rapidly interconverts from one chiral conformation to the other (Fig. 10).[54(b)] If the aliphatic link is substituted, however, asymmetry is created and the chelate may be fixed into a single, static chiral conformation by the requirement that the substituent be equatorially disposed (Fig. 11).[38(f),(g)]

Although free rotation of the acetylene in the active intermediate I_1 (Fig. 9) is possible, the orientation and the steric bulk of the substituents on the chiral bidentate phosphine ligands may restrict the rotation. Thus, asymmetrical induction of the chiral phosphine ligands on the reactive intermediate I_1 leads to a preference for forming one enantiomer of deltacyclene. Figure 12 shows the proposed asymmetrical complex with S,S-Chiraphos. The prediction of the absolute configuration of the product from this asymmetrical complex agrees with the experimental results.

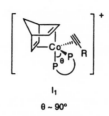

I_1

θ ~ 90°

Figure 9. The most likely active intermediate in the cobalt-catalyzed HDA reaction.

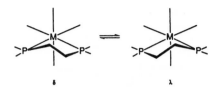

Figure 10. The chiral δ and λ conformations of a saturated five-membered chelate ring.

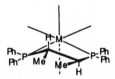

Figure 11. The preferred conformation of the *S,S*-Chiraphos chelate ring.

Brunner's group also independently investigated the enantioselective homo Diels–Alder reaction between NBD and phenylacetylene using the chiral bidentate phosphine Norphos.[21(b)] Up to 98.4% ee was obtained. They also tested this chiral Norphos ligand on the homo Diels–Alder reaction between NBD and acylonitrile. Unfortunately, only 12–15% ee (with *exo/endo* ratio 55/45) was achieved. Following these studies and our report on various chiral phosphines[21(a)] (Table 4), Brunner and Prester[21(c)] extended these studies using a variety of other bidentate phosphines (Fig. 13). The absolute stereochemistry of the chiral deltacyclenes was not determined but was predicted on the basis of GC retention times. High enantioselectivities were obtained with phenylacetylene (> 99.4% ee with Norphos or BDPP) and 1-hexyne (> 98% with Norphos).

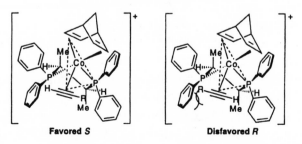

Favored *S*   Disfavored *R*

Figure 12. Proposed asymmetrical complex with *S,S*-Chiraphos.

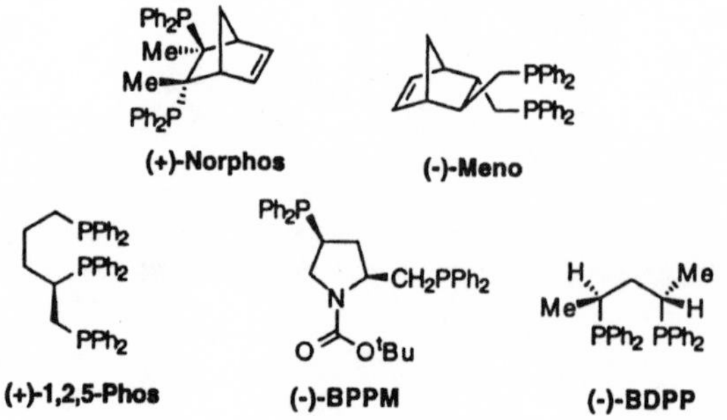

Figure 13.

More recently, Buono and Pardigon[21(d)] showed that the catalytic system used by Cheng and co-workers,[20(e)] [CoI$_2$(PPh$_3$)$_2$]/Zn, replacing PPh$_3$ with an amino acid based chiral phosphine (Fig. 14), gives a highly enantioselective homo Diels–Alder reaction (with phenylacetylene and 1-hexyne, up to 97% ee is achieved).

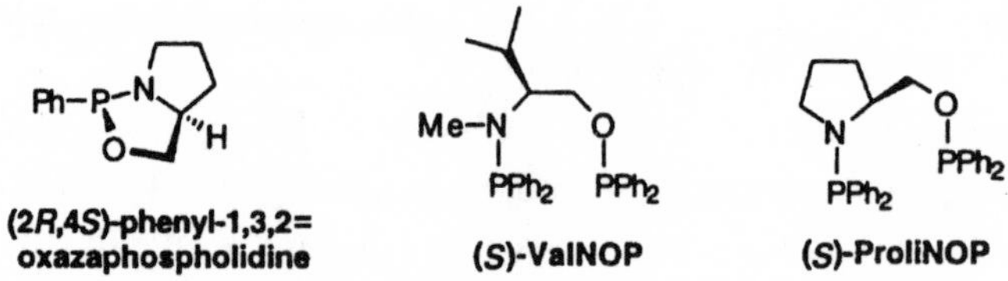

Figure 14.   Amino acid-based chiral phosphines.

D. Homo Diels–Alder Reaction with 2-Substituted Norbornadienes

Unlike the Diels–Alder reaction (where predictable and high regioselectivity is expected in a cycloaddition between an electron-rich diene and an electron-poor dienophile),[41] little is known about the regiochemical outcome of an analogous unsymmetrical HDA reaction. In fact, before our studies,[19(e)] reports of successful cycloadditions using substituted norbornadienes are rare in the literature.[22(i),23(d)] In one instance, TCNE, a symmetrical and highly reactive dienophile, reacts on the unsubstituted side of the 2-substituted NBD **17** to give the substituted cyclopropane derivative **18** (Eq. 12).[22(i)] Other attempts to promote the

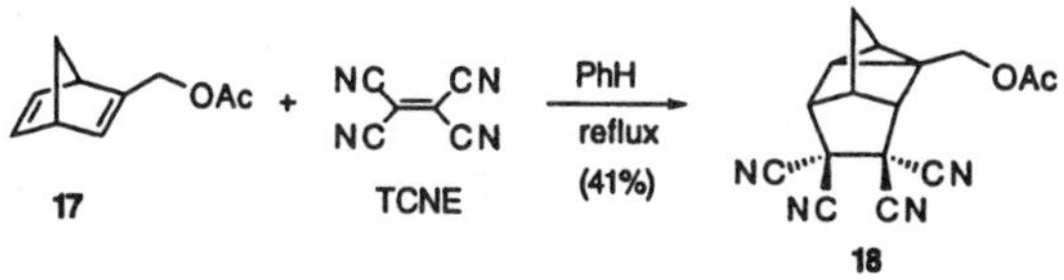

$$(12)$$

cycloaddition with substituted norbornadienes or other types of homo-conjugated dienes have been unsuccessful.[22(i),55]

With the development of active catalysts,[19(d),20(c),21(a)] we began our examination of the regioselectivity of the homo Diels–Alder reaction. We chose to investigate the reaction of 2-substituted norbornadienes bearing an electron-donating or electron-withdrawing group and various electron-deficient dienophiles.[19(e),(g),40] From the outset of this study, we were aware that as many as eight isomers could form when both the diene and dienophile are unsymmetrical. Thus, high levels of regioselectivity would be necessary before the reaction could become synthetically useful. The structures of the regioisomers, designated *ortho, meta, meta'* and *para*, are shown in Scheme 9. *Exo* and *endo* stereoisomers are possible for each cycloadduct.

Studies of the HDA reaction between electron-deficient, 2-substituted NBD **19** (Y = COOMe) with various dienophiles are shown in Table 6.[19(e)] The yields and the selectivities are highest when carrying out these cycloadditions at 80 °C, using Ni(COD)$_2$ (10–20%), PPh$_3$ (2 equiv. wrt Ni) in 1,2-dichloroethane. At lower temperatures, the reactions are not complete, even after two days. At a slightly higher temperature in toluene, the substituted NBD is completely consumed, but the yields are lower and the selectivities decrease slightly.[40]

The homo Diels–Alder reaction gives high levels of stereoselectivity, but which isomer would predominate is not easily predicted given the

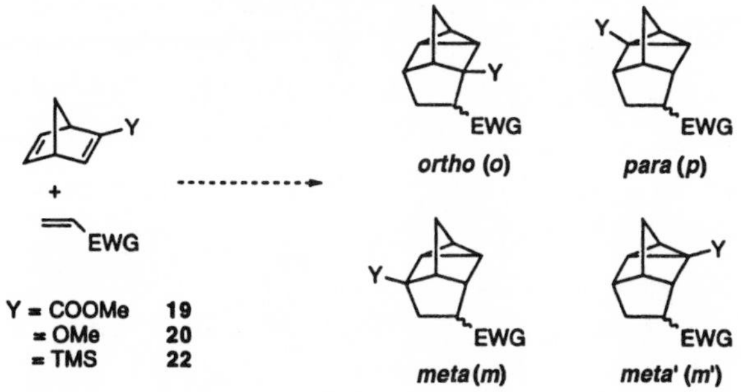

Scheme 9.

Table 6. HDA Reaction of **19** with Different Dienophiles
Catalyzed by Ni(COD)$_2$/PPh$_3$

			Ratio of Regioisomers (Exo:Endo)			
Entry	EWG	Yield (%)	Para	Meta'	Meta	Ortho
1	CN	94	100 (1:2.3)	–	–	–
2	SO$_2$Ph	75	66 (> 20:1)	33 (> 20:1)	–	–
3	COMe	84	70 (3:1)	10 (1:1.4)	–	20 (0:1)

current information. Acrylonitrile is not highly stereoselective with either the parent NBD **1** or 2-substituted NBD **19**. The selectivity reverses between those adducts, and the *endo* isomer predominates (2.3 : 1) in the *para* adducts from NBD **19** (Table 6, Entry 1) whereas the *exo* adduct is favored (4 : 1) from NBD **1** (Table 3, Entry 4). The sulphone cycloadditions are unselective with the parent NBD **1** (Table 3, Entry 5), whereas cycloaddition with the substituted NBD **19** is remarkably *exo* selective (Table 6, Entry 2).

The regioselectivity is less variable. The *para* isomer is favored for all of these dienophiles with electron-deficient NBD **19**. The substituent Y is connected to the cyclopropane in most adducts. The *ortho* and *meta'* isomers are minor products, but the *meta* adduct is not observed. The *para* orientation may be favored over the other for steric or electronic reasons.

Because of the stepwise nature of the reaction and the questions concerning the mechanism of this reaction, direct comparisons with the concerted Diels–Alder reaction may be invalid. These trends in reactivity may be complicated by a change in the rate-determining step with different factors controlling the various rates for each substrate.

Similar studies were carried out with electron-rich, 2-substituted NBD **20** (Y = OMe).[19(e),(g)] With methyl vinyl ketone (MVK) at 80 °C using Ni(COD)$_2$ (10–20%), PPh$_3$ (2 equiv. wrt Ni) in 1,2-dichloroethane, deltacyclane **21** is formed with moderate yield, accompanied by minor isomers (Eq. 13). The selectivity in this reaction is high, and the *ortho-*

$$\text{(13)}$$

endo HDA adduct **21** makes up > 80% of the total product (54% isolated yield). However, few other dienophiles undergo the HDA reaction with this substrate. In contrast to the reaction between MVK and NBD **1**, in which the *exo* isomer predominated, the *endo* isomer is formed as the major product in the reaction between MVK with NBD **20**. The origin of the regioselectivity may be electronic in nature because steric factors should result in an approach to the unsubstituted side of the NBD, as in electron-deficient NBD **19**.

To evaluate the relative reactivities of **1**, **19**, and **20**, two competition reactions were carried out: (1) between **1** and **19** with MVK and (2) between **1** and **20** with MVK. These experiments were carried out with equimolar amounts of each diene (2–3 equiv. each wrt MVK) to approach pseudo first-order conditions. The approximate value of the relative reactivity to cycloaddition with MVK can be calculated by comparing the ratios of adducts from each diene.[19(g),40] These competition studies showed that both electron-rich NBD **20** and electron-deficient NBD **19** are significantly less reactive than the parent, unsubstituted NBD **1** (Table 7).

When a silyl-2-substituted NBD (Y = SiR_3) is subjected to the previous cycloaddition conditions (using $Ni(COD)_2/2PPh_3$) with acrylonitrile, a complex mixture of HDA adducts and [2 + 2] adducts is obtained (Eq. 14).

$$(14)$$

By varying the cycloadditive conditions (including changing the reaction temperature, varying the ratio of the reactants and the ratio of Ni/phosphine, using different phosphine ligands, etc.), we found that

Table 7. Competition Results for 2-Substituted NBDs with MVK

Entry	NBD	*Reactivity Relative with* **20**
1	**1**	7.6
2	**19**	4.5
3	**20**	1.0

replacement of PPh_3 with $P(OPh)_3$ and performing the reaction at 80 °C with 10–15% of $Ni(COD)_2$, eliminates the [2 + 2] adducts and optimizes the selectivities in the HDA reactions (Scheme 10). Thus, *meta'* isomers predominate under these conditions with a silyl-2-substituted NBD. In most cases however the yields are lower than with other dienes.

Summarizing our studies of the Ni-catalyzed HDA reaction with 2-substituted norbornadienes, an electron-withdrawing group attached to the 2-position of the NBD favors the *para* regioisomer (Table 6), whereas an electron-donating group favors the *ortho* regioisomer (Eq. 13) and a silyl group favors the *meta'* regioisomer (Scheme 10).

Based on the proposed mechanism for the Ni-catalyzed HDA reaction described in Section IIB, reaction pathways leading to these different regioisomers are summarized as in Scheme 11. Unfortunately, we do not yet have a good explanation to rationalize our observations on the regioselectivities. Detailed mechanistic studies have to be carried out to understand these results fully.

Similar studies of the regioselectivity in the HDA reaction between 2-substituted norbornadienes and unactivated terminal acetylenes catalyzed by a cobalt catalyst were also undertaken.[56] Unfortunately, 2-substituted norbornadienes are much less reactive with unactivated acetylenes in the cobalt-catalyzed HDA reaction than with electron-deficient olefins in the Ni-catalyzed HDA cycloadditions. When an electron-withdrawing group (Y = COOMe) or an electron-donating group (Y = OMe) is attached to the 2-position of the NBD, no desired [2 + 2 + 2] cycloadduct is observed with 1-hexyne, even in refluxing toluene for three days under the usual cobalt-catalyzed conditions (Scheme 12). Instead, some NBD dimers and acetylene trimers are detected. With a

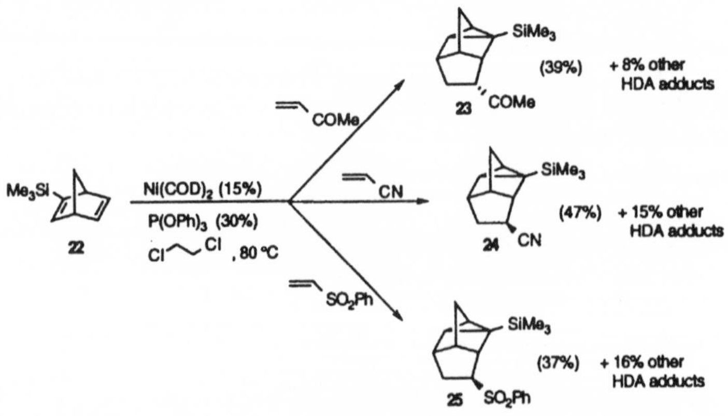

Scheme 10.

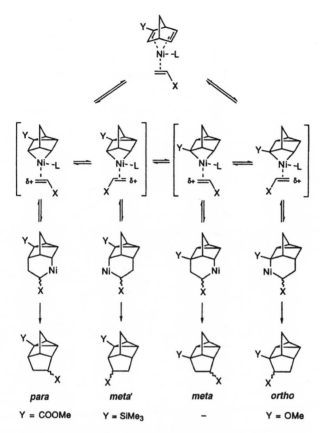

para   meta'   meta   ortho

Y = COOMe   Y = SiMe₃   –   Y = OMe

Scheme 11. Mechanistic pathways leading to different regioisomers in the Ni-catalyzed HDA reaction of 2-substituted NBDs.

silyl group for NBD **22** ($Y = SiMe_3$), the HDA reaction with 1-hexyne occurs, but the yields are low and none of the cycloadditions go to completion (Table 8).

The best regioselectivity obtained is 11.3 : 1 (Table 8, Entry 4), but the isolated yield is only 30%. The major regioisomer (**C**) was identified as the *ortho* isomer and the minor isomer (**B**) was identified as the *meta* isomer by using ^{13}C (APT) NMR and ^1H NMR decoupling techniques. Reactions in a 1:1 mixture of THF and benzene lead to an increased proportion of the *meta* isomer (**B**) (Table 8, Entries 3 and 6). Other studies using phenylacetylene instead of 1-hexyne give similar results, but the yields are never greater than 30%.

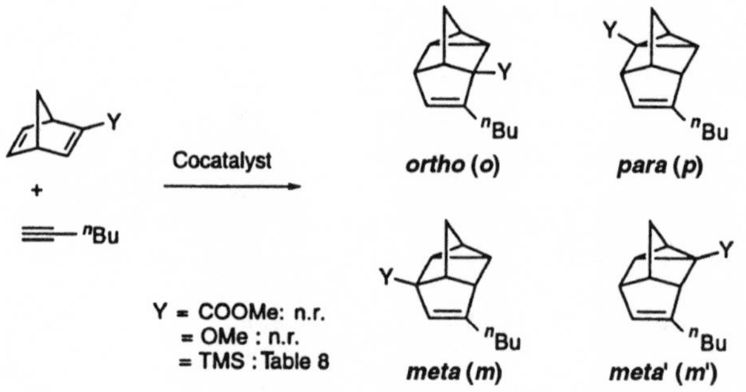

Y = COOMe: n.r.
  = OMe : n.r.
  = TMS : Table 8

Scheme 12.

Table 8. HDA Reaction between 2-TMS-NBD with 1-Hexyne

Entry	Cobalt-Catalyst[a]	Temp.	% Recovered 2-TMS-NBD	Total Isolated Yield of HDA Adducts (%)	Ratio[b] A : B : C
1	Co(acac)$_2$	r.t.	28	42	1 : 3.5 : 23.3
2		70 °C	30	50	1.7 : 1 : 3.7
3		r.t.[c]	60	< 10	0 : 2.1 : 1
4	Co(acac)$_3$	r.t.	28	30	0 : 1 : 11.3
5		70 °C	47	21	1 : 7 : 11
6		r.t.[c]	60	< 10	0 : 4.6 : 1
7	Co(acac)$_3$[d]	r.t.	42	37	1 : 1.2 : 8.3

Notes: [a]5% of cobalt-catalyst (with 1 equiv. of dppe and 4 equiv. Et$_2$AlCl wrt Co).

[b]Isomers C and B were identified as *ortho* and *meta* adducts, respectively, but isomer A was not identified.

[c]Using (1:1) THF/PhH as solvent instead of PhH alone.

[d]Another 5% of cobalt-catalyst was added after 1 day.

E. Homo Diels–Alder Reaction with 7-Substituted Norbornadienes

As described in the introduction, our overall objective was to investigate the HDA reaction as a route to polycyclic natural products by a cycloaddition-fragmentation sequence. This strategy requires efficient routes to cycloadducts bearing leaving groups α to the cyclopropane bonds which undergo bond cleavage under appropriate conditions. Although remote substituents at the 7-position of the NBD might be expected to have less control over the selectivity compared to the substituent which is directly attached to the reacting center (at the

2-position of the NBD), the literature indicates that certain reactions of bicyclo[2.2.1]heptane systems exhibit moderate to high levels of stereo- and regiochemical control.[4(d,e),14(b),14(d),17(c)] These results encouraged us to investigate the regioselectivity of the HDA reaction between a 7-substituted NBD and a dienophile.[19(f)]

Four possible homo Diels–Alder adducts may be formed in this reaction, namely, *anti-exo* (ax), *anti-endo* (an), *syn-exo* (sx) and *syn-endo* (sn) (Scheme 13). We first examined the effect of the nature of the 7-substituent on the NBD to the Ni-catalyzed conditions with MVK as the dienophile (Scheme 13, Table 9).

The results of the HDA reactions between various 7-substituted nor-bornadienes and methyl vinyl ketone (MVK) are shown in Table 9. In general, high yields are obtained. Although four possible products could be formed in this reaction (Scheme 13), little of the *endo* products (an and sn) were detected in most cases. In other words, all 7-substituted norbornadienes that we examined were highly *exo* selective (always > 97 : 3) with MVK. Interestingly, the *anti* to *syn* ratio increases as the group electronegativity[57] increases, eventually yielding 95 : 5 in favor of the *anti* isomer. Varying the temperature (from r.t. to 80 °C) or changing the nature of the phosphine (from PPh$_3$ to P(OR)$_3$) has very little effect on either the yields or the selectivities of these cycloadditions.

The regio- and stereoselectivity of three different dienophiles in the HDA reaction with two different 7-substituted norbornadienes were investigated (Scheme 14). Both acrylonitrile (AN) and phenyl vinyl sulfone (PVS) give excellent yields of the HDA products. However, a modest drop in *anti* : *syn* ratios was observed whereas the *exo* : *endo* ratios decrease dramatically as the dienophile changes from MVK to AN or PVS (Table 10).

A substituent at the 7-position in a NBD derivative breaks one of the planes of symmetry and leads to differentiation of the two olefins. One of the double bonds is *syn* to the substituent (denoted as *syn-π*) and another is *anti* (denoted as *anti-π*). For a [2π + 2π + 2π] HDA reaction to occur, the dienophile has to approach the NBD from the *endo* face of the double bonds (Fig. 15). Although we do not know which step in the cycloaddition is rate-determining, the 7-substituent should clearly exert an electronic effect on the two alkenes.

A direct through-space interaction[2(a)] between the 7-substituent with the neighboring double bond is possible. This interaction could involve either orbital overlap[58] or a field effect. The substituent orbitals could also interact with the π orbitals through bonds rather than through space.

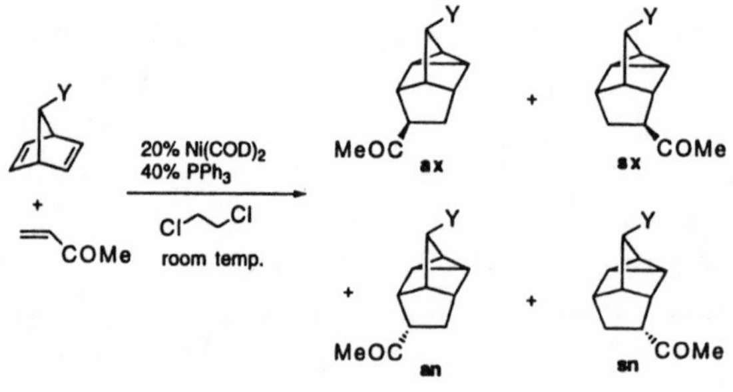

Scheme 13.

Table 9. Effect of the 7-Substituent on the Ni-Catalyzed HDA Reaction with MVK

Entry	Y	Yield (%)	ax : sx : an : sn	Anti:Syn	Exo:Endo
1	^{n}hexyl	83	40 : 58 : 1.6 : 0.4	42 : 58	98 : 2
2	Ph	84	54 : 45 : 0.8 : 0.2	55 : 45	99 : 1
3	Cl	60	71 : 28 : 0.8 : 0.2	72: 28	99 : 1
4	OCOPh	97	80 : 20 : 0 : 0	80 : 20	100 : 0
5	OTIPS	90	90 : 9 : 1 : 0	91 : 9	99 : 1
6	OMEM	89	88 : 9 : 3 : 0	91 : 9	97 : 3
7	O^{t}Bu	95	95 : 5 : 0 : 0	95 : 5	100 : 0

For example, the ability of the C_1–C_7–C_4 bridge of NBD to hyperconjugate with the two π orbitals could be altered by the 7-substituent.

From the *ab initio* calculations of various 7-substituted norbornadienes using the STO-3G basis set reported by Mazzocchi, Houk, and co-workers,[17(e)] there is a shift of electron density from the *anti*-π olefin to the *syn*-π olefin as the substituent group electronegativity increases (Fig. 16). Based on the mechanism of deltacyclane formation proposed by Noyori,[19(c)] metallacycles **B** are formed via complex **A** (Scheme 15). The larger the substituent group electronegativity, the greater the shift of electron density from the *anti*-π olefin to the *syn*-π olefin in the **A**. One can rationalize that the *syn*-σ metal-carbon has a greater electron density than the *anti*-σ metal-carbon in the metallacycles **B** as the group electronegativity of Y increases. Thus, the more reactive metallacycle **B₂** leads to the major *anti*-product.

The previous discussion is based on examining the electronic effect of the 7-substituents on the two olefins of the starting dienes. *Ab initio* calculations on the presumed nickel π and σ complexes are required to

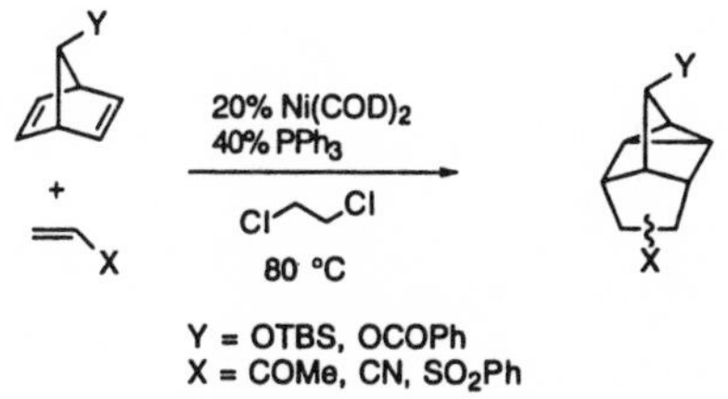
Y
+
X
20% Ni(COD)2
40% PPh3
Cl Cl
80 °C
Y
X
Y = OTBS, OCOPh
X = COMe, CN, SO2Ph

Scheme 14.

Table 10. Effect of Dienophiles in HDA Reactions with 7-Substituted NBDs

Entry	*Y*	*X*	*Yield (%)*	*Anti : Syn*	*Exo : Endo*
1	OTBS	COMe	93	90 : 10	98 : 2
2		CN	93	89 : 11	79 : 21
3		SO_2Ph	95	81 : 19	67 : 33
4	OCOPh	COMe	95	77 : 23	96 : 4
5		CN	95	75 : 25	75 : 25
6		SO_2Ph	97	70 : 30	70 : 30

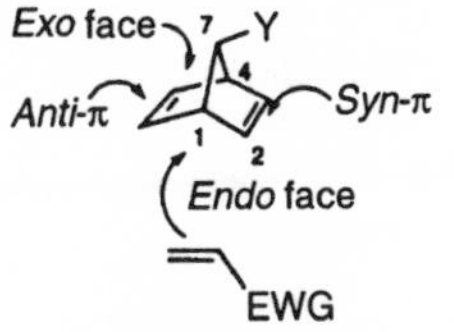
Exo face
7
Y
4
Anti-π
Syn-π
1
2
Endo face
EWG

Figure 15.

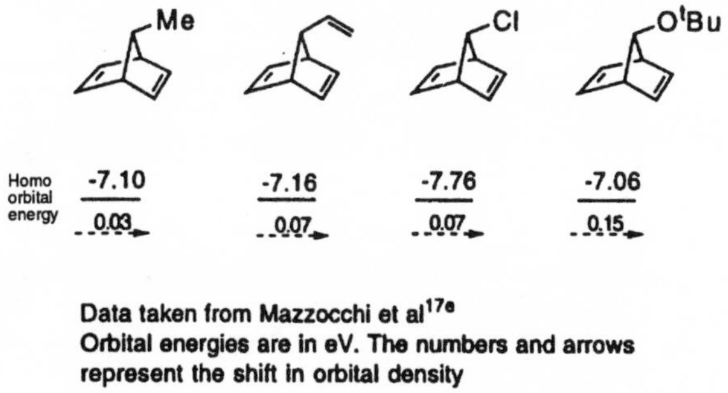
Me
Cl
O^tBu
Homo orbital energy
-7.10
0.03
-7.16
0.07
-7.76
0.07
-7.06
0.15
Data taken from Mazzocchi et al17e
Orbital energies are in eV. The numbers and arrows represent the shift in orbital density

Figure 16.

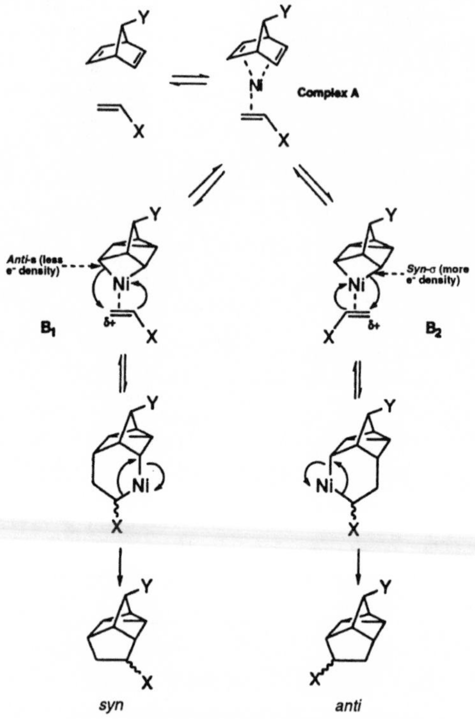

Scheme 15.

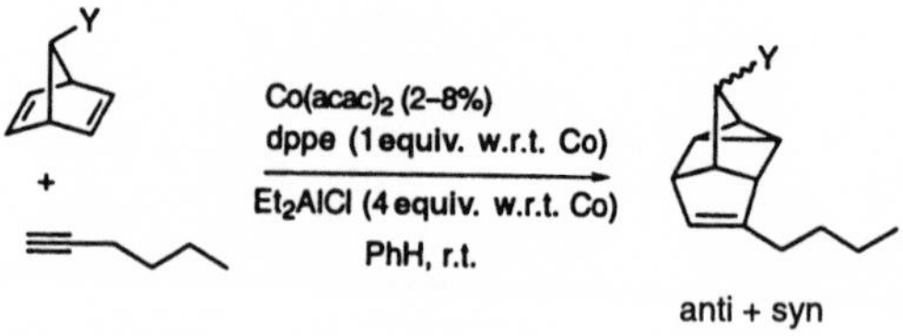

Scheme 16.

Table 11. Effect of the 7-Substituent on Cobalt-catalyzed HDA
Reactions with 1-Hexyne

Entry	Y	Yield (%)	Anti : Syn
1	^{n}hexyl	80	50 : 50
2	Ph	94	50 : 50
3	OCOPh	96	59 : 41
4	OTIPS	84	65 : 35

explore the relative contributions of steric and electronic effects in detail.[59]

Similar studies on the cobalt-catalyzed HDA reaction between various 7-substituted norbornadienes and an unactivated terminal acetylene, 1-hexyne, were also investigated (Scheme 16, Table 11).[56] Although the yields remain high, the *anti/syn* selectivities are poor compared with the corresponding Ni-catalyzed cycloadditions.

These poor selectivities, relative to the nickel catalyzed reactions, are perhaps not surprising. The difference in electron density between the two sp carbons in an unactivated alkyne is obviously much smaller compared to the two sp^2 carbons in an electron-deficient olefin. Thus, the B_2 type intermediate (Scheme 15) with an unactivated alkyne does not have much preference over the corresponding B_1 type intermediate.

F. Intramolecular Homo Diels–Alder Reactions

The additional ring which arises from an intramolecular HDA reaction may enable us to synthesize various kinds of triquinanes or other poly-cyclic compounds based on our proposed cycloaddition-fragmentation strategy (Scheme 3). The decrease in entropy associated with tethering the two reactive components suggests that the reaction would be significantly more facile than the intermolecular reaction.[60] This potential rate enhancement, however, is compromised by the dramatic decrease in rate associated with intermolecular cycloadditions with substituted norbor-nadienes (Section IID, Table 7). In fact, before our studies,[20(f)] only one example of an attempted intramolecular HDA reaction was reported in the literature (Eq. 15).[55] Instead of undergoing a [2 + 2 + 2] cycloaddition, an alternative intramolecular ene reaction occurs.[60(b)]

(15)

The failure of this substrate to undergo the HDA reaction is not difficult to rationalize given the results of our earlier studies. From our experience, the reactivity of a 2,3-disubstituted NBD in a HDA reaction is very low. Secondly, the ester functionality in the tether may prefer to adopt a conformation where a $[2 + 2 + 2]$ cycloaddition cannot occur (As shown in Fig. 17, a typical ester favors conformation **A** rather than **B**).[61]

In an intramolecular reaction, there are two possible modes of $[2 + 2 + 2]$ cycloaddition which have to be considered (Eq. 16). The dienophile

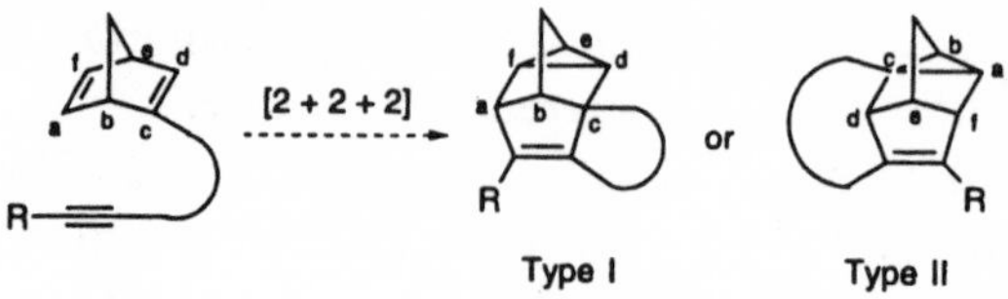

$$\text{(16)}$$

Type I   Type II

in the tether can cyclize on C^a–C^b–C^c to give a cycloadduct of type I, or it can cyclize on C^d–C^e–C^f to give a type II cycloadduct. Molecular models and MM2 calculations indicate that both products would be stable.

Dienynes **26** were prepared in two steps from NBD.[20(f)] Although the thermal cycloaddition of **26a** (at 170 °C in mesitylene for 24 h) gives only 3% of the desired HDA adduct **27a**, the cobalt-catalyzed HDA reaction at room temperature increases the yield of the cycloadduct five-fold to 15% (Scheme 17). Improved yields are obtained for dienyne **26b** under the same cobalt-catalyzed conditions to afford the corresponding HDA adduct **27b** with 22% yield. A *gem*-dialkyl effect may be responsible for the improvement.[62]

Although the yields of the cycloadditions are modest, these initial results were encouraging because as shown previously in Section IIA, intermolecular HDA reaction between NBD and propargyl ethers fails to give any cycloadduct, and in fact the propargyl ether poisons the catalyst.[20(c)] The Lewis acidic character of the reducing agent, Et_2AlCl, was also of concern because the starting materials **26** and the cycload-

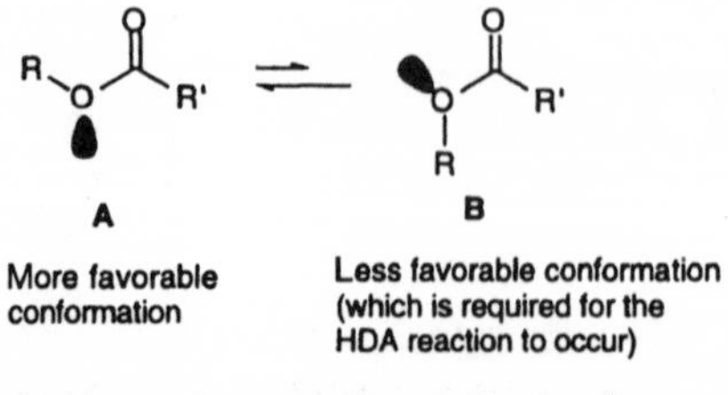

A   B

More favorable   Less favorable conformation
conformation   (which is required for the
                                      HDA reaction to occur)

Figure 17. Two conformations of an ester.

Co(acac)₃ (5–10%)
dppe (1 equiv. w.r.t. Co)
Et₂AlCl (4 equiv. w.r.t. Co)
PhH, r.t.

26a R = H
26b R = Me

27a : 15%
27b : 22%

Scheme 17.

ducts **27** contained sensitive functional groups (allylic ethers). Furthermore, the intermolecular reactions of 2-substituted norbornadienes were generally very slow. It appeared that a significant improvement in the yields of the cycloaddition would be realized if the oxygen within the tether were replaced by a methylene unit.

The all-carbon dienynes **28a–d** and **30a–c** were prepared and their intramolecular cycloadditions studied (Schemes 18 and 19). No reaction was observed upon treatment of **28a** at 150 °C in the absence of the catalyst. However, under cobalt-catalyzed conditions, cycloadditions occur smoothly.

All of the intramolecular HDA adducts observed were type I adducts (Eq. 16), and no type II adduct was detected in any case. Higher yields

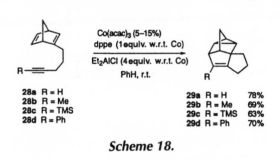

Co(acac)₃ (5–15%)
dppe (1 equiv. w.r.t. Co)
Et₂AlCl (4 equiv. w.r.t. Co)
PhH, r.t.

28a R = H
28b R = Me
28c R = TMS
28d R = Ph

29a R = H   78%
29b R = Me   69%
29c R = TMS   63%
29d R = Ph   70%

Scheme 18.

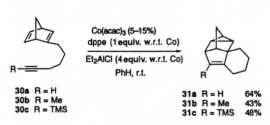

Co(acac)₃ (5–15%)
dppe (1 equiv. w.r.t. Co)
Et₂AlCl (4 equiv. w.r.t. Co)
PhH, r.t.

30a R = H
30b R = Me
30c R = TMS

31a R = H   64%
31b R = Me   43%
31c R = TMS   48%

Scheme 19.

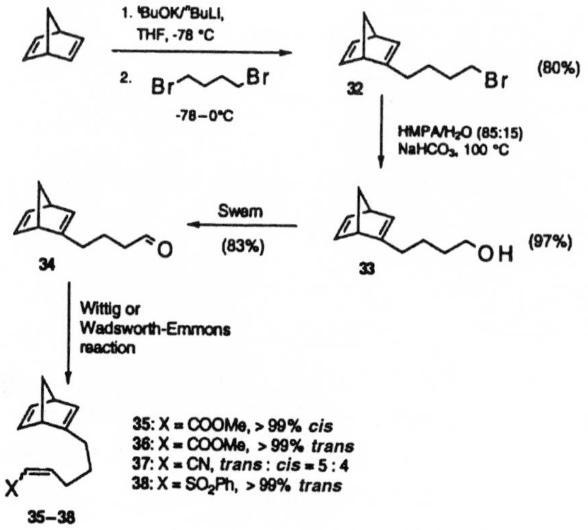

Scheme 20.

are obtained with the dienynes with a 3-carbon tether **28**, compared to the corresponding 4-carbon tether **30**. Attempted cycloadditions for the corresponding 2-carbon and 5-carbon tether dienynes were unsuccessful. Although Co(acac)$_2$ instead of Co(acac)$_3$ under the same conditions gives slightly lower yields, an increase in reaction temperature to 80 °C does not improve yields.

Similar studies were attempted on the intramolecular Ni-catalyzed HDA reaction.[56] Trienes **35–38** were prepared from NBD using a route similar to that used to prepare **28a–d** and **30a–c** (Scheme 20). Surprisingly, none of these trienes give any cycloadduct under the nickel-catalyzed conditions. Varying the temperature (from r.t. to 110 °C), using other phosphine ligands (e.g., PPh$_3$, P(OPh)$_3$, P(OiPr)$_3$, dppe, etc.), varying the amount of catalysts, changing the solvent system, etc. does not afford any cycloadduct and only starting materials were recovered in all cases.

III. NICKEL-CATALYZED [2π + 2π] CYCLOADDITION REACTIONS WITH NORBORNADIENES

During the studies of the nickel-catalyzed HDA reactions between substituted norbornadienes with various dienophiles, we noticed that some dienes undergo a [2 + 2] cycloaddition instead of a [2 + 2 + 2] cycloaddition.[40] As shown earlier (Section IID, Eq. 14), 2-TMS-substituted NBD reacts with acrylonitrile in the presence of Ni(COD)$_2$/2PPh$_3$, to afford a mixture of [2 + 2 + 2] and [2 + 2] adducts. In some cases, the

[2 + 2] cycloaddition occurs exclusively with high chemo- and regiose-lectivity and moderate levels of stereoselectivity. The [2 + 2] cycloaddi-tion between an unsymmetrical dienophile and a monosubstituted NBD produces up to sixteen products (Scheme 21). Two regioisomers are possible for the reaction of either olefin of the NBD. Four stereoisomers could result for each regioisomer by *exo* or *endo* attack, with *cis-syn-cis* or *cis-anti-cis* stereochemistry (Fig. 18) about the cyclobutane.

We have investigated the substituent effects of the olefin and enophile that lead to cyclobutane formation.[40] Reaction occurs exclusively with the unsubstituted double bond of the NBD irrespective of the electronic nature of the substituent (Y) giving type I [2 + 2] adducts (Scheme 21). In addition, only two of the remaining isomers are observed in most cases. We find that cyclobutane formation is particularly facile when unreactive dienes are reacted with very reactive alkenes. We also ob-served a clear trend between the *exo* versus *endo* mode of reaction and the electron density on the remote, unreacting alkene of the NBD. The specific examples are outlined following.

Cyclobutane formation was first observed in our studies of the cy-cloadditions between an electron-deficient NBD **19** (Y = COOMe) and dimethyl meleate.[19(e)] A small amount of [2 + 2] cycloadduct accompa-nies the expected [2 + 2 + 2] adduct (ratio 1:4.5). When the more reactive olefin N-phenylmaleimide (NPM) is used, the major product isolated is cyclobutane **exo-39** (34%) accompanied by *endo*-cyclobutane **endo-39** (17%) and two [2 + 2 + 2] adducts **40** (14%, *exo/endo* = 2.5:1) (Scheme 22). With electron-deficient 2,3-disubstituted NBD **41**, cycloaddition with NPM affords *exo*-cyclobutane **42** (Eq. 17). No other [2 + 2] or [2 +

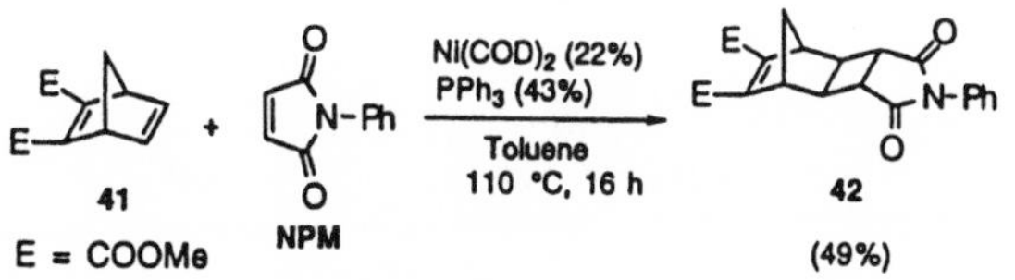

$$(17)$$

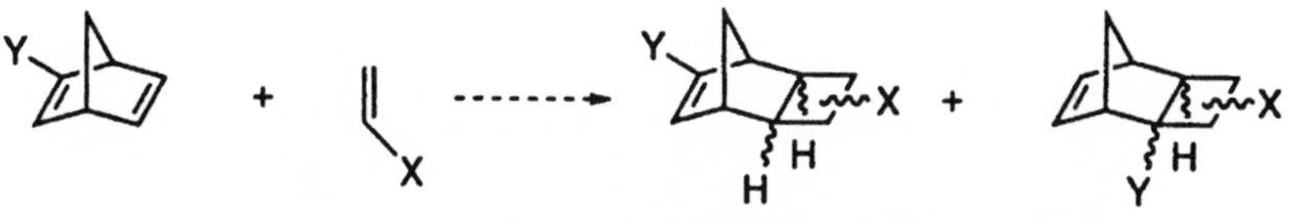

Type I   **Type II**

Scheme 21.

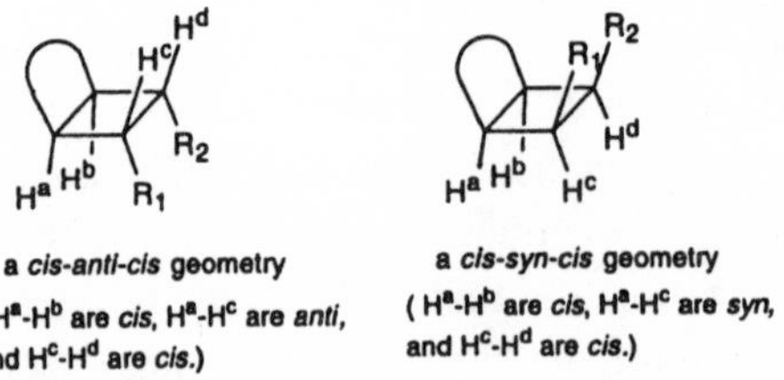

a *cis-anti-cis* geometry
(H^a-H^b are *cis*, H^a-H^c are *anti*,
and H^c-H^d are *cis*.)

a *cis-syn-cis* geometry
(H^a-H^b are *cis*, H^a-H^c are *syn*,
and H^c-H^d are *cis*.)

Figure 18. Geometries of cyclobutane.

2 + 2] cycloadducts were detected. The structure of **42** was confirmed by X-ray crystallography (Fig. 19).

Cycloaddition between the electron-rich, 2-substituted NBD **20** and acrylonitrile (AN) or N-phenylmaleimide (NPM) exclusively forms [2 + 2] adducts (Scheme 23). Unlike the cycloadditions with electron-deficient norbornadienes (**19** and **41**), which give the major adduct with an *exo-(cis-anti-cis)*-geometry (*exo*-**39** and **42**), cycloaddition with electron-deficient NBD **20** affords the *endo-(cis-anti-cis)*-adducts (**43** and **44**) as the major product. As shown in Scheme 23, improvement in the *endo* stereoselectivity to 12 : 1 is achieved by changing the phosphine from PPh$_3$ to P(OPh)$_3$, but unfortunately, this increase occurs at the expense of the yield.

The structure of **43** was confirmed by X-ray crystallography of the corresponding ketone **45** (Eq. 18, Fig. 20), and the structure of **44** was

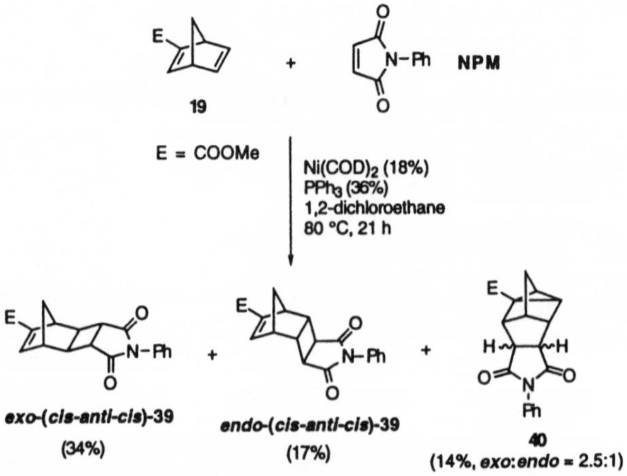

Scheme 22.

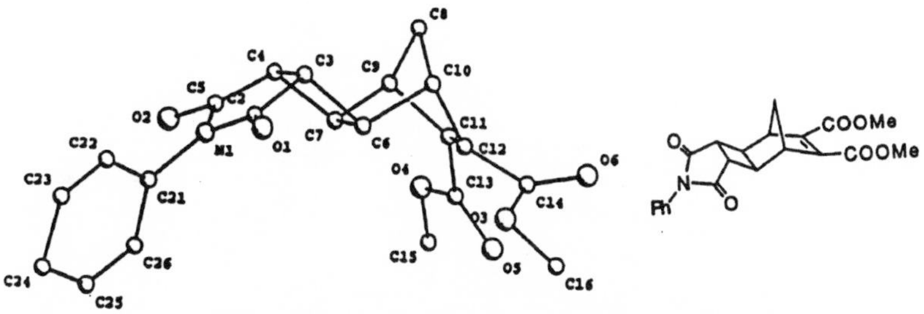

Figure 19. X-ray crystal structure **42.**

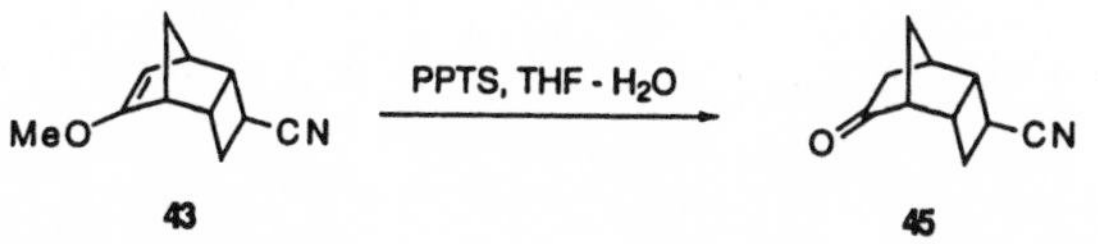

(18)

confirmed by using ^{1}H NMR techniques (NOE and decouplings experiments) on the corresponding tricyclic diester **46** (Eq. 19).

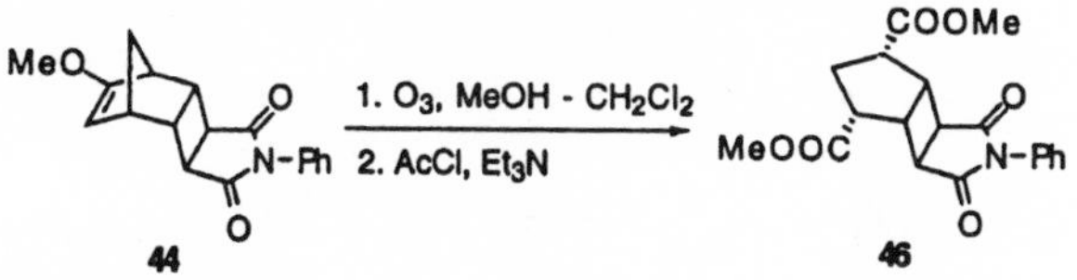

(19)

2-Silyl-substituted NBD **22** also undergoes a [2 + 2] cycloaddition with NPM, giving cyclobutanes **47** and **48** (Eq. 20). This reaction is more complicated because the stereochemistry varies as a function of the

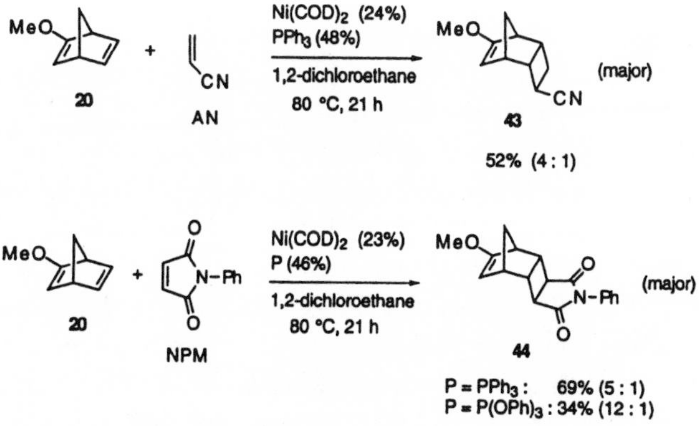

Scheme 23.

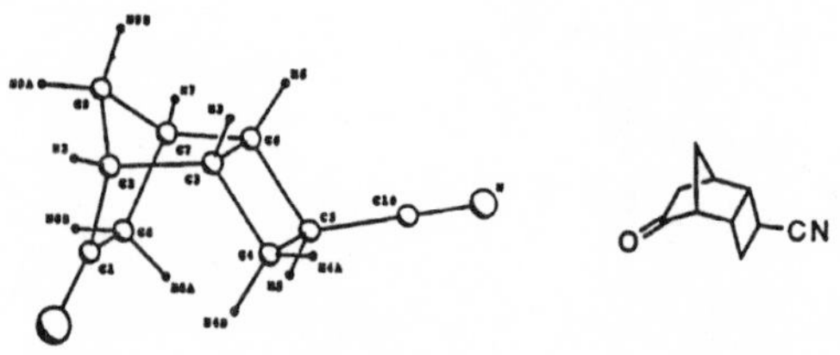

48

Figure 20. X-ray crystal structure of ketonitrile **45**.

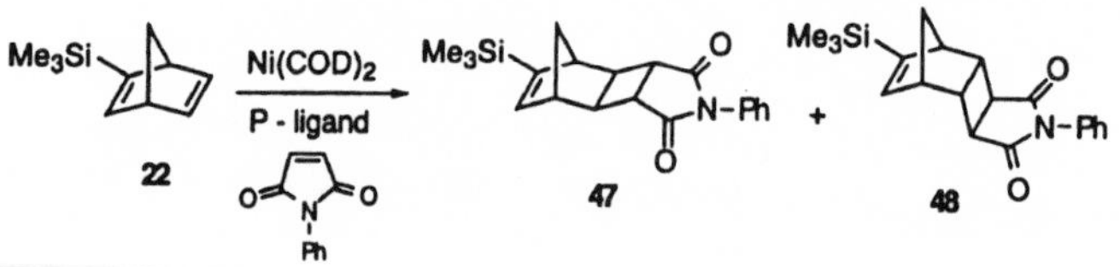

$$(20)$$

Table 12. Effect of Temperature and Phosphines on [2 + 2] Cycloaddition with NBD **22**

Entry	P-Ligand	Temperature	Yield (%)	**47** (Exo) : **48** (Endo)
1	2 PPh$_3$	110 °C	69^a	1 : 4.7
2	2 P(OPh)$_3$	110 °C	71	1 : 3.7
3	2 P(OPh)$_3$	80 °C	42	1.8 : 1
4	1 P(OPh)$_3$	80 °C	62	1 : 1

Note: ^{a}An additional 7–10% of [2 + 2 + 2] HDA adducts complicated the mixture with this ligand.

temperature and ligand (Table 12). At higher temperatures (110 °C) the *exo* adduct predominates regardless of the phosphine ligand used, but the chemoselectivity for the [2 + 2] vs. [2 + 2 + 2] reaction is increased by using P(OPh)$_3$ instead of PPh$_3$. Lower temperatures lead to reduced selectivity with a slight preference for the *endo* adduct.

Thus, although as many as sixteen possible [2 + 2] cycloadducts could be formed in these cycloadditions, only two of these isomers are observed in most cases. In all of the [2 + 2] adducts isolated, the reaction occurs on the less substituted olefin of the norbornadienes, and the stereochemistry of the cyclobutanes formed favors the less hindered *cis-anti-cis* geometry. Electron-deficient norbornadienes (**19** and **41**) favors *exo-(cis-anti-cis)*-adducts (*exo-39* and **42**), whereas the *endo-(cis-anti-cis)*-adducts (**43** and **44**) are preferred with electron-deficient NBD **20**.

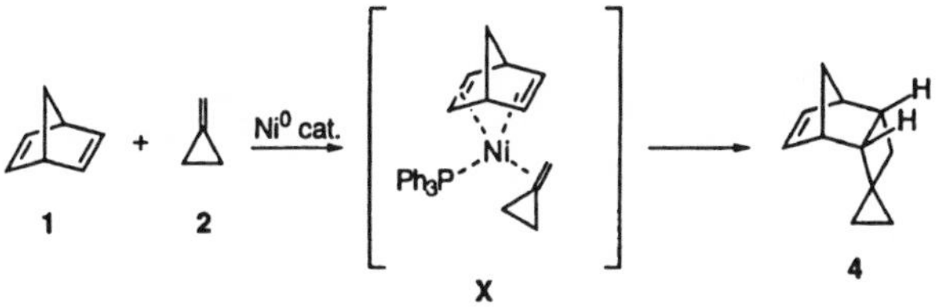

Scheme 24.

2-Silyl-substituted NBD **22** affords both *exo-* and *endo-(cis-anti-cis)*-adducts, and their ratios vary with conditions.

As described in the introduction, Noyori reported a nickel catalyzed [2 + 2] reaction for NBD **1** with methylenecyclopropane **2** (Scheme 24).[7(c)] This reaction is highly selective for addition to the *endo* face in the presence of Ni(COD)$_2$ and 1 equiv. of PPh$_3$ wrt Ni to give **4**. The addition of the phosphine ligand suppresses the formation of other dimers and codimers and increases the *endo* : *exo* selectivity.

The formation of the *endo* adduct suggests that complex **X** with **1** as an *endo* bidentate ligand is an intermediate. The order of coordination strength for each component, PPh$_3$ > > **1** (as bidentate) > **2** > **1** (as monodentate),[36] was used to rationalize the clean cross-coupling reaction in the presence of PPh$_3$.

The [2 + 2] reaction also occurs for stained olefins **49** and **50**, giving only the *exo* cycloadducts (Scheme 25).[9] It was thought that the reaction proceeds via the *exo* metallacyclopentane intermediate **Y** arising from monodentate coordination to the *exo* face of the bicyclic olefins. The similarity in the electronic spectra of nickel complexes of these olefins to those of the complex of norbornene, which was known to coordinate on the *exo* face,[36] supports this mechanism.

It is possible to rationalize the formation of the *endo* and *exo* [2 + 2] adducts of 2-substituted norbornadienes by the pathways shown in

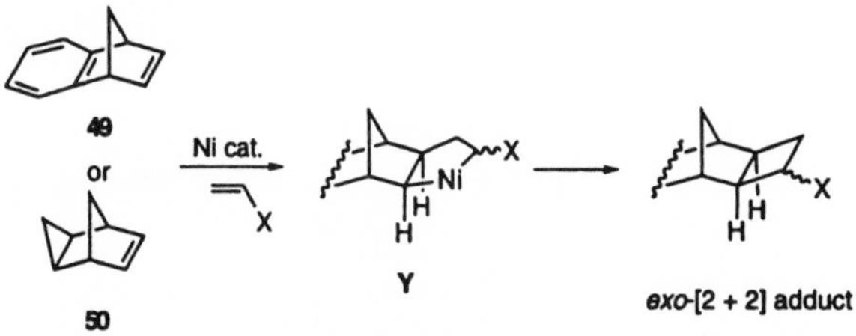

Scheme 25.

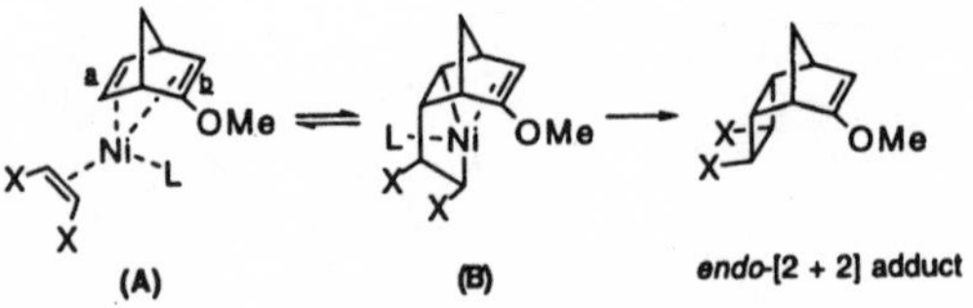

Scheme 26. Proposed mechanism for forming *endo* [2 + 2] adducts from NBD **20**.

Schemes 26 and 27. It is well known that Ni^0 complexes coordinate better with an electron-deficient olefin than with an electron-rich olefin.[36] This is caused by the more favorable overlap with the metal orbitals and the increase in the back donation from the metal to the olefin which is accompanied by partial oxidation of the metal.

Olefin $\underline{a}$ in complex (**A**) as shown in Scheme 26 is more electron-deficient than olefin $\underline{b}$. Thus, Ni coordinates more tightly with olefin $\underline{a}$ than olefin $\underline{b}$. Donation of electron density from olefin $\underline{a}$ to the electron-deficient dienophile leads to the formation of the *endo* metallacyclopentane (**B**). Reductive elimination of Ni leads to the observed *endo* [2 + 2] adduct.

Olefin $\underline{a}$ in complexes (**C**) and (**D**) is more electron-deficient than olefin $\underline{b}$, and therefore Ni might be expected to form a stronger π-complex. However, steric factors must also play an important role with a tetrasubstituted olefin such as $\underline{a}$. Our results suggest that olefin $\underline{b}$ coordinates to the Ni from the less hindered *exo* face, giving complex (**E**). Donation of electron density from olefin $\underline{b}$ to the dienophile can occur,

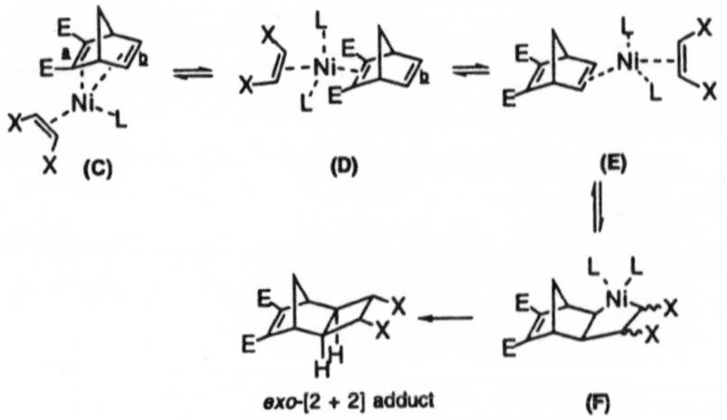

Scheme 27. Proposed mechanism for forming *exo* [2 + 2] adducts from NBD **41**.

affording *exo* metallacyclopentane (**F**), which on reductive elimination, leads to the observed *exo* [2 + 2] cycloadduct.

IV. [2 + 2] VERSUS [2 + 2 + 2] CYCLOADDITIONS— A MECHANISTIC CONSIDERATION

Although our studies have not provided a full explanation for the regioselectivity of the substituted dienes, several key observations have been made. The selectivity depends on the substituent on the diene, the dienophile, and the ligand used in the cycloaddition. As described earlier, the reactivity of 2-substituted norbornadienes in the HDA reaction is much lower than NBD itself. Thus, when these 2-substituted norbornadienes react with reactive dienophiles, an alternative [2 + 2] cycloaddition pathway may predominate.

The parent NBD **1** (Y = H) undergoes [2 + 2 + 2] cycloadditions with various dienophiles, as described in Section IIB. These cycloadditions may occur via pathway **I** → **II** → **IV** → deltacyclane, as proposed by Noyori (Scheme 28). An alternative pathway **I** → **III** → **IV** → deltacyclane is also possible. Evidence of the intermediate **III** is supported by the formation of the *endo*-[2 + 2] cycloadduct with methylenecyclopropane, as described in Scheme 24.

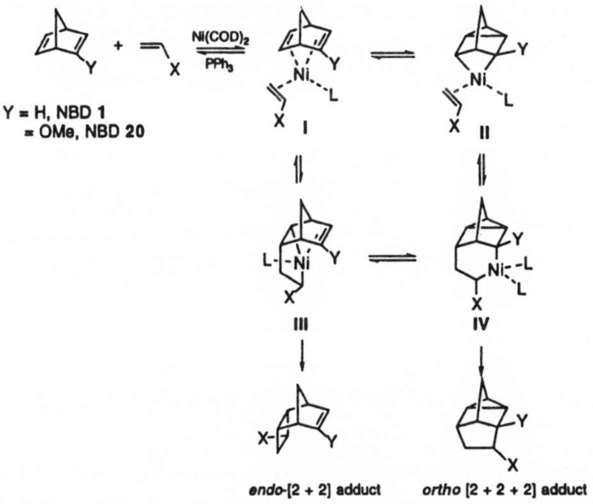

Scheme 28. Partitioning between [2 +2] and [2 + 2 + 2] cycloadditions.

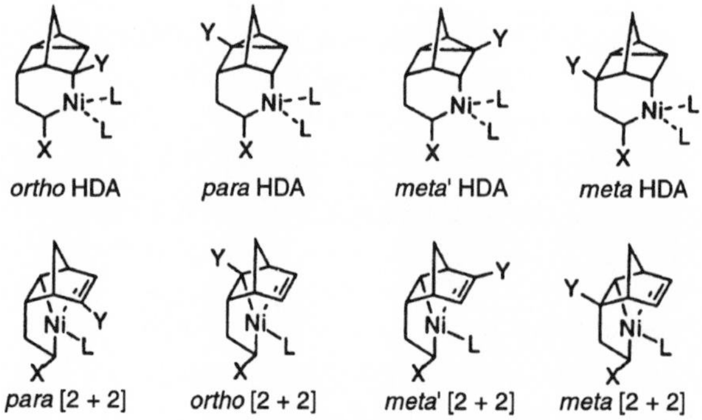

| *ortho* HDA | *para* HDA | *meta'* HDA | *meta* HDA |

| *para* [2 + 2] | *ortho* [2 + 2] | *meta'* [2 + 2] | *meta* [2 + 2] |

Figure 21.

NBD **20** (Y = OMe) gives an *ortho* [2 + 2 + 2] adduct when reacted with MVK (X = COMe) and affords a *para-anti, endo-*[2 + 2] adduct when reacted with AN (X = CN). Both of these adducts can be formed from the same intermediate **III**. This provides further support for the formation of **III** along the reaction pathway.

The other regioisomers are also formed from similar intermediates. The number of possible intermediates increases significantly with the introduction of the substituent on the diene. Four possible [2 + 2 + 2]-metallacycles and four possible [2 + 2]-metallacycles (Fig. 21) could lead to the formation of various regioisomers in the cycloadditions.

The present information is insufficient to provide a detailed mechanism for each of these reactions. The formation of [2 + 2] adducts with substituted dienes adds complexity and suggests that these reactions occur via intermediate **III** instead of intermediate **II**. However, this statement may not be true for all dienes or all dienophiles, and we cannot assume their presence in those cases where the [2 + 2] adducts were not observed.

V. CONCLUSIONS AND REMARKS

We have investigated the scope of the cobalt- and nickel-catalyzed [2 + 2 + 2] (homo Diels–Alder) and [2 + 2] cycloaddition reactions with norbornadienes. Excellent chemo-, regio- and stereoselectivities and high levels of enantioselectivities are achieved.

Many problems may arise in finding trends in reactivity or making generalizations for this reaction because of the multistep reaction mechanism. The factors affecting each step must be considered, and for each substrate, the rate-determining step of the reaction may be different, with different factors controlling these various rates. Nevertheless, these cycloadditions are very powerful and efficient for constructing highly strained polycyclic molecules. Our current investigations focus on the fragmentation of the cycloadducts and the applications of this cycloaddition-fragmentation protocol to synthesizing polycyclic natural products.[63,64]

ACKNOWLEDGMENTS

We express our sincere thanks to Dr. Louise G. Edwards who carried out many of the seminal experiments described in this review. Other important contributions were made by Cathleen M. Crudden, Julia C. Lautens, A. Catherine Smith, Christian Sood, and Marc Johnson.

Mark Lautens acknowledges NSERC(Canada), the Eli Lilly Grantee Program, the Alfred P. Sloan Fellowship, the University of Toronto, and the Bio-Mega Young Investigator Programme for their generous support of our programs. William Tam thanks the University of Toronto for a Simcoe Scholarship. Professor Mark Burk (Duke University) is thanked for generous gifts of various Duphos and BPE ligands, and Dr. Alan Lough (University of Toronto) for the X-ray structure determinations.

REFERENCES AND NOTES

1. Julius Hyman & Co., *Belgian Patent* **1951**, *498*, 176.
2. (a) Goldstein, M. J.; Natowsley, S.; Heilbronner, E.; Hornung, V. *Helv. Chim. Acta* **1973**, *56*, 294. (b) Bischof, P.; Heilbronner, E.; Prinzbach, H.; Martin, H. D. *Helv. Chim. Acta* **1971**, *54*, 1072. (c) Houk, K. N.; Domelsmith, L. N.; Mollerer, P. D.; Hahn, R. C.; Johnson, R. P. *J. Am. Chem. Soc.* **1978**, *100*, 2959.
3. Hall, H. K. Jr.; Smith, C. D.; Baldt, J. H. *J. Am. Chem. Soc.* **1973**, *95*, 3197.
4. For [2 + 1] and [2 + 2 + 1] cycloadditions of NBD with carbenes, see (a) Clark, S. C.; Johnson, B. L. *Tetrahedron* **1968**, *24*, 5067; (b) Jefford, C. W.; Kabengele, N. T.; Kovacs, J.; Burger, U. *Tetrahedron Lett.* **1974**, *15*, 257; (c) Jefford, C. W.; Kabengele, N. T.; Kovacs, J.; Burger, U. *Helv. Chim. Acta* **1974**, *57*, 104. (d) Jefford, C. W.; Mareda, J.; Gahret, J. C. E.; Kabengele, N. T.; Graham, W. D.; Burger, U. *J. Am. Chem. Soc.* **1976**, *98*, 2585; (e) Klumpp, G. W.; Kwantes, P. M. *Tetrahedron Lett.* **1976**, *17*, 707; (f) Dolgii, I. E.; Tomilov, Yu. V.; Shteinshneider, A. Yu.; Nefedov, O. M. *Izv. Akad. Nauk SSSR, Ser. Khim. (Engl. Transl.)* **1983**, *32*, 638.
5. For Lewis acids catalyzed [2 + 2] and [2 + 2 + 2] cycloadditions of NBD, see (a) Snider, B. B.; Rodini, D. J.; Conn, R. S. E.; Sealfon, S. *J. Am. Chem. Soc.* **1979**, *101*, 5283; (b) Hoffmann, H. M. R.; Fienemann, H. *J. Org. Chem.* **1979**, *44*, 2802.
6. For hetero-[2 + 2] and [2 + 2 + 2] cycloadditions of NBD, see (a) Cristol, S. J.; Alfred, E. L.; Wetzel, D. L. *J. Org. Chem.* **1962**, *27*, 4058; (b) Cookson, R. C.;

Gilani, S. S. H.; Stevens, I. D. R. *Tetrahedron Lett.* **1963**, *615*; (c) Moriarty, R. M. *J. Org. Chem.* **1963**, *28*, 2385; (d) Tufariello, J. J.; Mich, T. F.; Miller, P. S. *Tetrahedron Lett.* **1966**, 2293.

7. For metal-catalyzed [2 + 2] cycloadditions of NBD with enophiles, see (a) Schrauzer, G. N.; Glockner, P. *Chem. Ber.* **1964**, *97*, 2451; (b) Schrauzer, G. N. *Adv. Catal.* **1968**, *18*, 373; (c) Noyori, R.; Ishigumi, T.; Hayashi, N.; Takaya, H. *J. Am. Chem. Soc.* **1973**, *95*, 1674; (d) Mitsudo, T. A.; Kokuryo, K.; Takegami, Y. *J. Chem. Soc., Chem. Commun.* **1976**, 722; (e) Dolgii, I. E.; Tomilov, Yu. V.; Tsvetkova, N. M.; Bordakov, V. G.; Nefedov, O. M. *Izv. Akad. Nauk SSSR, Ser. Khim. (Engl. Transl.)* **1983**, *32*, 868; (f) Mitsudo, T.; Naruse, H.; Kondo, T.; Ozaki, Y.; Watanabe, Y. *Angew. Chem., Int. Ed. Engl.* **1994**, *33*, 580. See also refs. 9, 19(b) and 26(a, b).

8. For thermal [2 + 2] cycloadditions of NBD with enophiles, see (a) Heaney, H.; Jablonski, J. M. *Tetrahedron Lett.* **1967**, *8*, 2733; (b) Butler, D. N.; Barrette, A.; Snow, R. A. *Syn. Commun.* **1975**, *5*, 101. See also refs. 22(d, h, k) and 23(e).

9. Noyori, R.; Takaya, H.; Yamakawa, M. *Bull. Chem. Soc. Jpn.* **1982**, *55*, 852.

10. For dimerization and trimerization of NBD, see (a) Vallarino, L. *J. Chem. Soc.* **1957**, 2287; (b) Bennett, M. A.; Wilkinson, G. *J. Chem. Soc.* **1961**, 1418; (c) Chatt, J.; Shaw, B. L. *Chem. Ind. (London)*, 1961, 290; (d) Yoshikawa, S.; Kiji, J. Furukawa, J. *Bull. Chem. Soc. Jpn.* **1976**, *49*, 1093; (e) Rusina, A.; Vlcek, A. A. *Nature (London)*, **1965**, *206*, 295; (f) McCleverty, J. A.; Wilkinson, G. *Inorg. Syn.* **1966**, *8*, 214; (g) Schrauzer, G. N.; Bastian, B. N.; Fosselius, G. A. *J. Am. Chem. Soc.* **1966**, *88*, 4890; (h) Cannell, L. G. *Tetrahedron Lett.* **1966**, *7*, 5967; (i) Mrowca, J. J.; Katz *J. Am. Chem. Soc.* **1966**, *88*, 4102 and the references therein; (j) Evans, D.; Osborn, J. A.; Wilkinson, G. *Inorg. Syn.*, **1968**, *11*, 99; (k) Montelatici, S.; van der Ent; Osborn, J. A.; Wilkinson, G. *J. Chem. Soc. A.* **1968**, 1054; (l) Wittig, G.; Otten, J. *Tetrahedron Lett.* **1963**, 601; (m) Schrauzer, G. N.; Ho, R. K. Y.; Schlesinger, G. *Tetrahedron Lett.* **1970**, 543; (n) Schrock, R. R.; Osborn, J. A. *J. Am. Chem. Soc.* **1971**, *93*, 3089; (o) Acton, N.; Roth, R. J.; Katz, T. J.; Frank, J. K.; Maier, C. A.; Paul, I. C *J. Am. Chem. Soc.* **1972**, *94*, 5446; (p) ref. 11.

11. Weissberger, E.; Mantzaris, J. *J. Am. Chem. Soc.* **1974**, *96*, 1873.

12. For Pauson–Khand type [2 + 2 + 1] cycloadditions of NBD, see (a) Khand, I. U.; Knox, G. R.; Pauson, P. L.; Watts, W. E.; Foreman, M. I. *J. Chem. Soc., Perkin Trans. 1* **1973**, 977; (b) Khand, I. U.; Pauson, P. L. *J. Chem. Soc., Perkin Trans. 1* **1976**, 30; (c) Schore, N. E. *Syn. Commun.* **1979**, *9*, 41; (d) Pauson, P. L. *Tetrahedron* **1985**, *41*, 5885; (e) Hanaoka, M.; Mukai, C.; Uschiyama, M. *J. Chem. Soc., Chem. Commun.* **1992**, 1014.

13. For metal-catalyzed [2 + 3] cycloadditions of NBD with TMM, see (a) Trost, B. M.; Chan, D. M. T. *J. Am. Chem. Soc.* **1983**, *105*, 2315; with methylene cyclopropane, see (b) Binger, P.; Schuchardt, U. *Chem. Ber.* **1980**, *113*, 3334; (c) Binger, P.; Buch, H. M. *Top. Curr. Chem.* **1987**, *135*, 77. See also ref. 7(c).

14. For hetero-[2 + 3] cycloadditions of NBD with azides, see (a) Huisgen, R.; Mobius, L.; Muller, G.; Stangl, H.; Szeimies, G.; Vernon, J. M. *Chem. Ber.* **1965**, *98*, 3992; (b) Klumpp, G. W.; Weefkind, A. H.; Graaf, W. L.; Bickelhaupt, F. *Justus Liebigs Ann. Chem.* **1967**, *706*, 47; (c) McLean, S.; Findlay, D. M. *Tetrahedron Lett.* **1969**, 2219; (d) Halton, B.; Woolhouse, A. D. *Aust. J. Chem.* **1973**, *26*, 619.

15. For hetero-[2 + 3] cycloadditions of NBD with diazomethanes, see (a) Filipescu, N.; DeMember, J. R. *Tetrahedron* **1969**, *24*, 5181; (b) Wilt, J. W.; Malloy, T. P. *J. Org. Chem.* **1973**, *38*, 277; (c) Franck-Neumann, M.; Sedrati, M. *Angew. Chem., Int. Ed. Engl.* **1974**, *13*, 606; (d) Wilt, J. W.; Sullivan, D. R. *J. Org. Chem.* **1975**, *40*, 1036; (e) Wilt, J. W.; Roberts, W. N. *J. Org. Chem.* **1978**, *43*, 170.

16. For hetero-[2 + 3] cycloadditions of NBD with nitrile oxides and other 1,3-dipoles, see (a) Taniguchi, H.; Ikeda, T.; Yoshi, Y.; Imoto, E. *Bull. Chem. Soc. Jpn.* **1977**, *50*, 2694; (b) Cristina, D.; Micheli, C. D.; Gandolfi, R. *J. Chem. Soc., Perkin Trans. 1* **1979**, 2891; (c) Micheli, C. D.; Gandolfi, R.; Oberti, R. *J. Org. Chem.* **1980**, *45*, 1209; (d) Umano, K.; Mizone, S.; Tokisato, K.; Inoue, H. *Tetrahedron Lett.* **1981**, *22*, 73; (e) Motoki, S.; Saito, T. *J. Org. Chem.* **1979**, *44*, 2493.

17. For [2 + 4] Diels–Alder reactions of NBD, see (a) Byrne, L. T.; Rye, A. R.; Wege, D. *Aust J. Chem.* **1974**, *27*, 1961; (b) Mackenzie, K.; Astin, K. B. *J. Chem. Soc., Perkin Trans. 2* **1975**, 1004; (c) Battise, M. A.; Timberlake, J. F.; Malkus, H. *Tetrahedron Lett.* **1976**, *17*, 2529; (d) Mackenzie, K. *Tetrahedron Lett.* **1976**, *17*, 1203; (e) Mazzocchi, P. H.; Stahky, B.; Dodd, J.; Rondan, N. G.; Domelsmith, L. N.; Rozeboom, M. D.; Caramella, P.; Houk, K. N. *J. Am. Chem. Soc.* **1980**, *102*, 6482.

18. For isomerization of norbornadienes to quadricyclanes, see (a) Hammond, G. S.; Turro, N. J. *J. Am. Chem. Soc.* **1961**, *83*, 4674; (b) Dauben, W. G.; Cargill, R. L. *Tetrahedron* **1961**, *15*, 197; (c) Hammond, G. S.; Wyatt, P.; Deboer, C. D.; Turro, N. J. *J. Am. Chem. Soc.* **1964**, *86*, 2532; (d) Frey, H. M. *J. Chem. Soc.* **1964**, 365; (e) Prinzbach, H.; Rivier, J. *Helv. Chim. Acta* **1970**, *53*, 2201.

19. For nickel-catalyzed [2 + 2 + 2] homo Diels–Alder reactions of norbornadienes, see (a) Schrauzer, G. N.; Eichler, S. *Chem. Ber.* **1962**, *95*, 2764; (b) Yoshikawa, S.; Aoki, K.; Kiji, J.; Furukawa, J. *Bull. Chem. Soc. Jpn.* **1975**, *48*, 3239; (c) Noyori, R.; Umeda, I.; Kawauchi, H.; Takaya, H. *J. Am. Chem. Soc.* **1975**, *97*, 812; (d) Lautens, M.; Edwards, L. G. *Tetrahedron Lett.* **1989**, *30*, 6813; (e) Lautens, M.; Edwards, L. G. *J. Org. Chem.* **1991**, *56*, 3762; (f) Lautens, M.; Tam, W.; Edwards, L. G. *J. Chem. Soc., Perkin Trans. 1* **1994**, 2143; (g) Lautens, M.; Edwards, L. G.; Tam, W.; Lough, A. J. *J. Am. Chem. Soc.* **1995**, *117*, 10276. See also refs. 7(a,c) and 21(b).

20. For cobalt-catalyzed [2 + 2 + 2] homo Diels–Alder reactions of norbornadienes, see (a) Lyons, J. E.; Myers. H. K.; Schneider, A. *J. Chem. Soc., Chem. Commun.* **1978**, *636*, 638; (b) Lyons, J. E.; Myers. H. K.; Schneider, A. *Ann. N.Y. Acad. Sci.* **1980**, *333*, 273; (c) Lautens, M.; Crudden, C. M. *Organometallics* **1989**, *8*, 2733; (d) Lautens, M.; Crudden, C. M. *Tetrahedron Lett.* **1989**, *30*, 4803; (e) Duan, I. F.; Cheng, C. H.; Shaw, J. S.; Cheng, S. S.; Liou, K. F. *J. Chem. Soc., Chem. Commun.* **1991**, 1347; (f) Lautens, M.; Tam, W.; Edwards, L. G. *J. Org. Chem.* **1992**, *57*, 8. See also refs. 7(a–c), 21(a–d), 26(a),(b), and 27.

21. For asymmetric [2 + 2 + 2] homo Diels–Alder reactions, see (a) Lautens, M.; Lautens, J. C.; Smith, A. C. *J. Am. Chem. Soc.* **1990**, *112*, 5627; (b) Brunner, H.; Muschiol, M.; Prester, F. *Angew. Chem., Int. Ed. Engl.* **1990**, *29*, 652; (c) Brunner, H.; Prester, F. *J. Organomet. Chem.* **1991**, *414*, 401; (d) Buono, G.; Pardigon, O. *Tetrahedron: Asymmetry* **1993**, *4*, 1977; (e) Lautens, M.; Tam, W.; Lautens, J. C.; Edwards, L. G.; Crudden, C. M.; Smith, A. C. *J. Am. Chem. Soc.* **1995**, *117*, 6863.

22. For thermal [2 + 2 + 2] homo Diels–Alder reactions of norbornadienes with electron-deficient olefins, see (a) Ullman, E. F. *Chem. Ind.* **1958**, 1173; (b) Blomquist, A. T.; Meinwald, Y. C. *J. Am. Chem. Soc.* **1959**, *81*, 667; (c) Hall, H. K., Jr. *J. Org. Chem.* **1960**, *25*, 42; (d) Tabushi, I.; Yamamura, K.; Yoshida, Z.; Togashi, A. *Bull. Chem. Soc. Jpn.* **1975**, *48*, 2922; (e) Cookson, R. C.; Crundwell, E.; Hill, R. R.; Hudec, J. *J. Chem. Soc.* **1962**, 3062; (f) Cookson, R. C.; Dance, J.; Hudec, J. *J. Chem. Soc.* **1964**, 5416; (g) Furukawa, J.; Kobuke, Y.; Sugimoto, T.; Fueno, T. *J. Am. Chem. Soc.* **1972**, *94*, 3633; (h) Nickon, A.; Kwasnik, H. R.; Mothew, C. T.; Swartz, T. D.; Williams, R. O.; DiGiorgio, J. B. *J. Org. Chem.* **1978**, *43*, 3904; (i) Fickes, G. N.; Metz, T. E. *J. Org. Chem.* **1978**, *43*, 4057; (j) Jenner, G.; Papadopoulos, M. *Tetrahedron Lett.* **1982**, *23*, 4333; (k) Lucchi, O. D.; Licini, G.; Pasquato, L.; Senta, M. *Tetrahedron Lett.* **1988**, *29*, 831.

23. For thermal [2 + 2 + 2] homo Diels–Alder reactions of norbornadienes with activated acetylenes, see (a) Krespan, C. G.; McKusick, B. C.; Cairns, T. L. *J. Am. Chem. Soc.* **1961**, *83*, 3428; (b) Cookson, R. C.; Dance, J. *Tetrahedron Lett.* **1962**, 879; (c) Huebner, C. F.; Donoghue, E.; Dorfman, L.; Stuber, F. A.; Danieli, N.; Wenkert, E. *Tetrahedron Lett.* **1966**, 1185; (d) Solo, A. J.; Singh, B.; Kapoor, J. N. *Tetrahedron* **1969**, *25*, 4579; (e) Sasaki, T.; Eguchi, S.; Sugimoto, M.; Hibi, F. *J. Org. Chem.* **1972**, *37*, 2317; (f) Martin, H. D.; Forster, D. *Angew. Chem., Int. Ed. Engl.* **1972**, *11*, 54; (g) Jenner, G. *Tetrahedron Lett.* **1987**, *28*, 3927; (h) Schleyer, P. R.; Leone, R. E. *J. Am. Chem. Soc.* **1968**, *90*, 4164. See also refs. 8(a) and 22(j).

24. For [2 + 2 + 2] cycloadditions of dienes other than norbornadienes, see (a) Zimmerman, H. E.; Grunewald, G. L. *J. Am. Chem. Soc.* **1964**, *86*, 1434; (b) Grant, F. W.; Gleason, R. W.; Bushweller, C. H. *J. Org. Chem.* **1965**, *30*, 290; (c) Berson, J. A.; Olin, S. S. *J. Am. Chem. Soc.* **1969**, *91*, 777; (d) Spielmann, W.; Fick, H. H.; Meyer, L. U.; Meijere, A. *Tetrahedron Lett.* **1976**, *17*, 4057; (e) ref. 22(i); (f) Kaufmann, D.; Fick, H. H.; Schallner, O.; Spielmann, W.; Meyer, L. U.; Golitz, P.; Meijere, A. *Chem. Ber.* **1983**, *116*, 587; (g) Yamaguchi, R.; Ban, M.; Kawanisi M.; Osawa, E.; Jaime, C.; Buda, A. B. *J. Am. Chem. Soc.* **1984**, *106*, 1512; (h) Trost, B. M.; Imi, K.; Indolese, A. F. *J. Am. Chem. Soc.* **1993**, *115*, 8831.

25. For [σ² + σ² + π²] cycloadditions of quadricyclanes, see (a) Smith, C. D. *J. Am Chem. Soc.* **1966**, *88*, 4273; (b) Prinzbach, H. *Angew. Chem., Int. Ed. Engl.* **1967** *6*, 1068; (c) Kaupp, G. *Chem. Ber.* **1971**, *104*, 182; (d) Tabushi, I.; Yamamura, K. Yoshida, Z. *J. Am. Chem. Soc.* **1972**, *94*, 787. See also refs. 19(b), 22(k), and 23(g).

26. For [2 + 2 + 4] cycloadditions with NBD, see (a) Greco, A.; Carbonaro, A. Dall'Asta, G. *J. Org. Chem.* **1970**, *35*, 271; (b) Carbonaro, A.; Cambisi, F. Dall'Asta, G. *J. Org. Chem.* **1971**, *36*, 1443; (c) ref. 20(b); (d) Lautens, M.; Tam W.; Sood, C. *J. Org. Chem.* **1993**, *58*, 4513; (e) Inukai, T.; Takahashi, A. *J. Chem Soc., Chem. Commun.* **1970**, 1473.

27. Turecek, F.; Mach, K.; Antropiusova, H.; Petrusova, L.; Hanus, V.; Sedmera, P *Tetrahedron* **1984**, *40*, 3295.

28. (a) Sauer, J. *Angew. Chem., Int. Ed. Engl.* **1966**, *5*, 211; (b) Ciganik, E. *Org. React* **1984**, *32*, 1; (c) Fallis, A. G. *Can. J. Chem.* **1984**, *62*, 183.

29. (a) DeLucchi, O.; Modena, G. *Tetrahedron* **1984**, *40*, 2585; (b) Nakayama, J. Nakamura, Y.; Hoshino, M. *Heterocycles* **1985**, *23*, 1119; (c) DeLucchi, O. Lucchini, V.; Pasquato, L.; Modena, G. *J. Org. Chem.* **1984**, *49*, 596; (d) Carr, R C. V.; Paquette, L. A. *J. Am. Chem. Soc.* **1980**, *102*, 853; (e) Davis, A. P.; Whitham

G. H. *J. Chem. Soc., Chem. Commun.* **1980**, 639; (f) Paquette, L. A.; Moerck, R. E.; Harirchian, B.; Magnus, P. D. *J. Am. Chem. Soc.* **1978**, *100*, 1597; (g) Westberg, H. H.; Dauben, H. J. *Tetrahedron Lett.* **1968**, 5123.

30. For intramolecular reactions, see (a) Wender, P. A.; Jenkins, T. E. *J. Am. Chem. Soc.* **1989**, *111*, 6432; (b) Jolly, R. S.; Luedtke, G.; Sheehan, D.; Livinghouse, T. *J. Am. Chem. Soc.* **1990**, *112*, 4965 and references therein for earlier reports of intermolecular metal-catalyzed Diels–Alder reactions.

31. For preparation of Ni(COD)$_2$, see (a) Schunn, R. A.; Ittel, S. D.; Cushing, M. A. *Inorg. Synth.* **1990**, *28*, 94; (b) Schunn, R. A. *Inorg. Synth.* **1974**, *15*, 5.

32. (a) Vollhardt, K. P. C. *Angew. Chem., Int. Ed. Engl.* **1984**, *8*, 2733; (b) Hillard, R. L. III.; Parnell, C. A.; Vollhardt, K.P.C. *Tetrahedron* **1983**, *39*, 905 and references cited therein.

33. Wilkinson, G.; Stone, F. G. A.; Abel, E. W., *Comprehensive Organometallic Chemistry* (Kemmitt, R. D. W. and Russell, D. R., eds.), Pergamon Press, 1982, Vol. 5, Chapter 34.

34. (a) Swaminathan, S.; Lessinger, L. *Cryst. Struct. Commun.*, **1978**, *7*, 621; (b) Stephens, F. S. *J. Chem. Soc., Dalton Trans.*, **1972**, 1754; (c) Boer, F. P.; Flynn, J. J. *J. Am. Chem. Soc.* **1971**, *93*, 6495; (d) Langenbach, H. J.; Keller, E.; Vaherenkamp, H. *J. Organomet. Chem.* **1979**, *171*, 259; (e) Ng, Y. S.; Penfold, B. R. *Acta Cryst., Sect. B*, **1978**, *34*, 1978.

35. Wilkinson, G.; Stone, F. G. A.; Abel, E. W., *Comprehensive Organometallic Chemistry* (Jolly, P. W., ed.), Pergamon Press, 1982, Vol. 6, Chapter 37.

36. Jolly, P. W.; Wilke, G., *The Organic Chemistry of Nickel*, Academic Press, 1974, Vol. I.

37. McAuliffe, C. A.; Levason, W. *Phosphine, Arsine and Stibine Complexes of the Transition Elements*, Elsevier Scientific, 1979.

38. For studies and uses of bidentate ligands, see (a) Casey, C. P.; Whiteker, G. T.; Melville, M. G.; Petrovich, L. M.; Gavney, J. A.; Powell, D. R. *J. Am. Chem. Soc.* **1992**, *114*, 5535; (b) Burk, M. J.; Feaster, Harlow, R. L. *Organometallics* **1990**, *9*, 2653. (c) Burk, M. J. *J. Am. Chem. Soc.* **1991**, *113*, 8518; (d) Burk, M. J.; Feaster, J. E. *J. Am. Chem. Soc.* **1992**, *114*, 6266; (e) Auburn, P. R.; Mackenzie, P. B.; Bosnich, B. *J. Am. Chem. Soc.* **1985**, *107*, 2033; (f) Fryzuk, M. D.; Bosnich, B. *J. Am. Chem. Soc.* **1981**, *103*, 6262; (g) MacNeil, P. A.; Roberts, N. L.; Bosnich, B. *J. Am. Chem. Soc.* **1981**, *103*, 2273; (h) Tani, K.; Yamagata, T.; Tatsuno, Y.; Yamagata, Y.; Tomita, K. I.; Autagawa, S.; Kumobayashi, H.; Otsuka, S. *Angew. Chem., Int. Ed. Engl.* **1985**, *24*, 217; (i) *Aldrichmica Acta* **1983**, *16*(4).

39. (a) Halpern, J.; Cassar, L. *J. Chem. Soc., Chem. Commun.* **1970**, 1082; (b) Fraser, A. R.; Bird, P. H.; Beyman, S. A.; Shapley, J. R.; White, R.; Osborn, J. A. *J. Am. Chem. Soc.* **1973**, *95*, 597; (c) Halpern, J. *Org. Syn. Metal Carbonyls* **1977**, *2*, 705; (d) Evans, L.; Kemmit, R.; Kimura, B.; Russel, D. *J. Chem. Soc., Chem. Commun.* **1972**, 509.

40. Louise G. Edwards, Ph.D. Thesis, **1992**, University of Toronto.

41. March, J. *Advanced Organic Chemistry*, 4th ed., Wiley Interscience, New York, 1992.

42. (a) Tolman, C. A. *J. Am. Chem. Soc.* **1970**, *92*, 2953, 1956; (b) Tolman, C. A.; Seidel, W. C.; Gosser, L.W. *J. Am. Chem. Soc.* **1974**, *96*, 53.

43. (a) Morrison, J. D., *Asymmetric Synthesis* Academic Press, 1985, Vols. 1–5. (b) Evans, D. A. *Top. Stereochem.* **1982**, *13*, 1.

44. For two recent books on this subject, see (a) Noyori, R., *Asymmetric Catalysis in Organic Synthesis*, Wiley Interscience: New York 1994; (b) Ojima, I. *Catalytic Asymmetric Synthesis*, VCH: New York, 1993. Other reviews include (c) Ojima, I.; Clos, N.; Bastos, C. *Tetrahedron* **1989**, *45*, 6901; (d) Bosnich, B., Ed. *Asymmetric Catalysis*, NATO ASI Series E, Applied Sciences No. 103; Martinus Nijhoff Publishers: Dordrecht, 1986; (e) Koskinen, A., *Asymmetric Synthesis of Natural Products*, John Wiley & Sons: New York, 1993.

45. Reactions where catalytic asymmetric induction has been achieved include the Sharpless epoxidation: (a) Morrison, J. D., *Asymmetric Synthesis* (Finn, M. G. and Sharpless, K. B., eds.), Academic Press: New York, 1985, Vol. 5, Chap. 8. (b) Morrison, J. D., *Asymmetric Synthesis* (Rossiter, B. E., ed.) Academic Press: New York, 1985 Vol. 5, Chap. 7; the Jacobsen epoxidation: (c) Zhang, W.; Loebach, J. L.; Wilson, S. R.; Jacobsen E. N. *J. Am. Chem. Soc.* **1990**, *112*, 2801; (d) Jacobsen E. N.; Zhang, W.; Muci, A. R.; Ecker, J. R.; Deng, L. *J. Am. Chem. Soc.* **1991**, *113*, 7063; the asymmetric hydrogenation; (e) Knowles, W. S. *Acc. Chem. Res.* **1983**, *16*, 106; (f) Halterman, R. L.; Vollhardt, K. P. C.; Welker, M.E.; Blaser, D.; Boese, R. *J. Am. Chem. Soc.* **1987**, *109*, 8105; (g) Conticello, V. P.; Brard, L.; Giardello, M. A.; Tsuji, Y.; Sabat, M.; Stern, C.; Marks, T. J. *J. Am. Chem. Soc.* **1992**, *114*, 2761; (h) Pfaltz, A. *Acc. Chem. Res.* **1993**, *26*, 339; (i) Pfaltz, A. *Helv. Chim. Acta* **1991**, *74*, 232; (j) Evans, D. A.; Nelson, S. G.; Gagne, M. R.; Muci, A. R. *J. Am. Chem. Soc.* **1993**, *115*, 9800; asymmetric isomerizations: (k) Noyori, R.; Otsuka, S. *J. Am. Chem. Soc.* **1984**, *106*, 5208; (l) Chen, Z.; Halterman, R. L. *J. Am. Chem. Soc.* **1992**, *114*, 2276 and the dihydroxylation of olefins; (m) Jacobsen, E. N.; Marko, I.; Mungall, W. S.; Schroder, G.; Sharpless, K. B. *J. Am. Chem. Soc.* **1988**, *110*, 1968.

46. Some representative examples in catalytic asymmetric cyclopropanation include (a) Aratani, T. *Pure Appl. Chem.* **1985**, *57*, 1839; (b) Pfaltz, A. *Acc. Chem. Res.* **1993**, *26*, 339; (c) Nakamura, A.; Konishi, A.; Tsujitani, R.; Kudo, M.; Otsuka, S. *J. Am. Chem. Soc.* **1978**, *100*, 3449; (d) Evans, D. A.; Woerpel, K. A.; Hinman, M. M. *J. Am. Chem. Soc.* **1991**, *113*, 726; (e) Doyle, M. P.; Pieters, R. J.; Martin, S. F.; Austin, R. E.; Oalmann, C. J.; Muller, P. *J. Am. Chem. Soc.* **1991**, *113*, 1423; (f) Nishiyama, N. *J. Am. Chem. Soc.* **1994**, *116*, 2223.

47. For leading references in catalytic asymmetric Diels–Alder reactions, see (a) Evans, D. A.; Miller, S. J.; Lectka T. *J. Am. Chem. Soc.* **1993**, *115*, 6460; (b) Corey, E. J.; Loh, T. P. *Tetrahedron Lett.* **1993**, *34*, 3979; (c) Narasaka, K.; Iwasawa, N.; Inoue, M.; Yamada, T.; Nakashima, M.; Sugimori, J. *J. Am. Chem. Soc.* **1989**, *111*, 5340; (d) Narasaka, K. *Pure Appl. Chem.* **1992**, *64*, 1889; (e) Yamamoto, H.; Ishihara, K. *J. Am. Chem. Soc.* **1994**, *116*, 1561.

48. Some of the data taken from ref. 21(a), and some data from unpublished results of Marc Johnson and William Tam.

49. From a detailed study of phosphine ligands carried out previously in our group, it is clear that the maximum number of carbons from P to P must not exceed 4 (unpublished work of Crudden, C. M.)

50. For the use of the Mosher ester, see (a) Dale, J. E.; Mosher, H. S. *J. Org. Chem* **1969**, *34*, 2543; (b) Dale, J. E.; Mosher, H. S. *J. Am. Chem. Soc.* **1973**, *95*, 512.

51. For the use of the O-methyl mandelate ester, see (a) Trost, B. M.; Belletire, J. L.; Godleski, S.; McDougal, P. G.; Balkovec, J. M.; Baldwin, J. J.; Christy, M. E.; Ponticello, G. S.; Varga, S. L.; Springer, J. P. *J. Org. Chem.* **1986**, *51*, 2370; (b) Roy, B.; Deslongchamps, P. *Can. J. Chem.* **1985**, *63*, 651.

52. Lautens, J. C.; Lautens, M.; Lough, A. J. *Acta Cryst.* **1991**, *C47*, 2725. This is the only report to confirm the absolute stereochemistry by X-ray crystallography of a deltacyclene from the enantioselective HDA reaction using a chiral phosphine.

53. (a) Stalick, J. K.; Corfield, P. W. R.; Meek, D. W. *Inorg. Chem.* **1973**, *12*, 1668; (b) Toriumi, K.; Ito, T.; Takaya, H.; Souchi, T.; Noyori, R. *Acta Cryst.* **1982**, *B38*, 807; (c) Balavoine, G.; Brunie, S.; Kagan, H. B. *J. Organomet. Chem.* **1980**, *187*, 125; (d) Brunner, H. et al. *Inorg. Chim. Acta* **1984**, *83*, L93; (e) Brunner, H. et al. *Inorg. Chim. Acta* **1985**, *96*, 67; (f) Ball, R. G.; Payne, N. C. *Inorganic Chemistry* **1977**, *16*, 1187.

54. (a) Corey, E. J.; Bailar, J. C. *J. Am. Chem. Soc.* **1959**, *81*, 2620; (b) Beattie, J. K. *Acc. Chem. Res.* **1971**, *4*, 253.

55. Kelly, T. R. *Tetrahedron Lett.* **1973**, *14*, 437.

56. Unpublished work of William Tam.

57. Huheey, J. E. *J. Org. Chem.* **1966**, *31*, 2356.

58. (a) Liotta, C. L. *Tetrahedron Lett.* **1975**, 523; (b) Klein, J. *Tetrahedron* **1974**, *30*, 3349; (c) Eisenstein, O.; Lefour, J. M.; Klein, J. *Tetrahedron* **1979**, *35*, 225.

59. In collaboration with Professor M. Gugelchuk, University of Waterloo, Ontario, Canada.

60. For reviews on the intramolecular Diels–Alder reaction, see (a) Ciganek, E. *Org. React.* **1984**, *32*, 1; (b) Taber, D. F., *Intramolecular Diels–Alder and Alder Ene Reactions*; Springer-Verlag: Berlin, 1984; (c) Fallis, A. G. *Can. J. Chem.* **1984**, *62*, 183; (d) Oppolzer, W. *Angew. Chem., Int. Ed. Engl.* **1977**, *16*, 10.

61. Deslongchamps, P., *Stereoelectronic Effects in Organic Chemistry*, 1st ed., Pergamon Press: New York, 1983, Chap. 3.

62. Jung, M. E.; Gervay, J. *J. Am. Chem. Soc.* **1991**, *113*, 224 and references cited therein.

63. Lautens, M.; Tam, W.; Blackwell, J. *J. Am. Chem. Soc.* **1997**, *119*, 623.

64. Lautens, M.; Blackwell, J. *Synthesis* **1998**, 537.

STATE OF THE ART IN SELECTIVE HETERO- AND CARBOCYCLIC SYNTHESES MEDIATED BY CYCLOMETALLATED COMPLEXES

John Spencer and Michel Pfeffer

Advances in Metal-Organic Chemistry
Volume 6, pages 103–144.
Copyright © 1998 by JAI Press Inc.
All rights of reproduction in any form reserved.
ISBN: 0-7623-0206-2

I. INTRODUCTION

The chemistry of heterocycles occupies a central position in organic chemistry. Given their wide occurrence in nature, industrial importance, and applications in myriad disciplines, including the pharmaceutical industry, dyestuffs, and organic conductors, one goal of the synthetic chemist is to find new selective, expeditious, and economical routes to these ring systems.[1]

Over the last few decades, transition-metal complexes have emerged as extremely useful synthetic tools for forming heterocycles. They often impart high selectivities (chemo, regio and stereo) seldom observed in 'classical' synthetic methods. Moreover, the efficacy of these reactions, the possibility of employing mild conditions, and in many cases their use in catalytic quantities, all place metals as attractive precursors for orchestrating these transformations.[2]

Herein, we review nonexhaustively our contribution to the field of transition-metal-mediated heterocyclic synthesis. This chemistry is based mainly on using cyclopalladated complexes and their reactions with disubstituted alkynes that in many cases, lead to heterocyclic products by the selective intramolecular formation of carbon–carbon and carbon–heteroatom (C–N, C–O and C–S) bonds.[3] In some instances these reactions also lead to interesting carbocyclic derivatives. Emphasis is placed on the transformations of the alkynes. When they are allowed to react with the metallated ligands, they lead in several instances to heterocyclic or carbocyclic final products. We present in particular some of the more recent results emanating from our laboratory and comment briefly on some similarities of this chemistry to other, selected and related transition-metal-mediated reactions, thus demonstrating that this field of research remains in vogue in many different research groups.

II. SYNTHESIS OF CYCLOMETALLATED COMPLEXES

Cyclometallated complexes share the common feature of a carbon–metal σ bond and an intramolecular stabilizing dative bond between a donor group, such as an amine, phosphine, or thioether, and the metal.[4] A large number of cyclometallated complexes are particularly well documented for palladium (II) and tertiary amine ligands. Many are obtained by a C–H activation reaction. These organometallic reagents have the added advantage of being often easy to handle and air stable. Examples of C–H activation processes that lead to synthetically useful complexes (see later) are shown in Scheme 1.[5–8] Commercially available palladium (II) acetate is often used as metallating agent, as in the synthesis of **3**,[5] **7**,[5] **8**,[7] and **10**.[8] An indirect metallation technique using **4** as a palladium source can be employed[10] to synthesize the pyridine derivative **12**[9] (for clarity, metallated C is emphasized in Schemes 1 and 2).

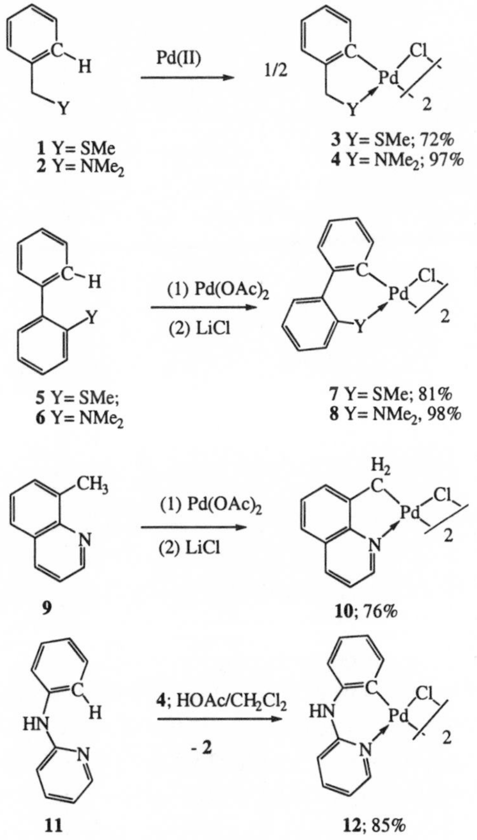

Scheme 1.

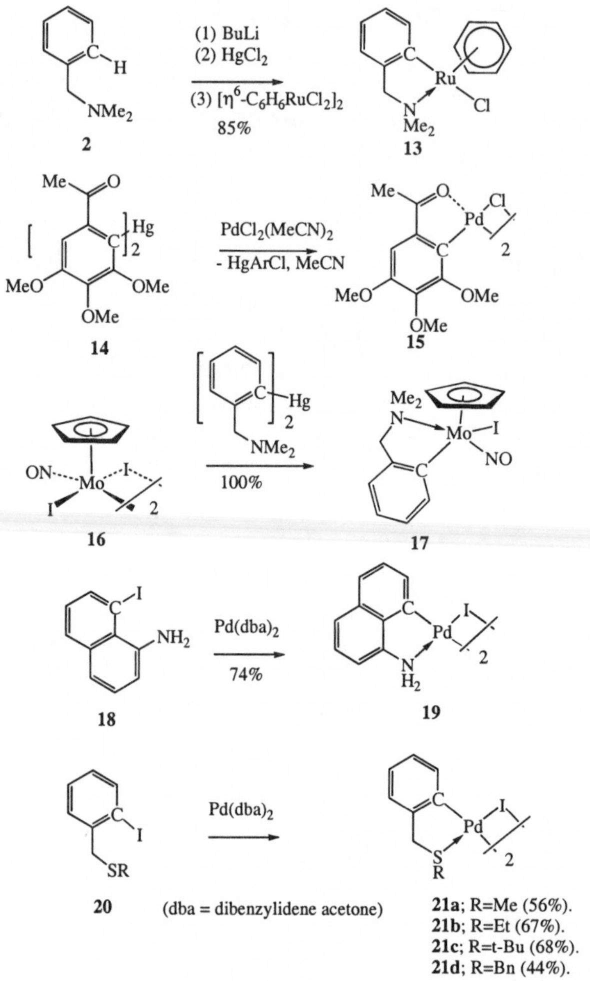

Scheme 2.

When the C–H activation process is inefficient or difficult to carry out, one can often resort to alternative methods such as transmetallation reactions, as in the synthesis of the interesting cycloruthenated complex **13**. The corresponding C–H activation process for **2** gives only a 38% yield.[11(a),(b)] A similar method was employed for the synthetically useful cyclopalladated ketone derivative **15**[12] and for the organomolybdenum complex **17**.[13] Oxidative additions of suitably substituted iodo derivatives on Pd(0) were employed to synthesize the primary 8-napthylamine complex **19**[14] and the complexes **21** with a variety of substitution patterns on the thioether group.[15]

Synthesis of [PdC$_6$H$_4$CH$_2$N(CH$_3$)$_2$(μ-Cl)]$_2$ **4** *by a C–H Activation Route*[6(b)]. To a well-stirred methanolic solution (300 mL) of Li$_2$[PdCl$_4$] (13.12 g, 50 mmol) at rt was added **2** (7.42 g, 55 mmol). After 5 min a cream precipitate began to form, and then a methanolic solution (50 mL) of NEt$_3$ (5.0 g, 50 mmol) was slowly added over a 1 h period. After 5 h stirring, the yellow precipitate of **4** was removed by filtration, washed with MeOH (3 × 50 mL), Et$_2$O (2 × 50 mL) and then dried, yielding 12.95 g, 97%.

Oxidative Addition Route to [C$_6$H$_4$CH$_2$-2-StBuPdI]$_2$ (**21c**). **20c** (R= t-Bu) (1.29 g, 4.2 mmol) and "Pd(dba)$_2$" (2.03 g, 3.5 mmol) were stirred overnight in toluene (50 mL) at rt. After filtration over Celite to remove traces of metallic palladium formed, CH$_2$Cl$_2$ (100 mL) was added to wash the product through, and the combined filtrates were evaporated. The resulting yellow solid was placed in a Soxhlet apparatus. Overnight extraction with hexane removed the dba liberated during the reaction, which was discarded from the filtrate. The yellow solid remaining in the Soxhlet apparatus was washed with ether (40 mL), then dried, yielding 0.990 g (68%).

III. MONOINSERTIONS OF DISUBSTITUTED ALKYNES IN THE Pd–C BOND

A. Organometallic Aspects

For some time our laboratory has been actively involved in studying the reactivity of mainly cyclopalladated amines and thioethers with disubstituted alkynes.[3(a),(b)] In many cases these reactions afford stable organometallic compounds in which the alkyne has inserted one to three times in the Pd–C bond (Scheme 3). Triple insertions are not discussed herein as they have already been covered in detail elsewhere.[3(b)]

Alkyne insertions in the Pd–C bond often occur stepwise although in certain cases the monoinserted complexes cannot be isolated and multiple insertions occur, even when a default of alkyne is used. Chelation of the heteroatom-containing group to palladium is conserved for each of these insertions.

In this section we restrict our discussion to monoinsertion reactions, a few actual examples of which are illustrated in Scheme 4. Whereas the chemistry of double- and triple- inserted complexes is not new and reminiscent of earlier work from Maitlis' group,[16] which often involved PdCl$_2$(MeCN)$_2$ as a palladium-containing species, that of the monoinserted complexes is quite unique and allows isolating a large range of

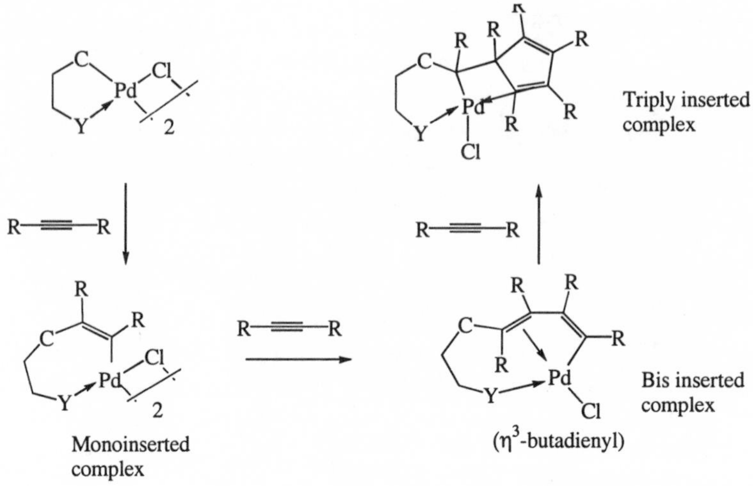

Scheme 3.

metallacyclic complexes for structural and reactivity studies. A few representative examples include the five-membered chelated metallacycles **10** and **21c** that react with alkynes to afford the seven-membered metallacycles **22** and **23**, respectively.[17,15] Similarly, the six-membered chelated metallacycles **7** and **25** are converted into the eight-membered metallacycles **24** and **26**, respectively, after alkyne insertion.[7,18]

In some cases alkyne insertion is accompanied by a skeletal rearrangement, as with **27** whereby the bulky trimethylsilyl group migrates via a 1,3 sigmatropic shift complex, affording the π-allyl complex **28** (Eq. 1).[17] Given the loss of the stereogenic center, this reaction was unhelpful for the important study of the stereoeselectivity of the C–C bond-forming alkyne insertion in the Pd–C bond (see Part D).[19]

B. Mechanistic Aspects

A recent theoretical study, using [PdCl(NH$_3$)(CH$_3$)] and C$_2$H$_2$ as models for the cyclopalladated and alkyne entities, has proposed a η^2-coordination of the symmetrically substituted alkyne to the metal prior to insertion in the Pd–C bond.[20] Related studies of inserting the isolobal metallacarbynes have suggested the possible involvement of a η^1-coordination mode of the alkyne before its insertion into the Pd–C bond.[21] Experimental kinetic studies have revealed that the rate-limiting step of this process involves a single bridge-splitting reaction of the

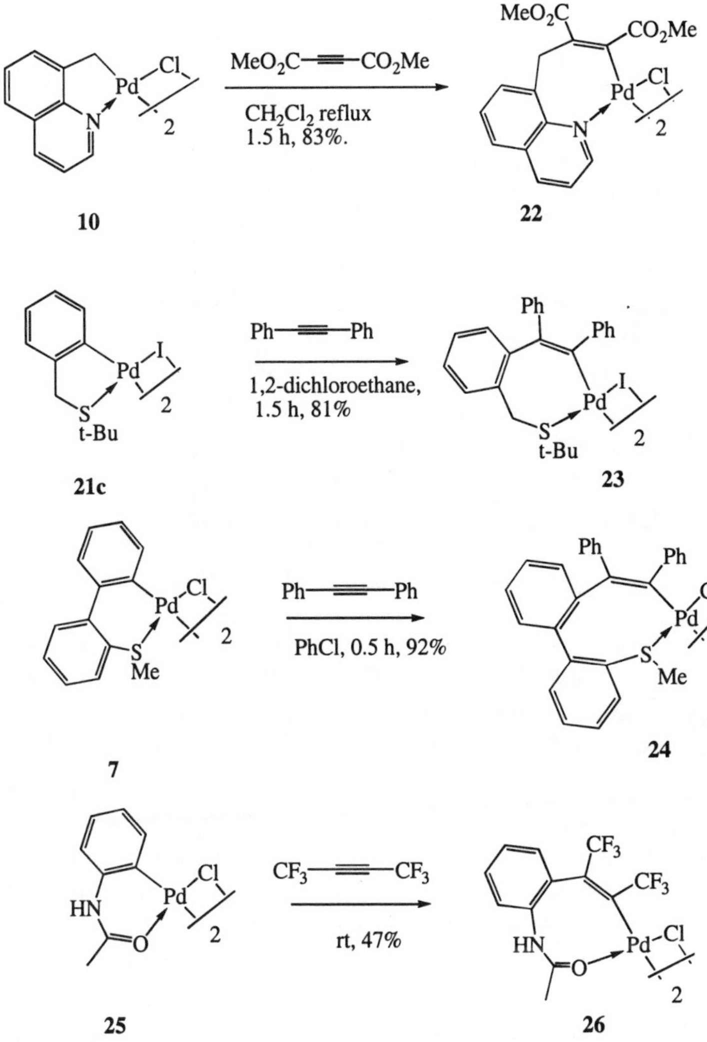
MeO2C—≡—CO2Me
CH2Cl2 reflux
1.5 h, 83%.
10
22
Ph—≡—Ph
1,2-dichloroethane,
1.5 h, 81%
21c
23
Ph—≡—Ph
PhCl, 0.5 h, 92%
7
24
CF3—≡—CF3
rt, 47%
25
26

Scheme 4.

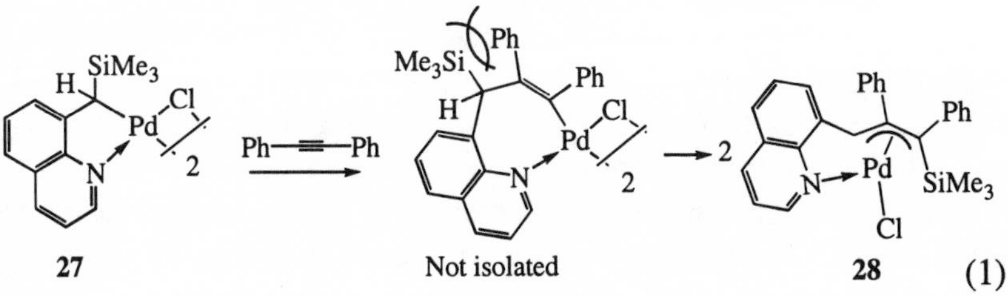
Ph—≡—Ph
Not isolated
27
28
(1)

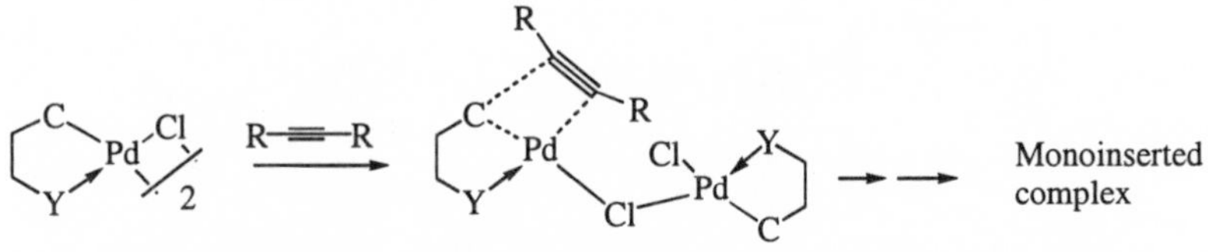

Scheme 5.

starting cyclopalladated dimeric species by the alkyne.[22(a)] This is followed by the usual *cis* migration-insertion in the Pd–C bond.[22(b)] Analogous kinetic studies carried out for the monoinsertion of internal alkynes into the Ni–C bond of monomeric cyclonickelated benzyldiphenylphosphine derivatives show that the alkyne is most likely to be coordinated at the fifth coordination site of the nickel atom before insertion leading to the seven-membered cyclonickelated ring.[23]

C. Regioselectivity of the Alkyne Insertion

The chemistry introduced so far is still relatively young, and we deemed it necessary to study some fundamental points that may be useful when the synthetic potential of such reactions becomes more widespread. With disymmetrical alkynes, for example, an interesting complication arises when inserting them in the Pd–C bond because this can theoretically give rise to two regioisomeric products. However, in practice, these reactions are often highly regioselective, and on the basis of the following observations, a steric, rather than electronic, controlled pathway is advocated to explain the high degree of selectivity.

An electronically biased insertion of phenyl propynoate and similar derivatives would be expected to afford a cyclopalladated complex in which the palladium is adjacent to the electron-withdrawing group (EWG = CO_2Et, CHO, COMe, SO_2p-tol...), a reaction akin to a 1,4-Michael addition of the palladated aryl group to the alkyne. In the majority of cases the regiochemistry of the insertion of these alkynes is such that the EWG ends up away from the palladium.[15,24] Here, to account for such regiochemistry, we tentatively propose that the oxygen atom of the EWG has an anchoring function via hydrogen bonding to the metallated aryl group before insertion in the Pd–C bond.

An atypical example exists whereby the cationic azobenzene cyclopalladated complex **30** reacts with methyl-3-phenylpropynoate to afford a cinnolium salt **31** which was assumed to result from a 1,4-Mi-

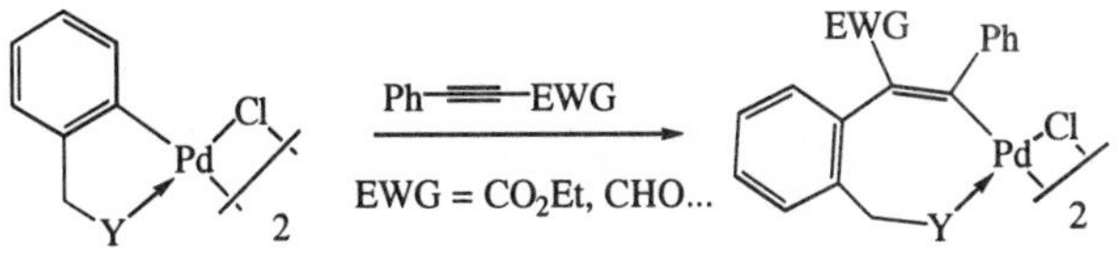

Scheme 6.

chael type addition of the metallated aryl group on the alkyne (Eq. 2).[25] In this particular example, the putative inserted metallacyclic intermediate is not isolable and a heterocyclization reaction occurs, which is discussed in the next section.

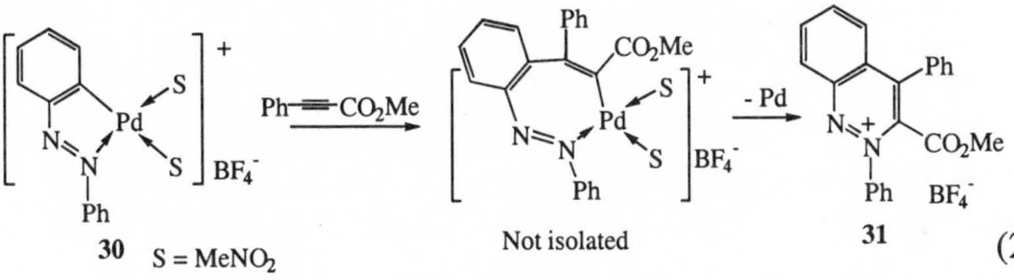

(2)

An electronically biased insertion of nonsymmetrical diaryl alkynes, in which one of the aryl groups is substituted by a strong electron-donating or electron-withdrawing group in the Pd–C bond would be expected to occur with high regioselectivity. However, it has been shown on many occasions that the insertion reactions of such alkynes are nonregioselective.[26,27]

To test to what extent steric factors influence the regiochemistry of these insertions, we recently studied the reactions of 4,4-dimethyl-2-pentyne, an alkyne that bears a 'large' *t*-Bu and 'small' Me group, with cyclopalladated complexes. These reactions are highly regioselective, and in the case of the strongly chelating complexes **3**, **7**, and **34**, the products obtained have the larger alkyl substituent on the same carbon as the palladium atom, a result confirmed by both [1]H nOe spectroscopy and X-ray structural studies.[28] Here, pyridine is added merely to facilitate the spectroscopic and analytic characterization of the complexes.

For these reactions a preinsertion scenario was proposed involving η^2-coordination of the alkyne to palladium in a *trans* position with

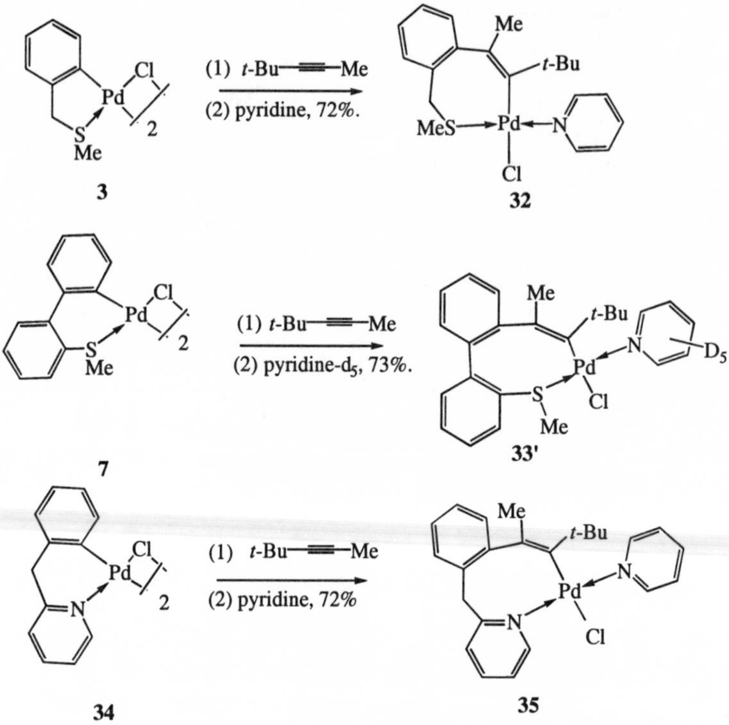

Scheme 7.

respect to the donor group Y. Given the relatively strong coordination of Y (Y = pyr, SMe) to palladium, the insertion of the alkyne has to occur in the same plane as the aryl group bonded to palladium. Of the two orientations possible, (b) in Scheme 8 should be favored because there is steric hindrance in (a) between the t-Bu group of the alkyne and the

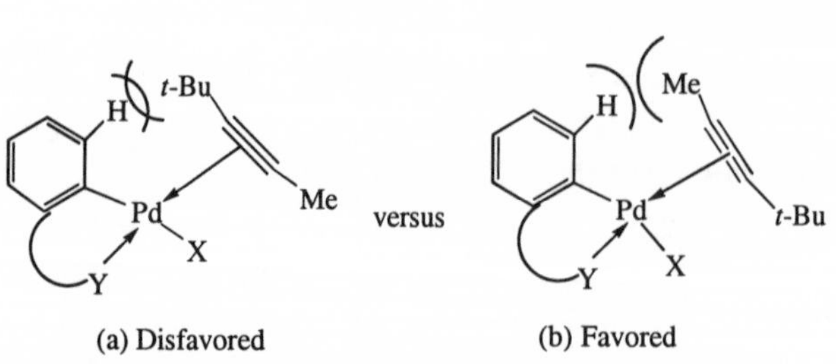

(a) Disfavored   (b) Favored

Scheme 8.

coplanar aromatic protons situated in the proximity of the palladium atom. This "chelate effect" places the *t*-Bu group next to the relatively large palladium atom, a regiochemistry opposite to the one classically observed by Maitlis.[16]

D. Stereoselectivity of the Alkyne Insertion

To conclude this section on the mechanistic and organometallic aspects of alkyne insertions in the Pd–C bond, we were recently interested in studying the stereochemistry of the C–C bond forming insertion of an alkyne in the Pd–C bond. First of all it was necessary to synthesize a cyclopalladated complex containing a stereogenic center directly bound to the metal. For this purpose we partially resolved a cyclopalladated 4-methyl-8-ethylquinoline derivative **36** employing (*S*)-leucine as chiral auxiliary. **37** could be formed with de's around 90% by a series of fractional precipitations. One of the diastereomers [the (*S*, *S*)-] was characterized by a single-crystal X-ray study. For practical purposes, **37** with ee's around 30–50% was readily available and **36** could be recov-

$$(3)$$

ered with virtually the same enantiomeric enrichment and absolute configuration.[29]

Next we required a reaction whereby we could find the stereogenic center of the starting material in the final product. Reaction of racemic **36** with DMAD as the alkyne afforded the racemic heterocyclic product **38** as a single diasteroisomer, to which we tentatively assigned the *anti* relative configuration. Given that this reaction proceeds via initial alkyne insertion in the Pd–C bond (see Scheme 32 for general mechanism), now we were in a position to relate the enantiomeric enrichment of the starting material with that of product **38**. Reaction of **36** (ee = 45%) with the same alkyne afforded **38** with almost the same ee. This result suggests a concerted type mechanism of the alkyne in the Pd–C bond, and we assume a retention of configuration for this reaction.[19]

(4)

Formation of {[Pd(MeO$_2$C)C=C(CO$_2$Me)CH$_2$C$_9$H$_6$N(μ-Cl)]$_2$} (22). A CH$_2$Cl$_2$ solution (60 mL) of dimethyl acetylenedicarboxylate (DMAD) (1g, 7.0 mmol) was added dropwise to a suspension of 10 (2 g, 7.0 mmol) in CH$_2$Cl$_2$ (400 mL) over 1.5 h. After 1.5 h reflux, the orange solution was concentrated to ca. 50 mL, and 22 was collected by filtration, washed with Et$_2$O, and then dried. Evaporation of the filtrate, followed by washing with CH$_2$Cl$_2$/hexane (3 × 25 mL of a 1:1 mixture), gives further 22. The combined samples afforded yellow 22 with a yield of 2.51 g (83%).

Formation of [PdC(t-Bu)=C(Me)-2-C$_6$H$_4$-C$_6$H$_4$-2'-SMe (μ-Cl)]$_2$ (33). A suspension of 7 (340 mg, 1.0 mmol) and 4,4-dimethyl-2-pentyne (100 mg, 1.04 mmol) was refluxed for 0.5 h in chlorobenzene (30 mL). After cooling, the resulting orange solution was evaporated to dryness. Stirring in pentane (20 mL), afforded 33 as a yellow solid which was collected by filtration and dried (320 mg, 73%). 33' was readily made by adding a drop of pyridine d-$_5$ to a CDCl$_3$ suspension of 33 at rt.

IV. TRANSITION-METAL-MEDIATED HETEROANNULATIONS OF ALKYNES

A. Formation of N-Heterocycles

One important synthetic use of the above monoinserted complexes involves forming heterocyclic products, resulting from their thermally induced demetallation. One problem that often has to be overcome is the high stability of the chloro-bridged monoalkyne-inserted organometallic complexes, given the important chelate effect due to coordination of the heteroatom to palladium. Activation of these complexes is often achieved by using either the corresponding iodo-bridged complexes or by removing the chloro bridges with a silver salt in the presence of a weakly binding solvent, such as THF, nitromethane, or acetonitrile. These transformations have a twofold effect on the reaction of the cyclopalladated complex with the alkyne, increased reactivity to alkyne insertion in the

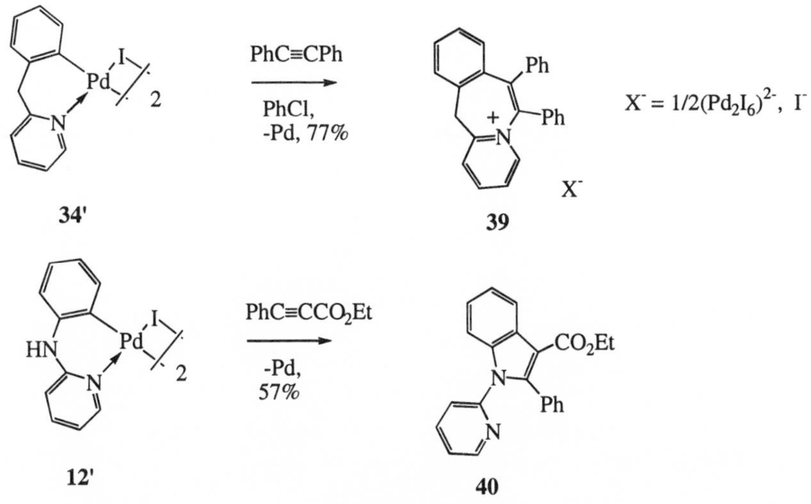

Scheme 9.

Pd–C bond due to the higher electrophilicity of the metal and greater thermal lability of the complex.

The iodo-bridged complexes **12′** and **34′** (formed by a simple halogen metathetic reaction of the corresponding chloro-bridged dimers with NaI in acetone, for example) both react with alkynes to afford heterocyclic products. **12′** leads to a range of indoles **40** in a 'one-pot' procedure,[9]

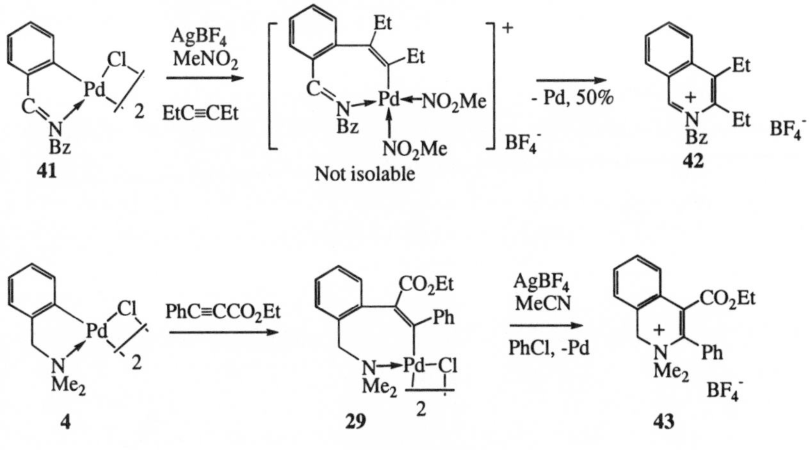

Scheme 10.

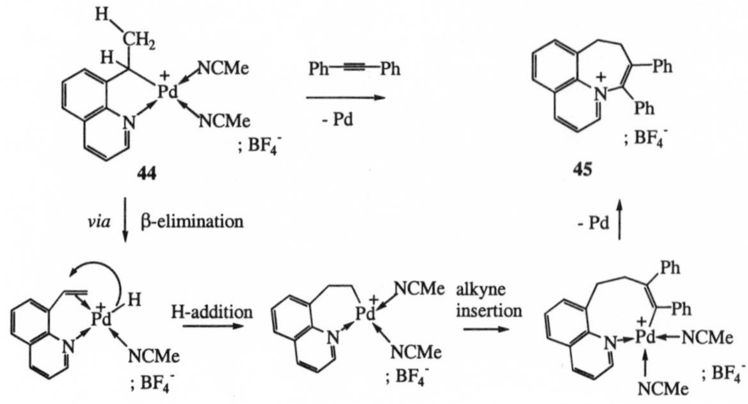

Scheme 11.

whereas **34′** reacts uniquely with diphenylacetylene to afford the cationic heterocyclic product **39**.[24(c)] Here, we see that incorporating a secondary amine function within the metallacyclic framework drastically changes reactivity, and instead of obtaining seven-membered heterocyclic products as with **34′**, five-membered derivatives are obtained exclusively by the preferential addition of the NH function in **12′** onto the alkyne.

Similarly, cationic complexes can either be used for a 'one-pot' procedure, such as in the synthesis of **42**,[30] or by previously isolating an enlarged metallacyclic complex, such as **29**, as in synthesizing **43**.[24(a)]

The cationic derivative **44** affords an interesting seven-membered heterocycle following skeletal rearrangement and loss of the original stereogenic center linked to the metal. A β-elimination of the metallated ethyl group, followed by intramolecular H-addition to the intermediate olefin-Pd complex, before alkyne insertion could explain the formation of **45**. Once again it is quite apparent why this reaction was not useful for the study of the stereoselectivity of the insertion of an alkyne in the Pd–C bond (see previous discussion).[19]

B. Use of Other Cyclometallated Complexes in N-Heterocyclic Synthesis

Cyclometallated complexes incorporating transition metals other than palladium react with alkynes and act as synthetic precursors to heterocycles. One early example is the synthesis of a 2-quinolinone derivative **48** from a cyclometallated cobalt derivative **46**. This process was somewhat limited in synthetic scope because it was only amenable to using hexafluorobutyne as the alkyne.[31]

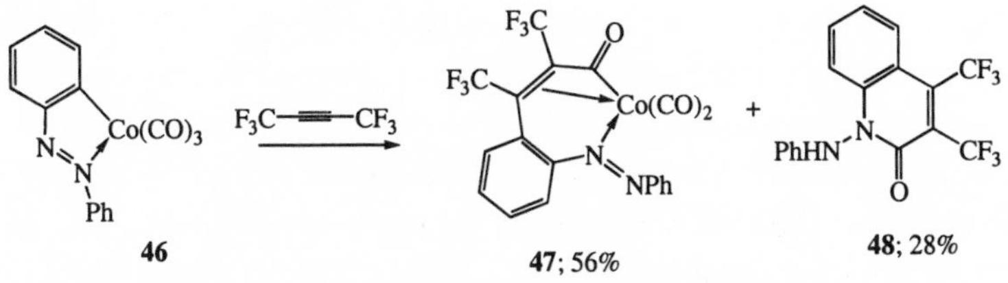

$$(5)$$

The cycloruthenated complex **13** reacts with disubstituted alkynes to afford the novel η^6-arene Ru(0) complexes **49** which were isolated, characterized by X-ray methods, and oxidatively demetallated by treatment with Cu(II) salts to afford metal-free(*1H*)-isoquinolium salts **43** (with PF_6^- as counterion).[11(a)] Kinetic studies have revealed that the reaction proceeds via insertion of the alkyne into the Ru–C bond and that for this process the rate-limiting step is very similar to that found for the analogous palladium compound.[11(c)] Irradiation or thermal treatment of the cycloruthenated complex **50** with diphenylacetylene yields diphenylindole **51** as product. A similar reaction is observed with the corresponding iron or molybdenum complexes.[32]

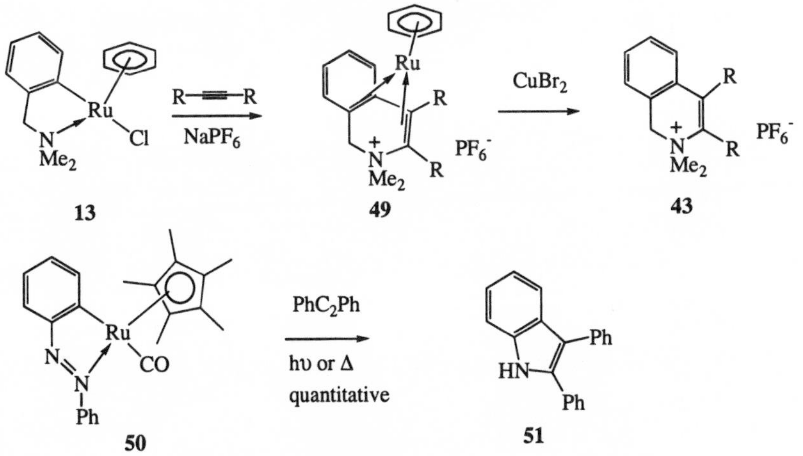

Scheme 12.

C. Formation of Sulfur-Containing Heterocycles

One challenge recently undertaken is to extend our synthetic method-ology presented in Schemes 9 and 10 to the formation of sulfur hetero-cycles. This was spurred by the greater availability of thioether containing cyclopalladated complexes made by either C–H activation or oxidative addition routes developed in our laboratory, and the fact that relatively few reactions are known in which a carbon–sulfur bond assisted by a transition metal is formed intramolecularly.[33] We found that a novel formal addition of a thioether group to a vinyl palladated group could be effected as in the synthesis of the *1H-2S*-benzothiopyrilium derivatives **52** and **53**.[15] Here, the methodology necessitates previously isolating inserted alkyne complexes (analogs of **23**). *In situ* generated cationic complexes, such as **54**, exclusively leads to neutral derivatives as the *t*-Bu group is lost as isobutene, which is easily detected by gas-phase chromatography.

Ruthenium-Mediated Synthesis of **43** *(R¹= R²= Ph).* A suspension of **13** (350 mg, 1 mmol), diphenylacetylene (210 mg, 1.2 mmol) and NaPF₆ (180 mg, 1.1 mmol) in MeOH was stirred for 1 h at rt. After adding CuBr₂ (470 mg, 2.1 mmol) and stirring overnight, the solvent was evaporated in vacuo, and the residue was extracted with CH₂Cl₂ (20 mL). An off-white solid was obtained by ether addition (yield 370 mg, 80%).

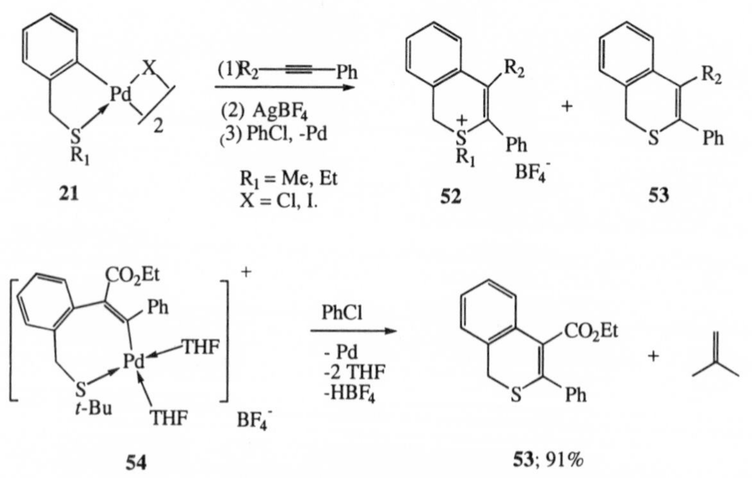

Scheme 13.

Synthesis of **52** (R^1 = *Et*, R^2= *COMe*). (Under N_2), a CH_2Cl_2 solution (20 mL) of **21b** (0.350 g, 0.91 mmol) and 4-phenyl-3-butyn-2-one (0.145 g, 1 mmol) was stirred at rt for 2.5 h. After solvent evaporation, an orange solid was obtained. To the latter, in THF (20 mL), was added $AgBF_4$ (0.210 g, 1.08 mmol). After 1 h stirring, removal of the AgI precipitate by filtration over Celite (the latter was washed with a further 20 mL CH_2Cl_2), an orange filtrate was obtained which was evaporated to dryness. The resulting orange solid was placed in PhCl (40 mL) and heated at reflux temperature for 0.5 h. After Celite filtration and solvent evaporation, a brown solid **52** was obtained from CH_2Cl_2/ pentane, collected by filtration, and dried to yield 0.240 g (69%).

Synthesis of **53** (R^2 = *CHO*). (Under N_2), **21c** (0.210 g, 0.51 mmol) and phenylpropargyl aldehyde (0.080 g, 0.61 mmol) were refluxed in 1,2-dichloroethane (20 mL) for 1 h. An orange solid was obtained after solvent concentration, hexane addition (30 mL), and evaporation of the combined solvents. Stirring a THF solution (40 mL) of this solid with $AgBF_4$ (0.120 g, 0.6 mmol) for 0.75 h was followed by AgI filtration and filtrate evaporation. Following 1 h PhCl reflux (40 mL), solvent evaporation, and CH_2Cl_2 (60 mL) extraction, Celite filtration afforded a dark solution. Concentration of this solution, followed by hexane addition, filtration to remove impurities, and reevaporation gave **53** as an orange oil with a yield of 0.095 g (78%) after solvent evaporation. Crystals suitable for X-ray diffraction were obtained from the slow evaporation of a $CHCl_3$/acetonitrile solution.

D. Catalytic Synthesis of Heterocycles

The previous reactions all involve the use of palladium in stoichiometric quantities although synthetic procedures using catalytic amounts of the metal are known now. We synthesized the potentially biologically interesting aporphine-related heterocycles **57** by a Heck-type reaction employing 1-iodo-8-dimethylaminonaphthalene **55**, which acts as a substrate and also reoxidizes *in situ* the palladium (0) liberated during the reaction.[26] The only side products of these reactions are methyl iodide and metallic palladium.

A similar reaction enables a direct route to precursors of the real aporphine skeleton, as shown in Scheme 15. The development of this reaction very much depended on finding an inexpensive, alternative route to the very expensive commercially available 9-phenanthrylamine, the precursor of 1-iodo-10-(N,N-dimethyl)phenanthrylamine **58** (by N-dimethylation, *ortho*-lithiation with *n*-BuLi, and iodination). This specific

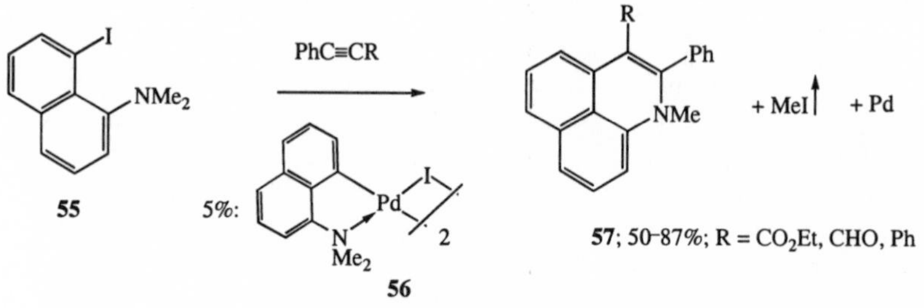

Scheme 14.

example represents a frequently encountered problem in this chemistry, in that the cost of a ligand or precursor significantly outweighs that of the palladium-containing compounds. This belies certain a priori beliefs held by chemists who avoid stoichiometric reactions involving palladium because of the high cost of the metal! Once 9-phenanthrylamine was obtained efficiently (using TMSCH$_2$N$_3$[34] and metallated 9-bromo-phenanthrylamine) we transformed a previously reported stoichiometric process[14] into a catalytic process, exemplified by the synthesis of **61**.[35]

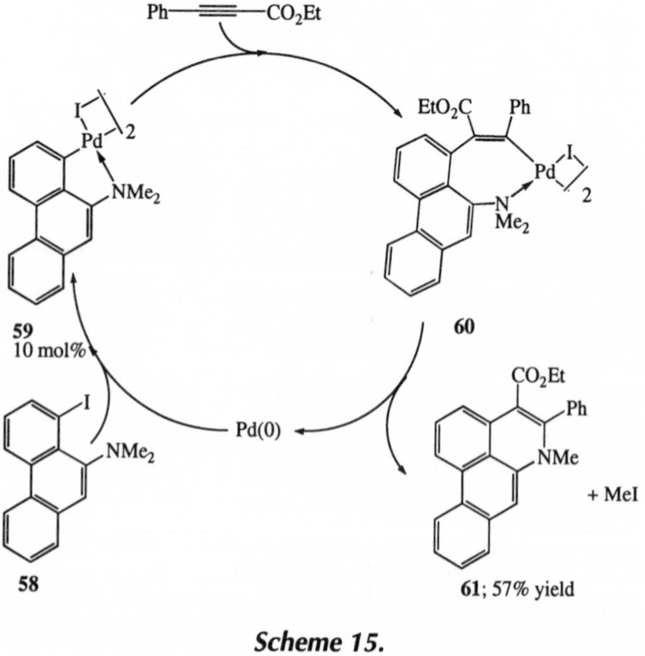

Scheme 15.

In similar Heck-type reactions, commercially available palladium acetate is used as a catalyst, providing a new route for synthesizing indoles, benzofurans, 1,2-dihydroisoquinolines and other interesting derivatives, quite gratifyingly with regioselectivities often matching those reported in our chemistry.[36]

General Procedure for Catalytically Synthesizing **57** *Using* **56** *as a Palladium Source.* A solution of the required alkyne (4.2 mmol) and **56** (80 mg, 0.2 mmol, 5 mol%) in PhCl (20 mL) was heated to reflux and a PhCl solution (20 mL) of 1-iodo-8-dimethylaminonaphthalene **55** was slowly added over a 1–2 h period. After further reflux for 2 h and solvent removal in vacuo, the residue was chromatographed over alumina. Pentane elution eliminated unreacted reagents, and elution with Et₂O afforded **57**.

Catalytic Synthesis of **61**. A PhCl solution of **59** (13.2 mg, 0.029 mmol in ca. 20 mL) was added in one go to a refluxing PhCl solution of ethyl-3-phenylpropynoate (54 mg, 0.311 mmol, in 20 mL). 1-Iodo-10-(*N,N*-dimethyl)phenanthrylamine **58** (101.4 mg, 0.292 mmol) in PhCl (20

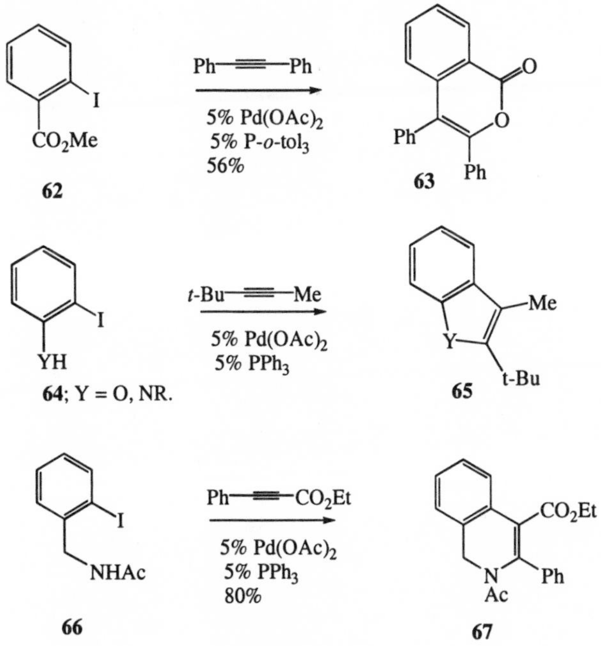

Scheme 16.

 mL) was added dropwise to the now dark solution over a period of 20 min. After complete addition of the latter, reflux was continued for 6 h. Cooling, filtration over Celite, and subsequent solvent evaporation in vacuo was followed by chromatographic purification over silica. Elution with ether:hexane (1:9) afforded unreacted **58** whereas increasing the proportion of ether gave **61** as yellow crystals after solvent evaporation (67 mg, 57%).

E. Formation of C–C, C–N, and C–S Bonds

One problem we have recently addressed was to try determining the mechanism of these heteroannulation processes that involve selectively forming both a carbon–carbon and carbon–heteroatom bond. One obvious hypothesis for the C–C bond formation involves a 'classic' insertion of the acetylene in the Pd–C bond, as in pathway (a), Scheme 17, after *trans* η^2-coordination with respect to the heteroatom linked to palladium. Then the resulting chelated metallacycle **A** undergoes a somewhat unprecedented reductive elimination process affording the heterocyclic product **C**. A second possibility involves pathway (b) whereby a zwitterionic intermediate **B**, resulting from nucleophilic attack of the palladium-bound heteroatom on the *cis*-bound acetylene, affords the heterocycle **C** by a more classic C–C bond-forming reductive elimination.[37]

So far, we have advocated pathway (a) for many reasons. The principal argument is that complexes, such as **A**, have already been characterized and lead to the heterocyclic products **C** in many instances, as in Scheme

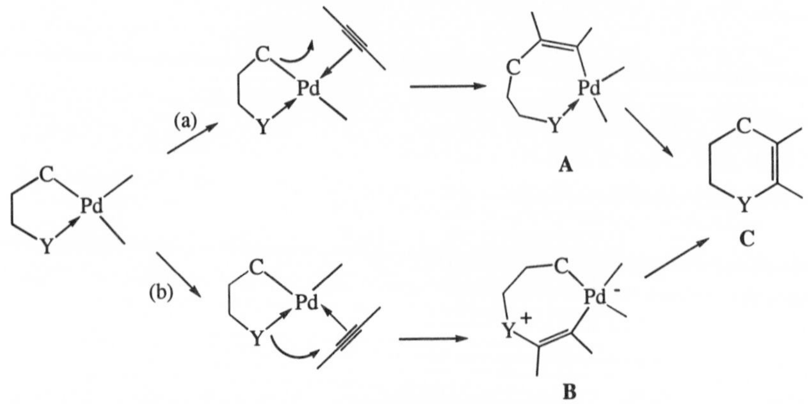

Scheme 17.

$$\text{24} \xrightarrow[\text{(iii) PhCl, -Pd}]{\substack{\text{(i) Ph}_2\text{C}_2 \\ \text{(ii) AgBF}_4/\text{THF}}} \text{68; 35\%} + \text{69; 22\%} \tag{6}$$

10 and Eq. 6, for example. Pathway (*b*) was proposed for one particular example, namely, for forming **40** from the masked cyclopalladated secondary amine function in **12'** (the NH group attacks the *cis*-bound acetylene to yield indole product finally), shown in Scheme 9.

Having favored pathway (a) for the majority of our reactions involving the formation of heterocyclic products from cyclopalladated complexes, now we need to explain the passage of **A** to **C** in Scheme 16, which appears to be the result of a reductive elimination process between the vinyl palladated group and the heteroatom coordinated to the metal.

The relative ease of the cyclization step from **A** to **C** may also be linked to the nucleophilic or coordinative ability of the heteroatom bound to the metal. The reaction of **7** with diphenylacetylene (Ph_2C_2) leads to the seven-membered derivatives **68** and **69** after prior isolation of the monoinsertion product **24**, treatment with a silver salt, followed by the usual thermolytic conditions. This is another rare example of an intramolecular formation of a C–S bond within the coordination sphere of a transition metal and a novel, albeit limited to one alkyne, route to the rare family of dibenzo[bd] thiepins.[15] With the closely related **8**, which differs from **7** only by the tertiary amine unit in the metallacyclic framework instead of a thioether function, a carbocyclic product **71** is obtained (see under carbocycle reactions, next section). The formation of the seven-membered S-heterocycles is attributed to the good coordinative ability of the thioether group in **7**. The S-atom remains close to the vinylic carbon function before the cyclization. With the poorly coordinating, readily displaced amine function in **8**, the N-atom is detached from the metal and ultimately affords a spirocyclic product (see Scheme 18).

A related observation involves the reaction of the cationic derivative **34''** (formed by treating **34** with $AgBF_4/MeCN$) with ethyl 3-phenyl-propynoate, which afforded a cationic heterocyclic product **70**, in which, surprisingly, no carboethoxy group could be found.[38] No analog of **39** was detected during this reaction. A reasonable explanation for this finding is a 1,4-Michael type nucleophilic addition of the pyridine function on the inserted alkyne of the minor regioisomer resulting from

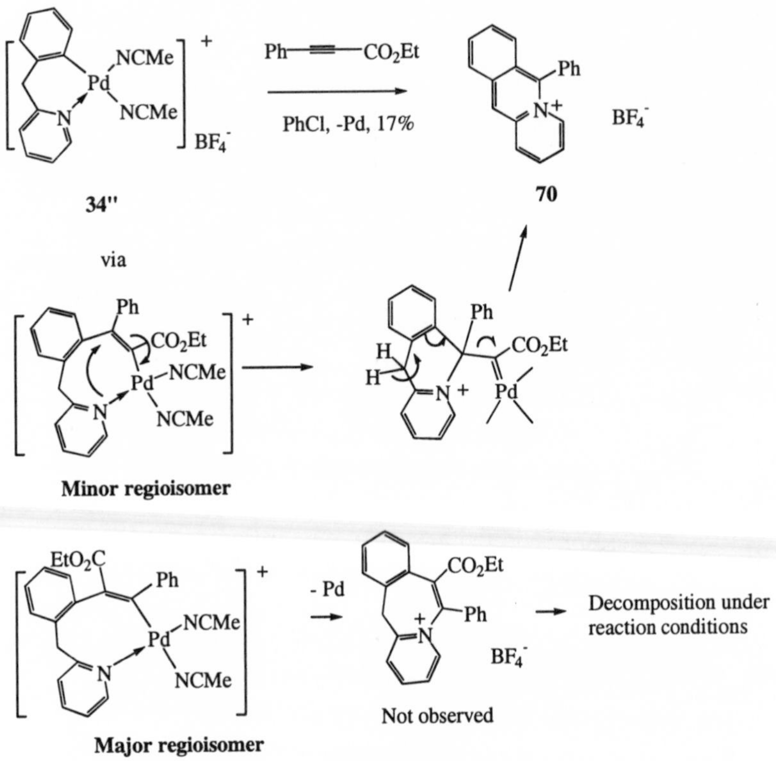

Scheme 18.

the nonregioselective alkyne insertion in the Pd–C bond of **34″**. Following proton loss and demetallation, the cationic heterocycle **70** is obtained. The major regioisomer of this reaction assumedly furnishes a seven-membered azepinium salt **39** which decomposes under the thermolytic conditions.

Synthesis of (**68**) *via a Monomeric Cyclopalladated Complex.* (Under N_2), **24** (0.190 g, 0.37 mmol) and $AgBF_4$ (0.090 mg, 0.46 mmol) were stirred in CH_2Cl_2/THF (20: 5 mL) for 1 h at rt. After AgCl filtration over Celite and evaporation of the filtrate, the resulting orange solid was heated in refluxing PhCl (30 mL) for 1.25 h. After cooling, Celite filtration to remove metallic palladium formed during the reaction, and solvent evaporation, a crude beige solid was obtained from CH_2Cl_2/hexane (2 mL: 20 mL), collected by filtration and dried. The filtrate was evaporated to afford an orange solid (0.060 g) which contained **69** and another unidentified

mechanism adapted for R_1=t-Bu; R_2=Me.

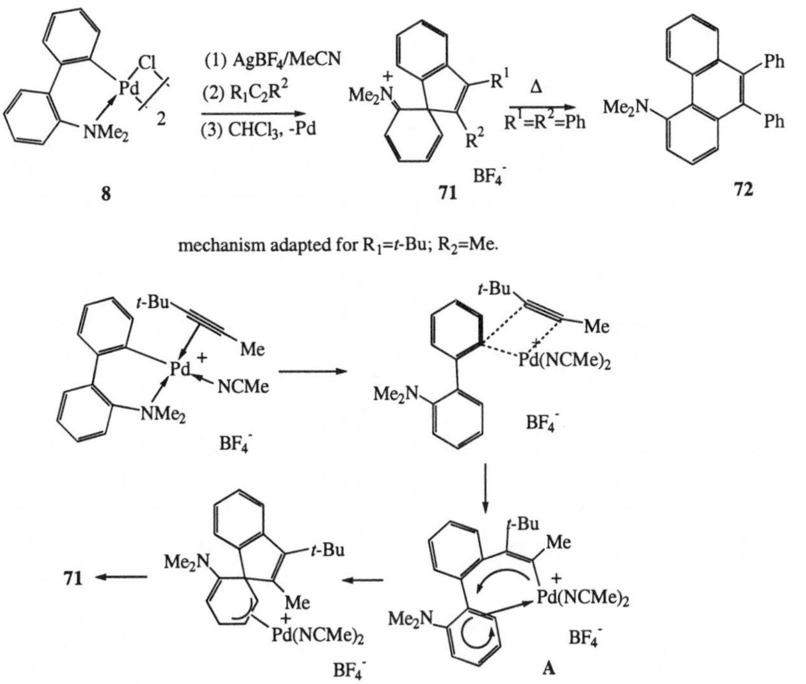

Scheme 19.

product. The beige solid was filtered over silica (acetone eluant) to yield an oil **68** with a yield of 0.060 g (35%). X-ray quality crystals were obtained by slow diffusion of hexane into a CH_2Cl_2 solution of **68**.

V. FORMATION OF CARBOCYCLES FOLLOWING ALKYNE MONOINSERTION IN THE Pd–C BOND

In some instances the monoinsertion of alkynes in the Pd–C bond leads to carbocyclic organic products. The cationic derivative of **8**, formed by silver-induced halide abstraction of the starting chloro dimer, reacts with a host of alkynes to yield the spirocyclic iminium salts **71**. It is thought that this reaction involves alkyne insertion in the Pd–C bond and decoordination of the weakly coordinating, readily displaced NMe_2 group. η^2-coordination of the aryl group with the electron-deficient metal precedes an intramolecular spirocyclization and then loss of palladium to furnish **71** finally.[7]

This reaction raises a few important central issues in this chemistry. We saw previously that a similar thioether-containing derivative (see Eq.

6) under similar conditions leads to an intramolecular C–S bond formation attributed to the greater coordinative ability of the SMe compared with an NMe_2 group. A second important feature of this spirocyclization reaction is that the spirocyclic derivative 71 ($R^1 = R^2 = Ph$) rearranges upon heating to yield the phenanthrene derivative 72, another result in favor of spirocyclic intermediates preceding annulated products as is encountered in Section VIIA (see later). Thirdly, another significant aspect is that the insertion of t-BuC_2Me in the Pd–C bond affords a product 71 whereby the t-Bu group is presumed to be away from the metal in the inserted vinyl palladated intermediate A (Scheme 19) that is opposite to the regiochemistry observed when an important chelate effect is present.[28] If we consider the poor chelating ability of the NMe_2 group with palladium, its decoordination enables the alkyne to insert in the Pd–C bond perpendicular to the plane of the aryl group linked to palladium, and the flat aryl face presents little steric hindrance to the t-Bu group, which is now oriented away from the palladium.

　　Reaction of the cyclometallated derivative of phenyl-2-pyridylketone 73 leads to indenol-chelated, palladium-containing derivatives 74.[9] Here, incorporating an electrophilic (CO) function in the starting palladacycle signifies that, following alkyne insertion in the Pd–C bond, an intramolecular attack of the vinyl palladated unit on the metal-bound, activated CO function occurs. This is in sharp contrast to the reaction described in Scheme 9 whereby incorporating a nucleophilic, masked, secondary amine function leads to indole derivatives 40 and to the azepinium synthesis from the metallated benzylpyridine complex 34'. Therefore, these reactions are rather sensitive to the nature of other potentially reactive functions within the metallacyclic framework.

$$(7)$$

73　　　　　　　　　Not isolable　　　　　　　74

Similar metal-mediated routes to indenols and indenones have been reported by other groups.[36(a),39] A cyclopalladated aldehyde 75 gives indenones 76 stoichiometrically and the related indenols catalytically in the presence of stoichiometric amounts of the corresponding organomer-

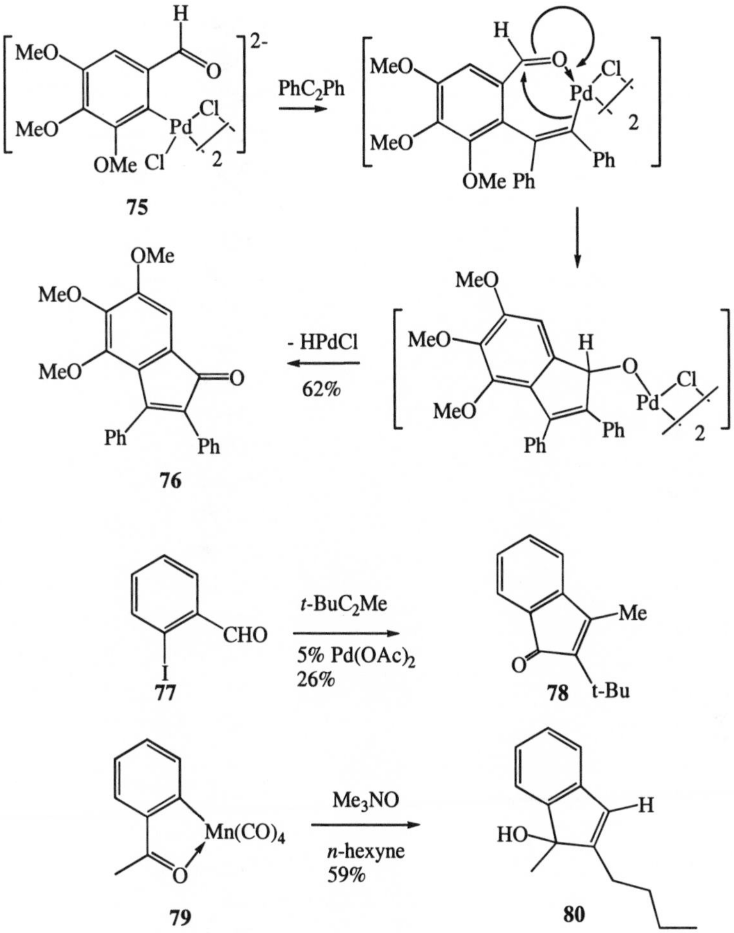

Scheme 20.

curial.[39(b)] Palladium-catalyzed routes[36(a),39(a),39(b)] and stoichiometric manganese-promoted[39(c),(d)] indenol syntheses have also been reported. The regiochemistry of these reactions often mirrors that observed following our chelate model (see Scheme 8).

VI. DOUBLE-INSERTION REACTIONS OF ALKYNES IN THE Pd–C BOND

Five-membered, nitrogen-containing chelates often react with disubstituted alkynes to yield chelated nine-membered metallacycles, as shown in Scheme 21.[36(a),40]

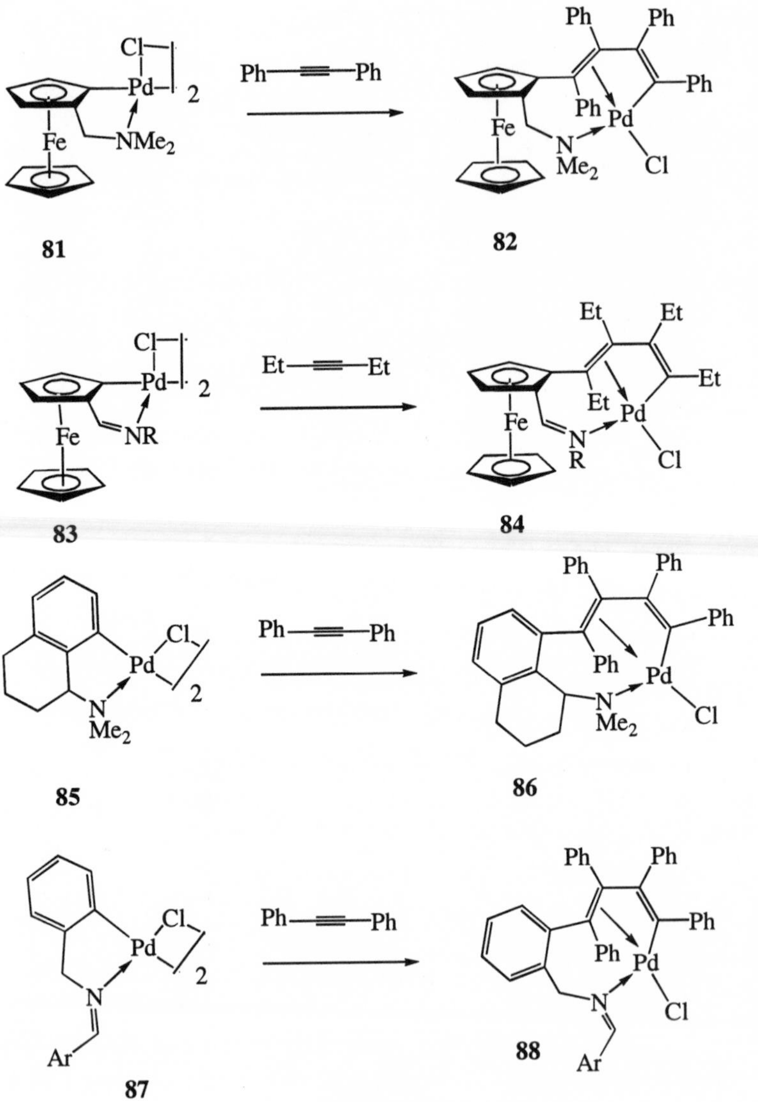

Scheme 21.

A common feature of the metallacyclic units, such as those found in **82** and **84**, is the *trans, cis* arrangement of the η^3-butadienyl segment. A proposition for such a stereochemistry has been rationalized by a *metallacyclic flip*.[18,41] Given that the *cis, cis* arrangement found in (a) Scheme 22 involves poor η^2-coordination of the olefin fragment to the metal

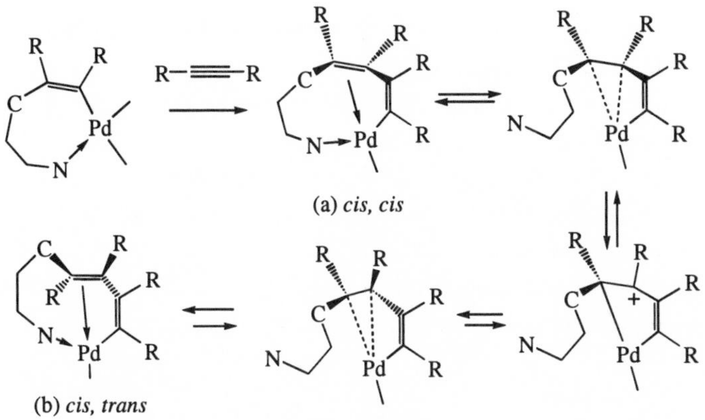

(a) *cis, cis*

(b) *cis, trans*

Scheme 22.

aused by inefficient overlap of the π-orbitals and the $d\pi$ orbitals of the netal, isomerization to a more stable *trans, cis* situation can occur, as roposed in Scheme 22. This process involves prior decoordination of he amine from the metal.

Synthesis of **82**[42]. Diphenylacetylene (216 mg, 1.2 mmol) and **81** (230 mg, 0.3 mmol of Pd) were stirred at rt for 5 h in CH_2Cl_2 (30 mL). After concentration of the solution to ca. 10 mL, chromatography over a silica gel column (10 × 2.5 cm, CH_2Cl_2 eluant) afforded **82** after solvent evaporation with a yield of 430 mg, 97%.

Synthesis of **86**[43]. Diphenylacetylene (358 mg, 2 mmol) and **85** (315 mg, 1 mmol of Pd) were refluxed for 4 h in CH_2Cl_2 (25 mL). Traces of metallic palladium were removed by filtration over a Celite column (4 cm long), and the filtrate was evaporated. The residue was washed with *n*-pentane (50 mL). **86** was precipitated from minimum CH_2Cl_2/pentane with a yield of 526 mg, 79%.

VII. SYNTHETIC APPLICATIONS OF η^3-BUTADIENYL FRAGMENTS

A. Deviation Reactions

As with their monoinserted counterparts, η^3-butadienyl-bound com-
plexes are often very stable, and activation methods are usually needed o employ them for synthesis. Moreover, we do not always observe heterocyclization processes and quite often carbocyclic derivatives, re-

sulting from annulation reactions, are formed. Before concentrating o
the quite distinct heterocyclization processes, it is convenient at this poin
to comment briefly on these carbocyclization reactions, as we see tha
mechanistically they are rather interesting and have led to some stimu
lating chemistry in so far as isolating intermediates is concerned. In th
following example, pyridine or maleic acid are employed as reactants t
destabilize the organometallic complex, and fulvene or naphthalen
carbocyclic products are obtained.[42,44] Similar reactions mediated b
silver-assisted halide abstraction have been reported by Heck.[45]

89   90; 38% isolated yield.   91; not isolated. Ratio of
                                                              90: 91 = 4:1 by NMR.

It is thought that the pyridine or maleic anhydride used in thes
reactions in large excess destabilizes the chelate in **89** by acting as
strong ligand that induces decoordination of the amine from palladium
This is followed by η^2-coordination of one of the phenyl rings originatin
from the starting acetylene (see **A** and **B**, Scheme 23) with the electron
deficient palladium atom. Now the palladium-bound phenyl group ca
be considered an activated olefin fragment, and ensuing nucleophili
attack by the vinyl palladated moeity, followed by spirocyclization
passing by **C** and **D**, eventually affords either the naphthyl or fulven
systems **90** and **91**, respectively.[42]

The mechanistic propositions in Scheme 23 are further supported b
a number of experimental findings, namely, the isolation and structura
characterization by X-ray methods of both organic and organometalli
spirocyclic derivatives in a number of reactions, such as **71**[7] and **93**,[24(c
the 'trapping' of an analog of **D** to yield **92** by an intramolecula
cyclization reaction in a triphenylphosphine-induced depalladation proc
ess,[42] and the structural characterization of some rare η^2-bound ary
palladium complexes, such as **94** found in Chart 1.[46]

Key intermediates were also isolable during the annulation reaction
shown in Scheme 24. Moreover, these reactions are very sensitive to th
nature of the heteroatom linked to palladium in the starting material. Wit

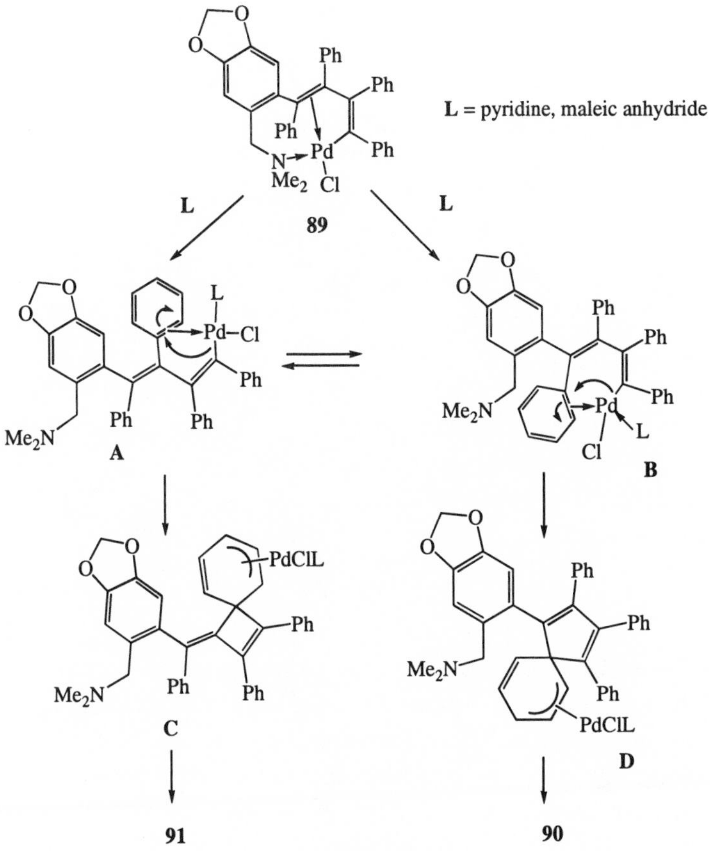

L = pyridine, maleic anhydride

Scheme 23.

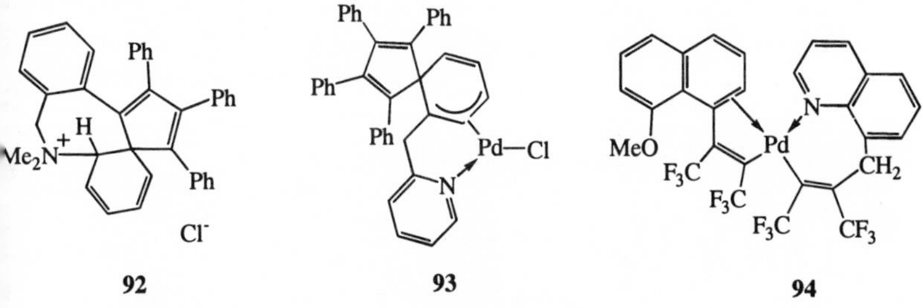

Chart 1.

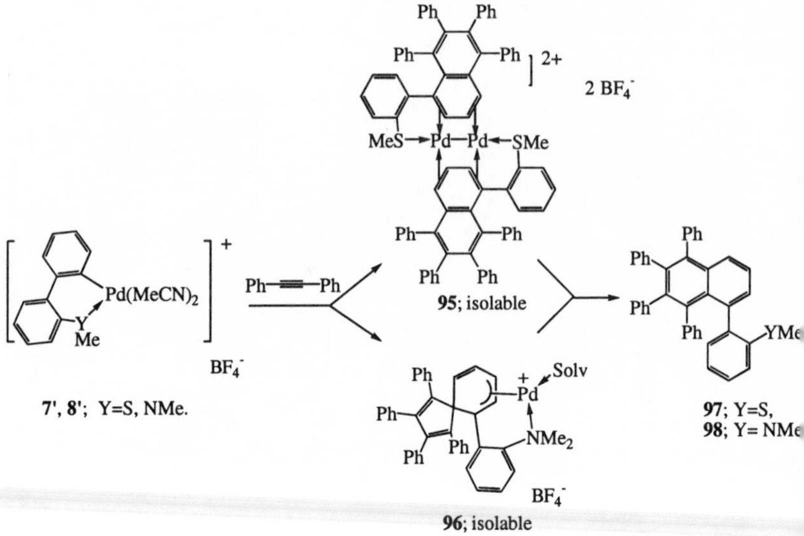

Scheme 24.

the sulfur-chelated derivative **8'**, a rare example of a η^4-bonded, neutra arene, dipalladium (I) complex **95** was isolated, characterized by X-ra methods and shown to afford the naphthyl thioether **97** upon therm: degradation. With the similar amine-containing derivative **7'**, a chelate spirocyclic palladium allyl complex **96** was isolated which was rathe unstable in solution and furnished the napthylamine **98**.[47] For the doubl annulation observed for the ferrocenyl derivative **82**, it was shown th: the second annulation yielding **100** results from the loss of the NMe group from **99**.[48]

Heck showed that the palladium-catalyzed annulation of variou halosubstituted aryl or vinyl derivatives could also lead to annulate products and, hence, that these reactions are not restricted to cyclopal

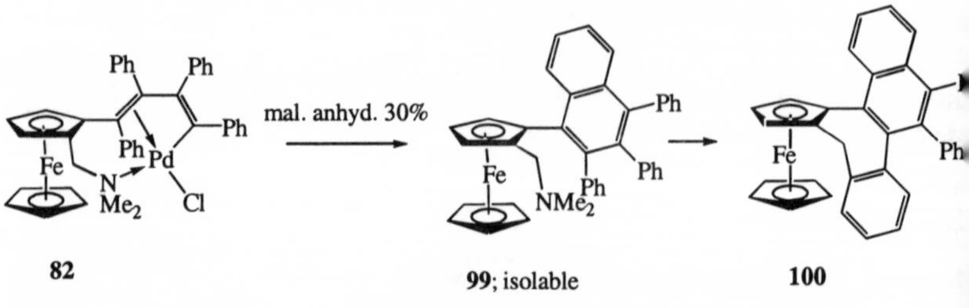

Scheme 25.

adated substrates.[45] Here, once again spirocyclic intermediates may be implicated to explain forming of the carbocyclic products.

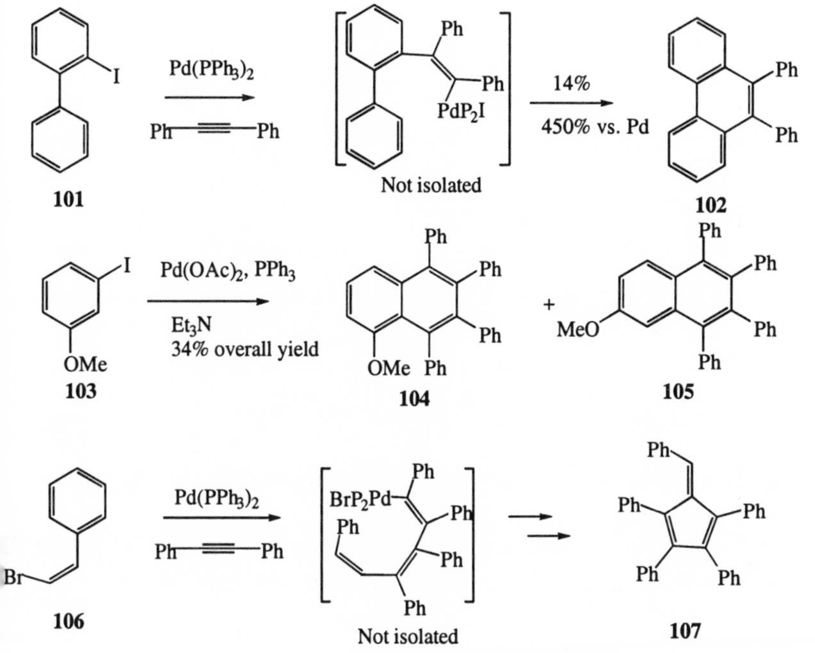

Scheme 26.

Once again the preponderance for forming spirocyclic carbocycles is illustrated by the formation of **109**. No doubt this reaction follows the mechanism shown for **90** and **91**, and in the example given a derivative of the π-allyl complex **108** was isolated and characterized by an X-ray study.[49]

C–C bond-forming reactions do not, however, always involve spirocyclic intermediates. The cyclopentadienyl product **111** is obtained during the depalladation of **110**, derived from a thermally induced 1,3-H shift of its isomer **114**. Here, PPh$_3$ plays a role similar to that of pyridine or maleic anhydride, in that it displaces the quinoline unit bound to palladium and, hence destabilizes the metallacyclic unit. The carbocyclic product **111** is assumedly formed as a result of an intramolecular C–H activation of the η^2-bound olefin unit assisted by the proximity of the

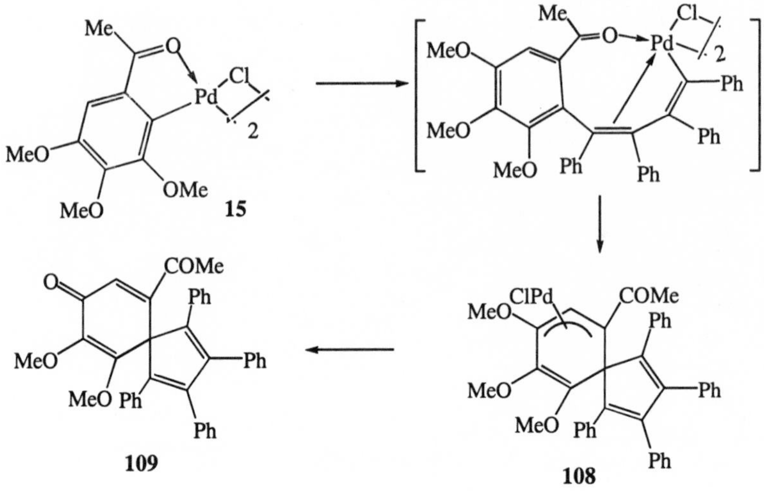

Scheme 27.

quinoline base. A C–C bond-forming reductive elimination process finally yields **111**.[44(b)]

A final deviation reaction is given by the O-heterocyclic synthesis shown in Scheme 29. Here, it is thought that the ester group of one of

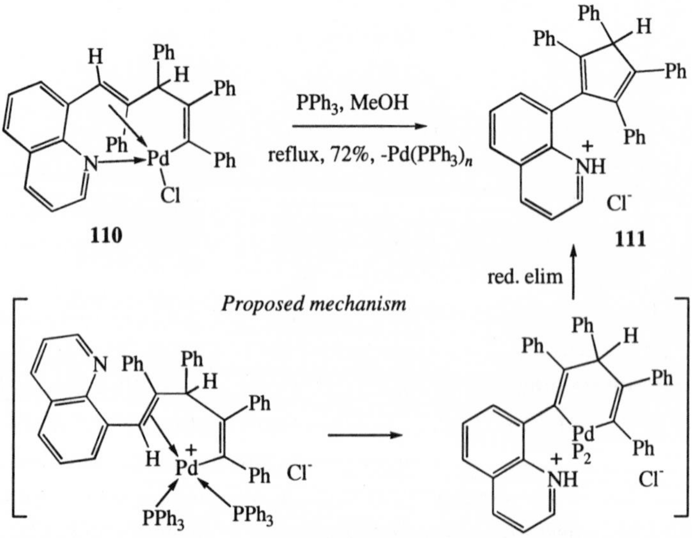

Proposed mechanism

Scheme 28.

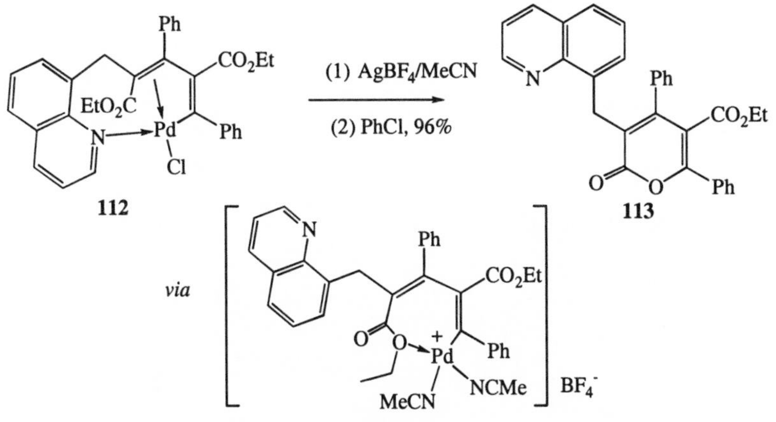

Scheme 29.

the inserted alkyne fragments is activated by coordination with palladium, dealkylated, and a pyrone **113** is obtained as product in near quantitative yield.[44(a)] Similar reactions can equally be carried out in the absence of silver salts, as with PPh$_3$ or maleic anhydride.[44(b)]

B. Heterocyclization Reactions Involving C–N Bond Formation

Triphenylphosphine has also recently been found useful as a depalladating agent for heterocyclization processes, and these reactions have the advantage of being very clean and employing mild conditions. Instead of obtaining palladium black, as in many of the previous reactions, a palladium phosphine complex is obtained. In the example given in Scheme 30, the phosphine is employed to destabilize the organometallic complex **114** by displacing the chloride linked to palladium before inducing the decoordination of the nitrogen atom. Nucleophilic attack of the liberated amine group on the butadienyl fragment leads to the C–N bond formation which is very likely to be followed by a C–C bond-forming reductive elimination step, affording the cationic heterocyclic product **115** with good yield.[44(b)]

The different reactivity of **114** compared to that of its isomer **110** in these demetallation reactions, that is, C–N versus C–C bond formation, is quite remarkable.

A situation involving both C–C and C–N bond formation was recently observed during the depalladation of the di-inserted alkyne complex **86**.[43] Whereas the formation of **116** and **117** (35% overall yield) is explained

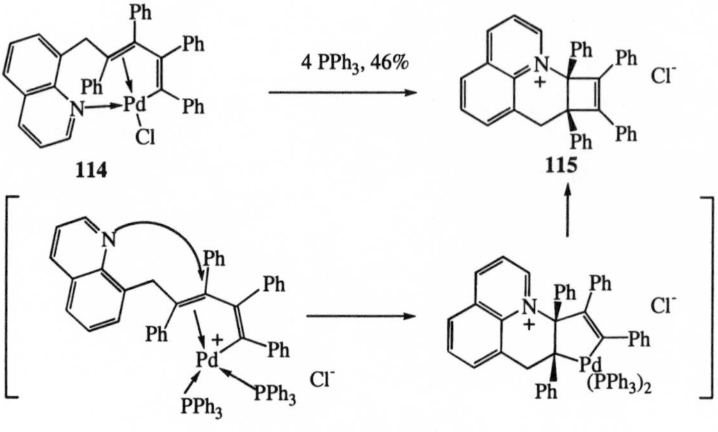

Proposed mechanism

Scheme 30.

by the annulation mechanism described earlier, that of **118** is less straightforward. A proposed pathway involves initial C–N bond formation by a nucleophilic attack of the amine on the η^2-bound olefin fragment of the bis-inserted complex, followed by a C–C bond formation via insertion of a palladacarbene into the C–H bond of one of the amine methyl groups.

Our final route to using η^3-butadienyl complexes is without doubt the most interesting from a synthetic standpoint. In the example given, activation of the cyclopalladated complex **119**, as in Eq. 8, is not

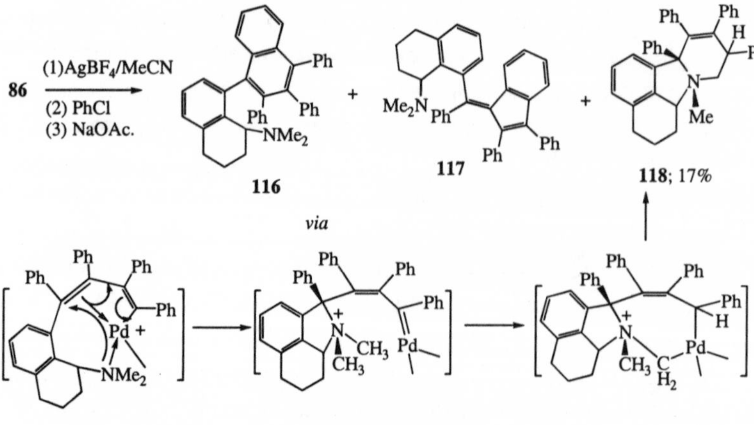

via

Scheme 31.

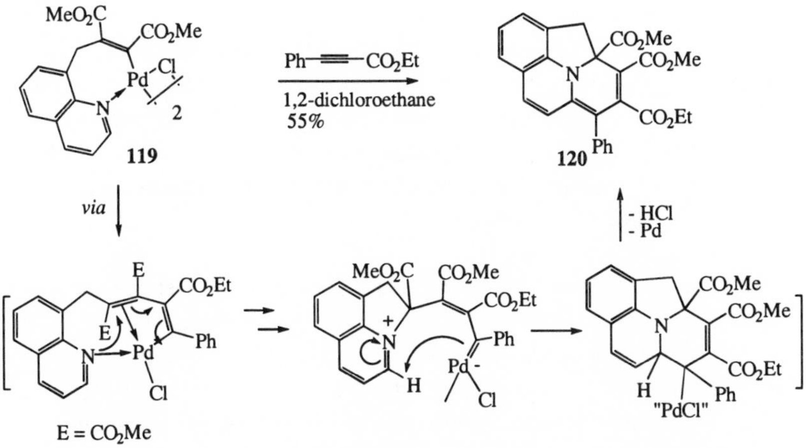

Scheme 32.

necessary. This versatile synthetic procedure is applicable to a whole host of alkynes, and its assets include the selective formation of both C–C and C–N bonds, the creation of a quaternary carbon center, and a high regioselectivity when using disymmetrical alkynes in the second step of the reaction.[50] A nucleophilic attack of the nitrogen atom on the activated η^2-butadienyl fragment is the key step in the C–N bond formation, and a C–H insertion of a palladacarbene was proposed to explain the final C–C bond formation.

The cycla[2.3.3] cyclazine heterocyclic products are new additions to the indolizine family, the skeleton of which is found in many naturally occurring alkaloids, such as crepidine and poranthericine (Chart 2).

Synthesis of **115** *by a Triphenylphosphine-Mediated Depalladation.* (Under N_2), **114** (640 mg, 1 mmol) and PPh_3 (1.50 g, 4 mmol) were

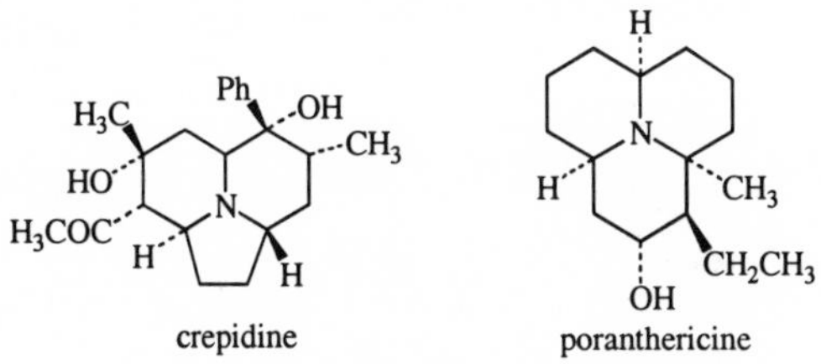

crepidine   poranthericine

Chart 2.

suspended in MeOH and refluxed for 1 h. After cooling and removal of the yellow [Pd(PPh$_3$)$_n$] precipitate, the filtrate was evaporated in vacuo, and the residue was extracted with MeOH (10 mL). The extract was filtered, and the residue dried in vacuo. From the latter a light yellow solid was obtained from CH$_2$Cl$_2$/hexane/ether (10 mL. 30 ml: 30 mL) with a yield of 245 mg, 46%. Analytically pure **115** was obtained from an ether-layered CH$_2$Cl$_2$ solution.

Synthesis of **120** *from* **119**.   **119** (1.23 g, 2.88 mmol of Pd) and ethyl-3-phenylpropynoate (0.49 mL, 2.96 mmol) were refluxed for 18 h in 1,2-dichloroethane (100 mL). After filtration to remove the metallic palladium formed, the red-purple filtrate was evaporated to dryness, washed with cold hexane (2 × 50 mL), and then passed over alumina. A 1:1 CH$_2$Cl$_2$/pentane mixture removed unreacted alkyne and impurities, and then CH$_2$Cl$_2$/acetone afforded purple **120** after solvent evaporation with a yield of 700 mg, 55%.

VIII.   HETEROANNULATION OF ALLENES AND 1,3-DIENES

In most of the heterocyclizations seen previously, alkynes are used as C-2 units linking the σ-bonded carbon atom and the nitrogen atom of the cyclopalladated starting materials. Dienes also play a similar role leading to a broader scope for heterocyclic synthesis.

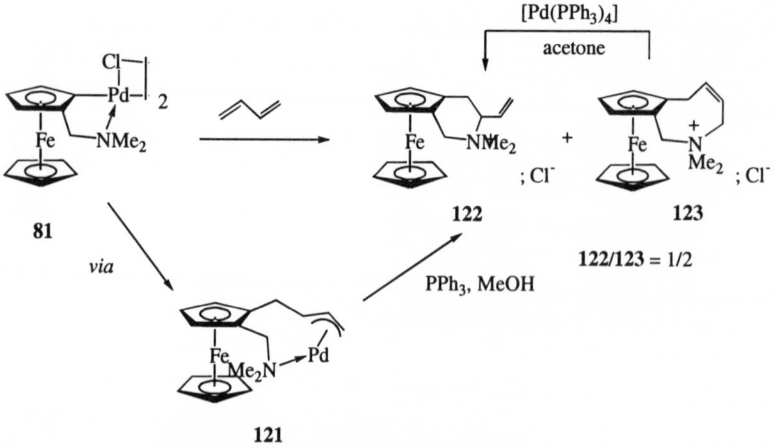

Scheme 33.

We have recently found, albeit in only one example derived from a cyclopalladated ferrocenyl starting material, that 1,3-dienes indeed insert into the Pd–C bond of these compounds and lead to stable η^3-allylpalladium compounds that may be depalladated in the presence of PPh_3 in methanol to afford six- or eight-membered rings.[51]

Several other dienes can be used for this reaction which is another example of a rare coupling between a tertiary nitrogen and a π-allyl Pd derivative.[52(a),(b)]

It is well known that the insertion of an allene into a Pd–C bond produces an allyl fragment π-bonded to Pd.[53] Having shown previously that the intramolecularly bonded tertiary nitrogen atom in such organopalladium complexes readily functions as a nucleophile,[52] we have confirmed that, using a series of N-containing cyclopalladated complexes, this process is an interesting way of building up the synthesis of a range of heterocycles with starting materials obtained from the intramolecular C–H activation process.[54]

It appears that with the nonsymmetrical 3-methyl-1,2-butadiene, the formation of the N–C bond is under kinetic control. Usually the most abundant regioisomer formed is the one having the most substituted carbon atom attached to N, a feature frequently encountered in related allylic derivatives.[55] This latter species may isomerize in solution in the presence of catalytic amounts of Pd(0) to the thermodynamically more stable isomer that has a CH_2 group bound to N. This reaction offers a large range

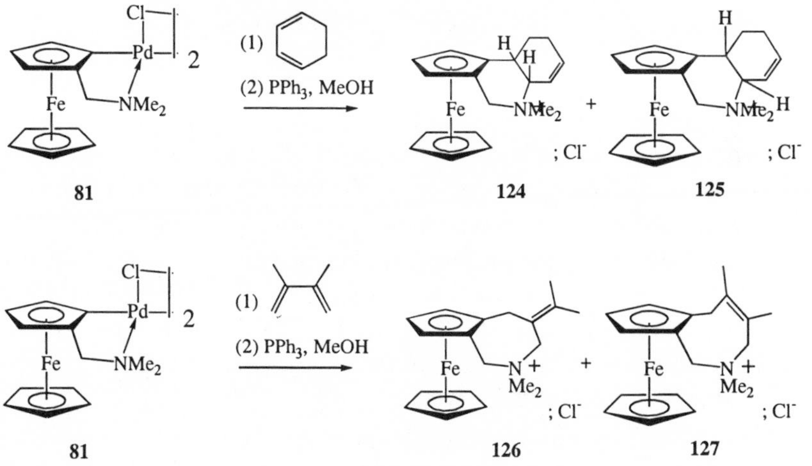

Scheme 34.

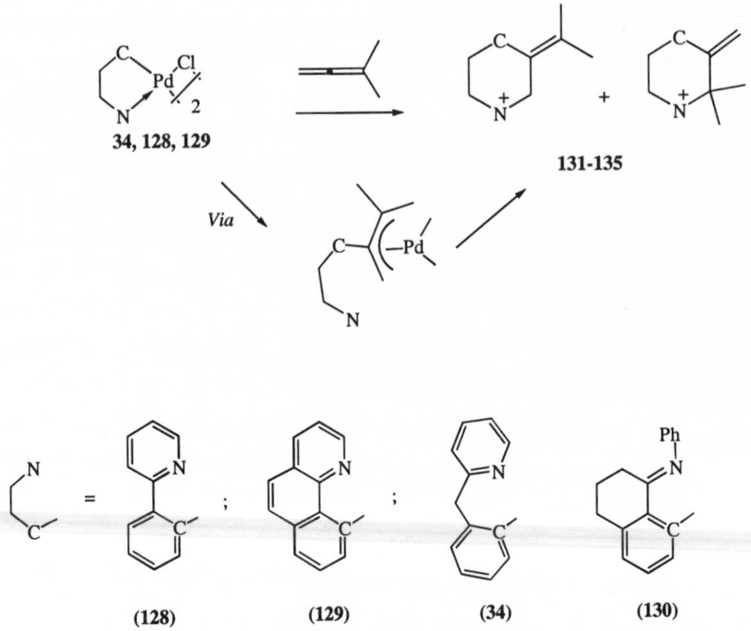

Scheme 35.

of possibilities, and it may well be an interesting alternative to the use of alkynes in the heterocyclization reactions we have studied so far.

IX. CONCLUSION

The aim of this account was to show that cyclometallated complexes, the majority being palladium-containing derivatives, in stoichiometric or catalytic quantities, can certainly be part of the arsenal of the chemist seeking to synthesize cyclic compounds selectively and efficiently. One of the major themes of our research in this respect has been to study these reactions from an organometallic chemist's point of view, with the particular goal of defining the precise role of the metal in the hetero- or carbocyclization process. This has led to some unusual and fruitful mechanistic findings and some useful synthetic approaches to many novel hetero- and carbocyclic systems.

Many of the reactions shown herein involve a formal reductive elimination process between a tertiary amine function or a thioether unit and a vinylic carbon atom, a reaction that has been rarely encountered before our work. Alternatively, heterocycles are obtained via nucleophilic addition of a tertiary amine to an activated olefin unit within the coordination sphere of the transition metal (mainly Pd). Such additions remain somewhat unexplored when compared to the wealth of literature existing for the corresponding reactions involving secondary or primary amine functions.[56]

Used in small scale preparations these reactions offer selective, expedient routes to hetero- and carbocyclic systems. Other salient features of these reactions include the unusual substitution patterns that can be impended onto the final heterocyclic product because of the availability of a large variety of substituents on the alkyne and the high level of regiocontrol achievable.

ACKNOWLEDGMENTS

The CNRS and the Commission of the European Communities are thanked for financial support. John Spencer was the grateful recipient of a thesis fellowship (ULP: 1990–94) from the latter (contract no. ERBSCC*CT000011). T. Heightman and Dr. P. Fowler (Vasella group, ETH-Zürich) are thanked for critical comments concerning this manuscript.

REFERENCES AND NOTES

1. (a) *Comprehensive Heterocyclic Chemistry*, Katrizky, A. R.; Rees, C. W.; Boulton, A. J.; McKillop, A., Eds., Pergamon: Oxford, 1984; (b) *Heterocyclic Chemistry*, Gilchrist, T. L., Longman Scientific and Technical: London, 1985; (c) *Heterocyclic Chemistry*, Joule, J. A.; Smith, G. F., 2nd Edition, Van Nostrand Reinhold International: New York, 1978.

2. Some pertinent reviews include (a) Sakamoto, T.; Kondo, Y; Yamanaka, H. *Heterocycles* **1988**, 27, 2225; (b) Colquhoun, H. M.; Holton, J.; Thompson, D. J.; Twigg, M. V., In *New Pathways for Organic Synthesis: Practical Applications of Transition Metals*, Plenum Press: New York and London, 1984; (c) Davidson, J. L.; Preston, P. N. *Advances in Heterocyclic Chemistry* **1982**, 30, 319 and references cited therein; (d) Hegedus, L. S. *J. Organomet. Chem.* **1993**, 457, 167; (e) Hegedus, L. S. *Angew. Chem., Int. Ed. Engl.* **1988**, 27, 1113; (f) Vollhardt, K. P. C. *Angew Chem., Int. Ed. Engl.* **1984**, 23, 539; (g) Bäckvall, J. E. *Pure Appl. Chem.* **1992**, 64, 429; (h) Knölker, H. J. *Synlett* **1992**, 371; (i) Trost, B. M. *Angew. Chem., Int. Ed. Engl.* **1989**, 28, 1173; (j) Rudler, H.; Audouin, M.; Chelain, E.; Goumont, R.; Massoud, A.; Parlier, A., Pacreau, A.; Rudler, M.; Yefsah, R.; Alvarez, C.; Degado-Reyes, F. *Chem. Soc. Rev.* **1991**, 20, 503.

3. (a) Pfeffer, M. *Pure Appl. Chem.* **1992**, 64, 335; (b) Pfeffer, M. *Recl. Trav. Chim. Pays-Bas* **1990**, 109, 567; (c) Ryabov, A. D. *Synthesis* **1985**, 233.

4. For more complete discussions of such complexes, recent examples include (a) Ryabov, A. D. *Chem. Rev.* **1990**, 90, 403; (b) Omae, I. *Organometallic Intra-*

molecular Coordination Compounds, Elsevier Publ.: Amsterdam, 1986; (c) Dunina, V. V.; Zalevskaya, O. A.; Potapov, V. M. *Russ. Chem. Rev.* **1988**, *57*, 250 and references cited therein.

5. Dupont, J.; Beydoun, N.; Pfeffer, M. *J. Chem. Soc., Dalton Trans.* **1989**, 1715.

6. (a) Cope, A. C.; Friedrich, E. C. *J. Am. Chem. Soc.* **1968**, *90*, 909; (b) Pfeffer, M. *Inorg. Synth.* **1989**, *26*, 211.

7. Dupont, J.; Pfeffer, M.; Theurel, L.; Rotteveel, M. A.; De Cian, A.; Fischer, J. *New J. Chem.* **1991**, *15*, 551.

8. (a) Deeming, A. J.; Rothwell, I. P. *J. Organomet. Chem.* **1981**, *205*, 117; (b) Pfeffer, M. *Inorg. Synth.* **1989**, *26*, 213.

9. Maassarani, F.; Pfeffer, M.; Spencer, J.; Wehman, E. *J. Organomet. Chem.* **1994**, *466*, 265.

10. Ryabov, A. D.; Kazankov, G. M. *J. Organomet. Chem.* **1984**, *268*, 284.

11. (a) Abbenhuis, H. C. L.; Pfeffer, M.; Sutter, J. P.; De Cian, A.; Fischer, J.; Li Ji, H.; Nelson, J. H. *Organometallics* **1993**, *12*, 4464; (b) Pfeffer, M.; Sutter, J. P.; Urriobabeitia, E. *Inorg. Chim. Acta* **1996**, *249*, 63; (c) Ferstl, W.; Sakodinskaya, I. K.; Beydoun-Sutter, N.; Le Borgne, G.; Pfeffer, M.; Ryabov, A. D. *Organometallics*, **1997**, *16*, 411; (d) Pfeffer, M.; Sutter, J. P.; Urriolabeitia, E. P. *Bull Soc. Chim. Fr.*, **1997**, *134*, 947.

12. Vicente, J.; Abad, J. A.; Gil-Rubio, J.; Jones, P. G.; Bembenek, E. *Organometallics* **1993**, *12*, 4151.

13. Pfeffer, M.; Urriolabeitia, E. P.; De Cian, A.; Fischer, J. *J. Organomet. Chem.* **1995**, *494*, 187.

14. Pfeffer, M., Beydoun-Sutter, N.; De Cian, A.; Fischer, J. *J. Organomet. Chem.* **1994**, *453* 139.

15. Spencer, J.; Pfeffer, M.; De Cian, A.; Fischer, J. *J. Org. Chem.* **1995**, *60*, 1005.

16. Maitlis, P. M. *J. Organomet. Chem.* **1980**, *200*, 161.

17. Pereira, M. T.; Pfeffer, M.; Rotteveel, M. A. *J. Organomet. Chem.* **1989**, *375*, 139.

18. Dupont, J.; Pfeffer, M.; Daran, J. C.; Gouteron, J. *J. Chem. Soc., Dalton Trans.* **1988**, 2421.

19. Spencer, J.; Pfeffer, M. *Tetrahedron: Asymmetry* **1995**, *6*, 419.

20. de Vaal, P.; Dedieu, A. *J. Organomet. Chem.* **1994**, *478*, 121.

21. Engel, P. E.; Pfeffer, M.; Dedieu, A. *Organometallics* **1995**, *14*, 3423.

22. (a) Ryabov, A. D.; van Eldik, R.; Le Borgne, G.; Pfeffer, M. *Organometallics* **1993**, *12*, 1386; (b) Samsel, E. G.; Norton, J. R. *J. Am. Chem. Soc.* **1984**, *106*, 5505.

23. Martinez, M.; Muller, G.; Panyella, D.; Rocamora, M.; Solans, X.; Font-Bardia, M. *Organometallics* **1995**, *14*, 552.

24. (a) Maassarani, F.; Pfeffer, M.; Le Borgne, G. *J. Chem. Soc., Chem. Commun.* **1987**, 565; (b) Maassarani, F. et al. *Organometallics* **1987**, *6*, 2029; (c) Maassarani, F. et al. *Organometallics* **1987**, 6, 2043.

25. Wu, G.; Rheingold, A. L.; Heck, R. F. *Organometallics* **1987**, *6*, 2386.

26. Beydoun, N.; Pfeffer, M. *Synthesis* **1990**, 729.

27. Huggins, J. M.; Bergman, R. G. *J. Am. Chem. Soc.* **1981**, *103*, 3002.

28. Spencer, J.; Pfeffer, M.; Kyritsakas, N.; Fischer, J. *Organomet.* **1995**, *14*, 2214.

29. Spencer, J.; Maassarani, F.; de Cian, A.; Fischer, J. *Tetrahedron: Asymmetry* **1994**, *5*, 321.

30. Wu, G.; Geib, S. J.; Rheingold, A. L.; Heck, R. F. *J. Org. Chem.* **1988**, *53*, 3238.

31. Bruce, M. I.; Goodall, B. L.; Stone, F. G. A. *J. Chem. Soc., Dalton Trans.* **1975**, 1651.

32. Garn, D.; Knoch, F.; Kisch, H. *J. Organomet. Chem.* **1993**, *444*, 155.

33. (a) Müller, E. *Synthesis* **1974**, *11*, 761; (b) Kozikowski, A. P.; Wetter, H. F. *Synthesis* **1976**, 586; (c) Meier-Brocks F.; Weiss, E. *J. Organomet. Chem.* **1993**, *453*, 33; (d) Fagan, P. J.; Nugent, W. A.; Calabrese, J. C. *J. Am. Chem. Soc.* **1994**, *116*, 1880; (e) Cámpora, J.; Guttiérrez, E.; Monge, A.; Palma, P.; Poveda, M. L.; Ruíz, C.; Carmona, E. *Organometallics* **1994**, *13*, 1728; (f) Buchwald, S.L.; Fang, Q. *J. Org. Chem.* **1989**, *54*, 2793; (g) Le Bozec, H. L.; Dixneuf, P. H. *J. Chem. Soc., Chem. Commun.* **1983**, 1462; (h) Bianchini, C.; Mealli, C.; Meli, A.; Sabat, M.; Silvestre, J.; Hoffmann, R. *Organometallics* **1986**, *5*, 1733; (i) Raseta, M. E.; Mishra, R. K.; Cawood, S. A.; Welker, M. E.; Rheingold, A. L. *Organometallics* **1991**, *10*, 2936; (j) Yih, K. H.; Lin, Y. C.; Cheng, M. C.; Wang, Y. *Organometallics* **1994**, *13*, 1561; (k) Choi, N.; Kabe, Y.; Ando, W. *Tetrahedron Lett.* **1991**, *35*, 4573; (l) Moody, C. J.; Taylor, R. J. *Tetrahedron* **1990**, *46*, 6501.

34. Nishiyama, K.; Tanaka, N. *J. Chem. Soc., Chem. Commun.* **1983**, 1322.

35. Gies, A. E.; Pfeffer, M.; Spencer, J. unpublished results, 1994.

36. (a) Tao, W.; Silverberg, L. J.; Rheingold, A. L.; Heck, R. F. *Organometallics* **1989**, *8*, 2550; (b) Larock, R. C.; Yum, E. K. *J. Am. Chem. Soc.* **1991**, *113*, 6689; (c) Larock, R. C.; Yum, E. K.; Doty, M. J.; Sham, K. K. C. *J. Org. Chem.* **1995**, *60*, 3270.

37. Brown, J. M.; Cooley, N. A. *Chem. Rev.* **1988**, *88*, 1031.

38. Maassarani, F.; Pfeffer, M.; Le Borgne, G. *Organometallics* **1990**, *9*, 3003.

39. (a) Larock, R. C.; Doty, M. J.; Cacchi, S. *J. Org. Chem.* **1993**, *58*, 4579; (b) Vicente, J.; Abad, J. A.; Gil-Rubio, J. *J. Organomet. Chem.* **1992**, *436*, C9; (c) Liebeskind, L. S.; Gasdaska, J. R.; McCallum, J. S.; Tremont, S. J. *J. Org. Chem.* **1989**, *54*, 669; (d) Grigsby, W. J.; Main, L.; Nicholson, B. K. *J. Organomet.* **1993**, *12*, 397; (e) Vicente, J.; Abad, J. A.; Gil-Rubio, J. *Inorg. Chim. Acta* **1994**, *222*, 1–4.

40. (a) Bahsoun, A.; Dehand, J.; Pfeffer, M., Zinsius, M.; Bouaoud, S. E.; Le Borgne, G. *J. Chem. Soc., Dalton Trans.* **1979**, 547; (b) Sutter, J. P.; Pfeffer, M.; De Cian, A.; Fischer, J. *Organomet.* **1992**, *11*, 386; (c) Albert, J.; Granell, J., Sales, J., Solans, X. *J. Organomet. Chem.* **1989**, *379*, 177; (d) Lopez, C.; Bosque, R.; Solans, X.; Font-Bardia, M.; Silver, J., Fern, G. *J. Chem. Soc., Dalton Trans.* **1995**, 1839; (e) Vicente, J.; Saura-Llamas, I.; Ramirez de Arellano, M. C. *J. Chem. Soc., Dalton Trans.* **1995**, 2529.

41. Taylor, S. H.; Maitlis, P. M. *J. Am. Chem. Soc.* **1978**, *100*, 4700.

42. Pfeffer, M.; Sutter, J. P.; DeCian, A.; Fischer, J. *Tetrahedron* **1992**, *48*, 2427.

43. Beydoun, N.; Pfeffer, M.; De Cian, A.; Fischer, J. *Organomet.* **1991**, *10*, 3693.

44. (a) Pfeffer, M.; Rotteveel, M. A.; De Cian, A.; Fischer, J.; Le Borgne, G. *J. Organomet. Chem.* **1991**, *413*, C15; (b) Pfeffer, M.; Sutter, J. P.; De Cian, A.; Fischer, J. *Organomet.* **1993**, *12*, 1167.

45. (a) Wu, G.; Rheingold, A. L.; Geib, S. J.; Heck, R. F. *Organometallics* **1987**, *6*, 1941; (b) Wu, G.; Rheingold, A. L.; Geib, S. J.; Heck, R. F. *Organometallics* **1986**, *5*, 1922.

46. Ossor, H., Pfeffer, M.; Jastrzebski, J. T. B. H.; Stam, C. H. *Inorg. Chem.* **1987**, *26*, 1169.

47. Dupont, J.; Rotteveel, M. A.; Pfeffer, M.; De Cian, A.; Fischer, J. *Organomet.* **1989**, *8*, 1116.
48. Pfeffer, M.; Rotteveel, M. A.; Sutter, J. P.; De Cian, A.; Fischer, J. *J. Organomet. Chem.* **1989**, *371*, C21.
49. (a) Vicente, J.; Abad, J. A.; Gil-Rubio, J.; Jones, P. G. *Organometallics* **1995**, *14*, 2677; (b) Vicente, Jo et al. *Inorg. Chim. Acta* **1994**, *222*, 1.
50. Pfeffer, M.; Rotteveel, M. A.; De Cian, A.; Fischer, J. *J. Org. Chem.* **1992**, *57*, 233.
51. Pfeffer, M.; Sutter, J. P.; De Cian, A.; Fischer, J. *Inorg. Chim. Acta* **1994**, *220*, 115.
52. (a) van der Schaaf, P. A.; Sutter, J. P.; Grellier, M.; van Mier, G. P. M.; Spek, A. L.; van Koten, G.; Pfeffer, M. *J. Am. Chem. Soc.* **1994**, *116*, 5134; (b) Grellier, M.; Pfeffer, M.; van Koten, G. *Tetrahedron Lett.* **1994**, *35*, 2877.
53. Rülke, R. E.; Kliphuis, D.; Elsevier, C. J.; Fraanje, J.; Goubitz, K.; van Leeuwen, P. W. N. M.; Vrieze, K. *J. Chem. Soc., Chem. Commun.* **1994**, 1877.
54. Chengebroyen, J.; Pfeffer, M.; Sirlin, C. *Tetrahedron Lett.* **1996**, *37*, 7263; Diederen, J. H.; Frühauf, H. W.; Hiemstra, H.; Vrieze, K.; Pfeffer, M. *Tetrahedron Lett.* **1998**, *39*, 4111.
55. Akermark, B.; Akermark, G.; Hegedus, L. S.; Zettenberg, K. *J. Am. Chem. Soc.* **1981**, *103*, 3037. Julia, M.; Nel, M.; Righini, A.; Uguen, D. *J. Organomet. Chem.* **1982**, *235*, 113. Larock R. C.; Varaprath, S.; Lau, H. H.; Fellows, C. A. *J. Am. Chem. Soc.* **1984**, *106*, 5274.
56. West, F. G.; Naidu, B. N. *J. Am. Chem. Soc.* **1993**, *115*, 1177; McKillop, A.; Stephenson, G. R.; Tinkl, M. *J. Chem. Soc., Perkin Trans. 1* **1993**, 1827.

SYNTHETIC APPLICATION OF CYCLOPENTADIENYL MOLYBDENUM(II)- AND TUNGSTEN(II)-ALLYL AND DIENE COMPOUNDS IN ORGANIC SYNTHESIS

Rai-Shung Liu

Advances in Metal-Organic Chemistry
Volume 6, pages 145–186.
ISBN: 0-7623-0206-2

146   RAI-SHUNG LIU

I. INTRODUCTION

With the stabilizing effect of a cyclopentadienyl group, molybdenum
compounds are chemically robust to sequential chemical reactions. Thus
one can elaborate multiple functions on their hydrocarbyl ligands, leading to unexpected chemical reactions that have both fundamental and
practical implications. Other multidentate ligands, such as hydrido(3,5-
dimethylpyrazolyl)borate(Tp′)[1-4] and 1,2-dicarbononaborane[5] anions
$C_2B_9H_{11}^{2-}$, similarly contribute to the chemical stability of their corresponding molybdenum(II) compounds, but possibly follow reaction
pathways distinct from their cyclopentadienyl analogs. Cyclopentadienyl molybdenum compounds, particularly those having a π-allyl or
π-diene group, have proved useful in organic synthesis. Besides their
stability to air and chemical reagents, the preparation of these allyl
compounds is inexpensive, and they are easily handled in preparations
on a multigram scale. Separating reaction products in this system can be
conveniently conducted chromatographically on a conventional silica or
alumina column, so it is not tedious.

Pearson and others have published review articles,[6-8] which are in part
concerned with the organic reactions of cyclopentadienyl molybdenum
π-allyl compounds. Because of limited work at that time, the scope of
this area made it difficult to comprehend its synthetic potential compared
with organoiron and chromium compounds.[6(a),9-15] Recent developments
in this field emerged rapidly, particularly in diastereo- and enantioselective organic synthesis via molybdenum π-allyl and π-diene compounds.
These new reports are significant and diverse, and it is beneficial to have
a current account of this area.

Cyclopentadienylmolybdenum π-allyl and π-diene compounds are
the main focus in this field due to the pioneering work of Faller.[16-21]
Subsequent utilization of these compounds by Pearson[6,7] is acknowledged. Not all reported π-allyl reagents produce useful organic compounds. Many π-allyl compounds exist that possess useful ligand
transformation, and they may enable special organic syntheses with
further effort. For example, in a recent paper, Green et al.[22] reported that
the molybdenum-η[3]-butadienyl compound **1** undergoes protonation to
give a η[4]-vinylketene cation **2** (Scheme 1). Because of an easily replace-

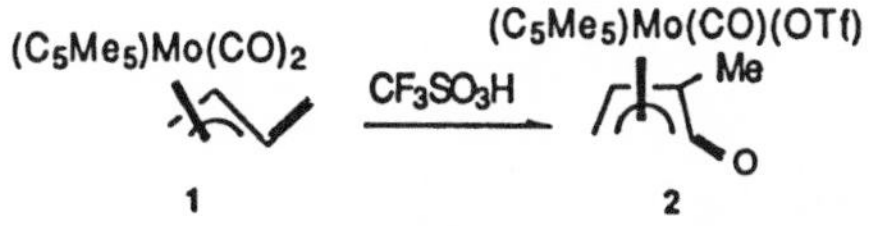

$(C_5Me_5)Mo(CO)_2$ $(C_5Me_5)Mo(CO)(OTf)$

    CF_3SO_3H

1   **2**

Scheme 1.

able $CF_3SO_3^-$ ligand in this compound, its chemistry may be compared to those of reported cobalt,[23] chromium,[24,25] and iron-η^4-vinyl ketene[26] species, which react further with alkynes to produce phenol, furan, and other organic compounds. This reaction will become more valuable if other electrophiles, such as aldehydes, acid halides, and unsaturated enones can be applied to generate the corresponding η^4-vinyl ketene cations.

This review is intended to familiarize readers with the organic chemistry of molybdenum π-allyl compounds and also to describe those π-allyl compounds that undergo fundamental and significant organometallic reactions, hopefully to stimulate further creative work in this field.

II. π-ALLYL AND π-DIENE CHEMISTRY

A. Synthesis of π-Allyl Compounds

From Allyl Halides, Acetates, Tosylates, and Diphenylphosphinates

Reactions of $CpMo(CO)_3Na$ and allylic halides or tosylates in THF afford $CpMo(CO)_3(\eta^1$-allyl), and subsequent decarbonylation of these products by anhydrous Me_3NO or by photolysis give π-allyl compounds.[27] By this method, allyl compounds $CpMo(CO)_2(\pi$-1-R-allyl) (R = H, alkyl, aryl, vinyl, aldehyde, COMe, COOMe)[28,29] are obtained with

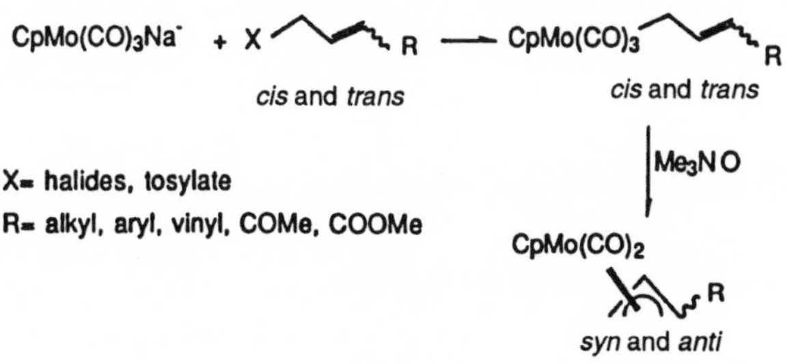

$CpMo(CO)_3Na^-$ + X ⟋⟍ R ⟶ $CpMo(CO)_3$ ⟋⟍ R

     cis and *trans*   *cis* and *trans*

X= halides, tosylate

R= alkyl, aryl, vinyl, COMe, COOMe

                 $\bigg\downarrow Me_3NO$

            $CpMo(CO)_2$

                  R

           syn and *anti*

Scheme 2.

reasonable yields (50–70%). Because of the only moderate thermal stability of the η^1-allyl compounds, the molybdenum π-allyls $CpMo(CO)_2(\pi$-allyl) are generally synthesized in one stage, that is, direct addition of Me_3NO to the η^1-allyl compound in CH_2Cl_2. The isomeric ratios of *syn(trans)* to *anti(cis)* of these η^3 products are generally consistent with the *cis* and *trans* composition of the starting allyl halides.[29(c)] The main restriction of this reaction is that it operates only with primary halides $XCH_2CR=CR_2$. In reactions involving secondary and tertiary halides, $CpMo(CO)_3X$ is formed exclusively.

Oxidative addition of $Mo(CO)_3(CH_3CN)_3$ with allyl halides followed by LiCp treatment is the most convenient and practical method of preparing molybdenum and tungsten π-allyl compounds $CpM(CO)_2(\pi$-allyl)(M=Mo, W) with high yields.[30] In contrast, allylic tosylates and methanesulfonates are too unstable to be generally useful. Allylic acetates have been successfully used in several cases,[31,32] and there is a report of allylic diphenylphosphinate as an efficient precursor.[33] The reactions are also efficient for synthesizing cyclic $CpMo(CO)_2(\pi$-allyl) compounds in which the intermediates $Mo(CO)_2(CH_3CN)_2(\pi$-allyl)X are generally present as well-defined yellow solids. Although no molybdenum *anti*-π-allyl compounds have been produced by this method, the reaction of *cis*-1-R-allylic halides, such as *cis*-crotyl halide, with a Mo(0) complex also gives the expected anti-1-R-allylic compound with negligible isomerization to its more stable syn-η^3-allyl isomer.[34] The molybdenum anti-η^3-allyl compounds are prepared either from $CpMo(CO)_3Na$ and *cis*-3-R-allylic halides, as mentioned before, or by $NaBH_4$ reduction of the $CpMo(CO)_2(\eta^4$-*cis*-diene)$^+$ cation. (see before)

The stereochemical course of oxidative addition of the chiral allylic acetates **3** with $Mo(CO)_3(CH_3CN)_3$ has been elucidated.[31(b)] The outcome is that formation of the Mo-allylic bond proceeds via retention of configuration with respect to the acetate–carbon bond cleavage, as shown in Scheme 4. Similarly, the reaction between the Mo(0) species and the chiral cyclic allylic acetate[32] **5** gives the retention product **6**. The

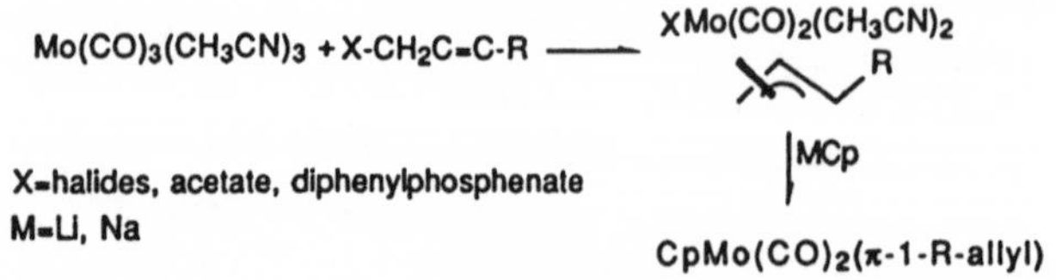

$$Mo(CO)_3(CH_3CN)_3 + X\text{-}CH_2C\text{=}C\text{-}R \longrightarrow \begin{array}{c} XMo(CO)_2(CH_3CN)_2 \\ \diagdown\diagup\diagdown\diagup^R \end{array}$$

X=halides, acetate, diphenylphosphenate
M=Li, Na

$$\Big|MCp$$

$$CpMo(CO)_2(\pi\text{-}1\text{-}R\text{-}allyl)$$

Scheme 3.

Scheme 4. $Mo(O) = Mo(CO)_3(CH_3CN)_3$

optical activities of the two allyl products are consistent with those of the starting chiral acetate compounds. The observed stereochemistry here is distinct from the corresponding reactions of Pd(0) complexes which proceed via inversion of configuration.[35] Because of the labile CH_3CN ligand of $Mo(CO)_3(CH_3CN)_3$, it is possible that before oxidative cleavage of the allylic-acetate bond, formation of a chelation intermediate through coordination of the allylic C=C bond and the carbonyl group with molybdenum accounts for the retention of stereochemistry.

In contrast, the reactions between $Mo(CO)_3(CH_3CN)_3$ and chiral allylic bromides proceed via inversion with respect to cleavage of the allylic carbon–halide bond. Examples are provided by the work of Liebeskind[36(a)] as shown in Scheme 5. The ring conformation is important for the reaction, and an interesting case is the chiral *trans* bromide isomer **7** whose pseudoaxial bromide group is displaced by $Mo(CO)_3(CH_3CN)_3$ presumably according to an S_N2 mechanism, whereas the *cis* isomer **9**

Scheme 5. $Mo(O) = Mo(CO)_3(CH_3CN)_3$

with bromide in the pseudoequatorial site fails to react with Mo(0) species to give the expected π-allyl product.[36(b)]

From Metal-η^1-Hydrocarbyl Compounds

Molybdenum-η^1-hydrocarbyl compounds $CpMo(CO)_3R$ are prone to carbonyl insertion to form an acyl group. Subsequent insertion of the acyl group into an olefin or an alkyne is feasible and well documented.[37] Several molybdenum π-allyl compounds have been prepared according to this method. Reactions between $CpMo(CO)_3Na$ and allene halides $Br(CH_2)_nC=C=CH_2$ (n = 1–4) produce allylic ketone compounds[22,38(a)] through insertion of the acyl group of the intermediate (A) into its allene C=C bond (Scheme 6). The reaction of $CpMo(CO)_3Na$ and 3-chloromethylpent-3-ene-2-one in cold THF gives the η^1-allyl compound 10. This compound undergoes a slow, skeletal rearrangement in diethyl ether at room temperatures to give a η^3-allyl butyrolactone (11) (yields 65%).[38(b)] Experimental evidence indicates that the product is derived from intramolecular insertion of an acyl group into the organic ketone

Scheme 6. $(M^- = CpMo(CO)_3)$

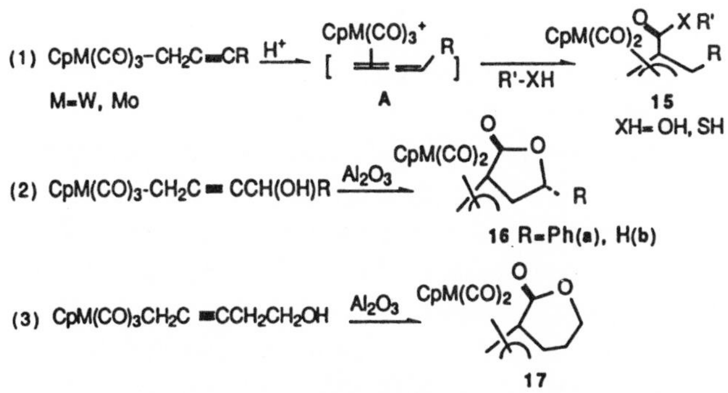

Scheme 7.

group rather than enol attack on the acyl group. Such an intramolecular insertion and its analogous η^3-butyrolactone allyl product **13** is also obtained from carbonylation of the η^1-enonyl compound **12** under high CO pressure (60 atm).[39] Further treatment of this enonyl complex with excess *t*-butylacetylene at room temperature gives a remarkable η^3-pyranyl complex **14**.[40] In the conversion of the vinyl ketone complex into the η^3-pyranyl complex **14**, it is proposed that the intermediate (**B**) forms and subsequently undergoes either a thermally allowed disrotatory ring closure or an insertion reaction to form a cyclopentadiene molybdenum species **C**.

Protonation of propargyl compounds $CpMo(CO)_3(\eta^1\text{-}CH_2C{\equiv}CR)(R{=}H, Me, Ph, CH{=}CH_2, CH{=}C{=}CH_2)$[41–43] in diethyl ether or other nonpolar solvents produces the η^2-allene cations **A** (Scheme 7), which further react with water, alcohol, and thiols to give the molybdenum and tungsten-η^3-2-carboxylated allyl complexes **15**. This reaction was extended to the hydroxyl-σ-alkynyl molybdenum compounds $CpMo(CO)_3[CH_2C{\equiv}C(CH_2)_nCR(OH)]$[44] (n = 0, R=H, Ph ; n = 1, R = H) that are activated by alumina to produce the γ- or δ-lactone π-allyl compounds **16–17** with 30–55% yields through an intramolecular cyclization. This reaction proceeds with high diastereoselectivity in the case of phenyl (R = Ph) in which only one diastereomer is obtained for the butyrolactone product **16a**.

B. [CpMo(CO)NO(π-allyl)]$^+$

The carbonyl group of $CpMo(CO)_2(\pi\text{-allyl})$ is easily replaced by the $NOBF_4$ salt to give the cationic salt $[CpMo(CO)(NO)(\pi\text{-allyl})]^+$,[16,17,19,45] which functions in synthetic equivalence as an allyl cation because of its

facile reactivity with numerous nucleophiles to give
Cpmo(CO)(NO)(Nu-CH$_2$CH=CH$_2$). The latter is easily oxidized by air
or Ce(IV), liberating α-functionalized olefins. Scheme 8 shows the
regio-controlled substitution of the CO group by NO for CpMo(CO)$_2$(π-
syn-crotyl) **18** in which NO and methyl groups are mutually *trans* in the
exo[46] conformer **19b** and *cis* in the *endo* conformer **19a**.[19] Nucleophilic
attack on the two conformers of this allyl cation[19] proceeds with high
stereospecificity, that is, attack at the allyl carbon *trans* to NO in the *endo*
isomer **19a** and *cis* to NO in the *exo* isomer **19b**. In this manner, the
resulting olefin complexes **20a** and **20b** have the most stable orientation
due to the electronically favorable arrangement of their olefin C=C bond
parallel to the carbonyl group.

Scheme 8 illustrates one instance[45] in which attack on the two conformers **19a** and **19b** by the enamine of isobutaldehyde gives only
η^2-olefin complexes **21** as a mixture of two conformers. To assert the
importance of controlling the nucleophilic stereochemistry by the asymmetric CpMo(CO)NO$^+$ fragment, the two diastereomers **22** and **24** were
independently prepared[47] and their alkylation reaction produced **23** and
25, respectively, consistent with the previous model. Notably, the sterically demanding phenyl group of **24** does not prevent *cis* addition to the
NO group in the *exo* isomer.

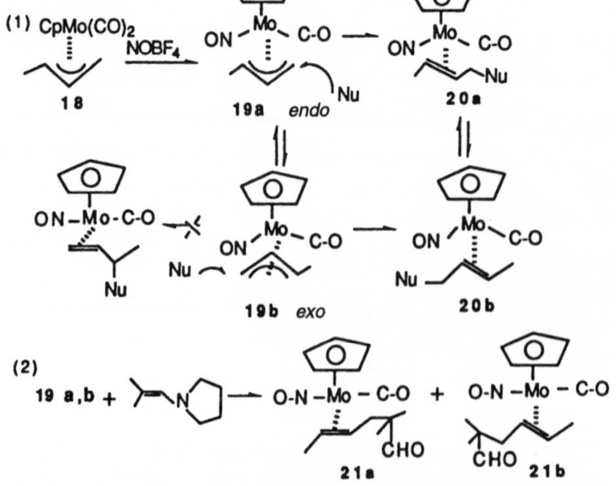

Scheme 8.

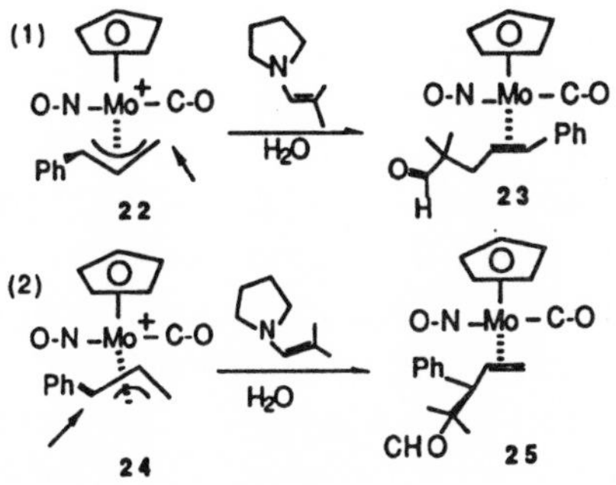

Scheme 9.

A few chiral CpMo(CO)(NO)(π-allyl)⁺ cationic compounds have been prepared. Introduction of a chiral neomenthyl (NM) group onto the cyclopentadienyl group, such as **26**,[45] allows simple resolution of its *S*-enantiomer, which affords 2,2,3-trimethylhex-4-enal **28** with ee > 96% after alkylation by the enamine of isobutyaldehyde. Chiral complex (+)-(NMCp)Mo(NO)Br(π-cyclooctenyl) **30** was resolved from diastereomeric mixtures of (NMCp)Mo(NO)Br(η³-C₈H₁₃).[48] Treating this compound with AgPF₆ in the presence of CO yields the chiral cation **31** with retention of configuration, which subsequently reacts with water with high selectivity to give (−)-(*R*)-3-hydroxylcyclooctene **32** with 93% ee.

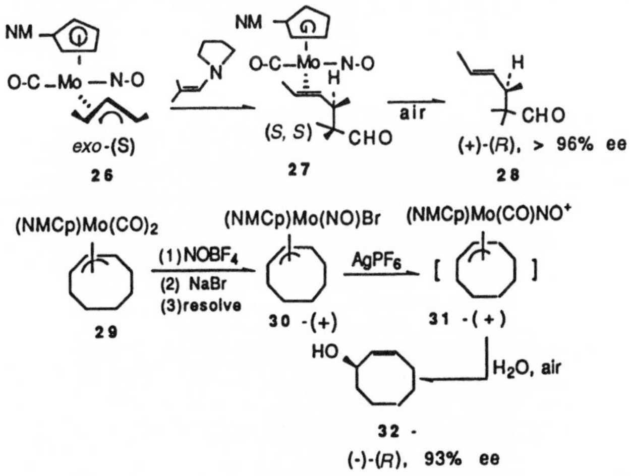

Scheme 10.

The functional equivalence of $CpMo(CO)NO(\eta^3\text{-allyl})^+$ to an allyl cation enables versatile decomplexation of the corresponding dicarbony allyl compounds. The reaction is generally conducted as a one-step synthesis without isolating the allyl cation. As shown in Scheme 11, treating **4** with $NOBF_4$ in equimolar proportions, followed by addition of $NaCH(COOMe)_2$ gives the chiral olefin **33**[31(b)] that maintains the optical purity of the original allyl compound **4**.

Intramolecular nucleophilic attack at the allyl carbon terminus via this allyl nitroso cationic species is feasible. Examples are represented by compounds **34**[49] and **36**,[50] whose side-chain carboxylate or hydroxyl termini undergo cyclization in basic conditions to give the cyclized products **35** and **37** in reasonable yields (> 50%).

Halogen-induced demetallation of the dicarbonyl π-allyl complexes $CpMo(CO)_2$ creates a reactive Mo(iv)-allyl cationic intermediate **A** (Scheme 12) which is subject to nucleophilic attack. This method is effectively used to decomplex dicarbonyl compounds, and its stereo-chemical outcome has been elucidated by Pearson et al.[51,52] In the lactonization reaction (Eq. 1), the carboxylate group of **38** attacks at the allyl group *trans* to the metal fragment to give **39** with good yields. If no other nucleophile is present, the iodide or bromide ion attacks the allyl group opposite the metal fragment to give *cis*-allylic halide derivatives, such as **41** and **42**. This reaction provides an alternative method for generating an allyl cationic species. According to some reports,[52] Ce(iv) ion oxidizes dicarbonyl molybdenum π-allyl compounds to create a highly electrophilic allyl center, but the reaction lacks chemoselectivity and works well only in limited cases.

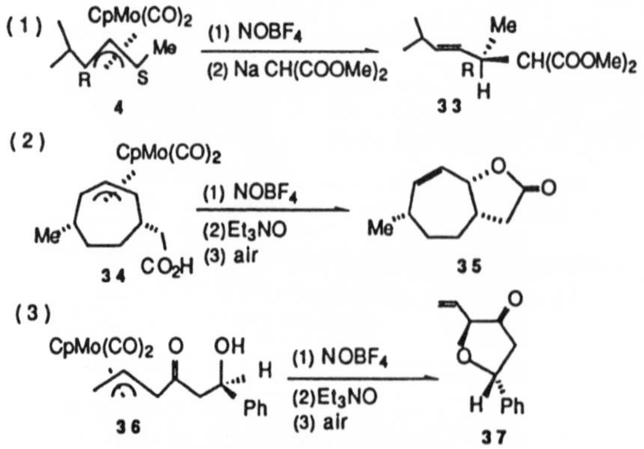

Scheme 11.

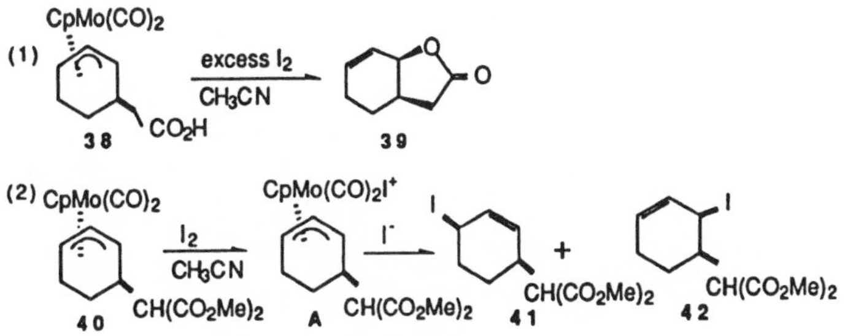

Scheme 12.

C. Cpmo(NO)X(π-allyl) (X = halides)

Stereoselective synthesis of a secondary homoallylic alcohol is an important topic in organic reactions. Utilizing an organometallic fragment as a stereotemplate for asymmetrical induction is a practical method. The well-established systems include the σ-allyl compounds of chromium, stannanes, aluminum, and boranes.[53-55] Cpmo(NO)X(π-allyl) (X = halides) compounds are easily prepared from LiX and Cpmo(NO)(CO)(π-allyl)$^+$.[56,57] Faller reported[58-60] that this allyl compound is also effective in giving a secondary homoallylic alcohol in condensation with aldehydes with a high degree of stereoselectivity. Because of the distinct electronic properties of halide versus nitrosyl, the metal-allyl bonding of compounds of this class is asymmetrical and prone to dissociation to η^1-allyl to leave a vacant site for aldehyde to coordinate, leading the allyl group to add to the aldehyde. This reaction is stereospecific, as manifested in the two instances following. Upon

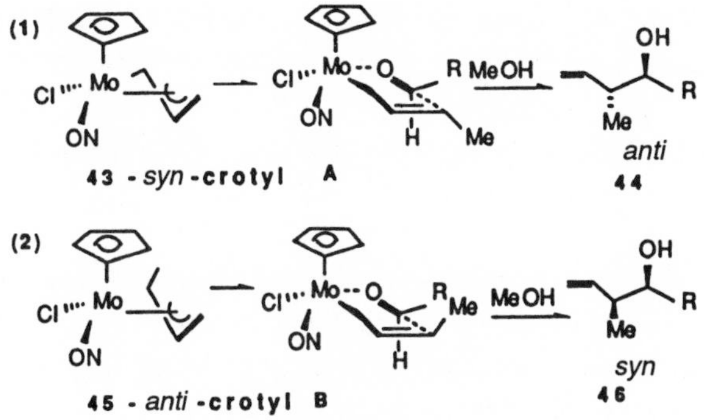

Scheme 13.

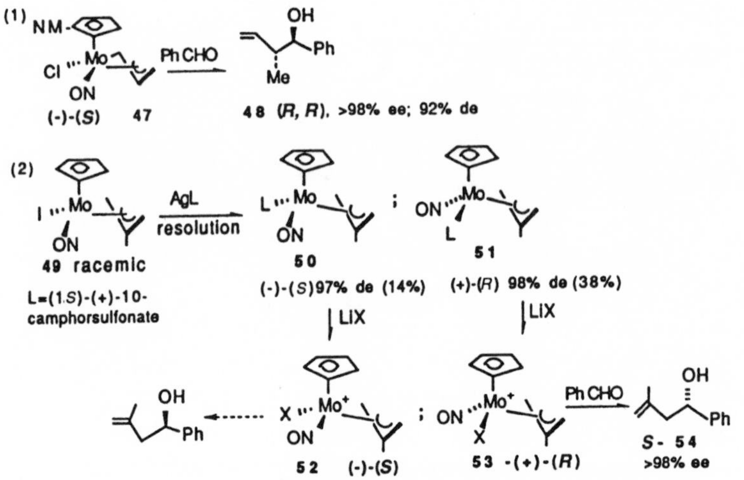

Scheme 14.

treatment with aldehydes, Mo-π-syn-crotyl complex **43** generates *anti* homoallylic alcohol **44** (de = 93%) whereas Mo-π-anti-crotyl compound **45** gives the *syn* alcohol **46** (de = 98%). Control of the product stereo-chemistry is attributed to a chairlike transition-state structure, repre-sented by **A** and **B**[21,59] in which the aldehyde R group occupies the equatorial site to minimize interligand steric hindrance. This reaction was extended to Mo-allyl compounds having functionality, such as 1-α-hydroxyl-allyl **136**, which condenses with aldehyde to give 1,3-diol[29(c)] with good diastereoselectivity (see Scheme 29) and this result reflects a useful extension of this reaction.

Chiral compounds of this class condense with aldehydes to give a reasonable degree of enantioselective synthesis of homoallylic alcohols. Reaction of (−)-(S)-(neomenthylcyclopentadienyl) Mo(NO)Cl(π-syn-crotyl) **47** with benzaldehyde affords (+)-(R, R)-2-methyl-1-phenyl-3-buten-1-ol **48** with >98% ee.[21] (+) and (−) chiral CpMo(NO)(π-2-methallyl)X (X = (S)-(+)-10-camphorsulfonate) com-plexes **50** and **51** were prepared[60] and separated from each other by resolving a diastereomeric mixture of CpMo(NO)(L)(π-2-methylallyl) **49** (L = (+)-1(S)-camphorsulfonate) through fractional crystallization. In a typical run the yields of (−)-(S)-**50** (97% de) and (+)-(R) **51** (de = 98%) were 14% and 38%, respectively. For these chiral compounds, replacing their camphorsulfonate group with halides allows generating enan-tiomerically pure CpMo(NO)X(π-methallyl) (X = halides)[(−)-(S)-**52**,

(+)-(*R*)-**53**] with configuration retained at the metal center. Treatment of benzaldehyde with (+)-(*R*)-CpMo(NO)(π-methallyl)X **53**(X = Cl, Br, I) yields the chiral homoallylic alcohol (*S*)-(–)-3-methyl-1-phenyl-3-buten-1-ol **54** with ee values between 90 and 98%.

D. $CpMo(CO)_2$(π-diene) Cations

The most important feature that makes $CpMo(CO)_2$(π-allyl) useful in organic synthesis is its facile transformation to the $CpMo(CO)_2$(π-diene)$^+$ cation. This reaction is particularly suitable for ring systems with six or seven members because their α-methylene protons adjacent to the Mo-π-allyl moiety are subject to Ph_3CBF_4-promoted hydride abstraction.[20,49,52] For an acyclic system, this method is successful in only a few cases, mainly on simple molybdenum (π-*anti*-crotyl) compounds (Eq. 2, Scheme 15). The molybdenum-π-diene cation is useful because of its reasonable reactivity to nucleophilic attack on carbanions and common

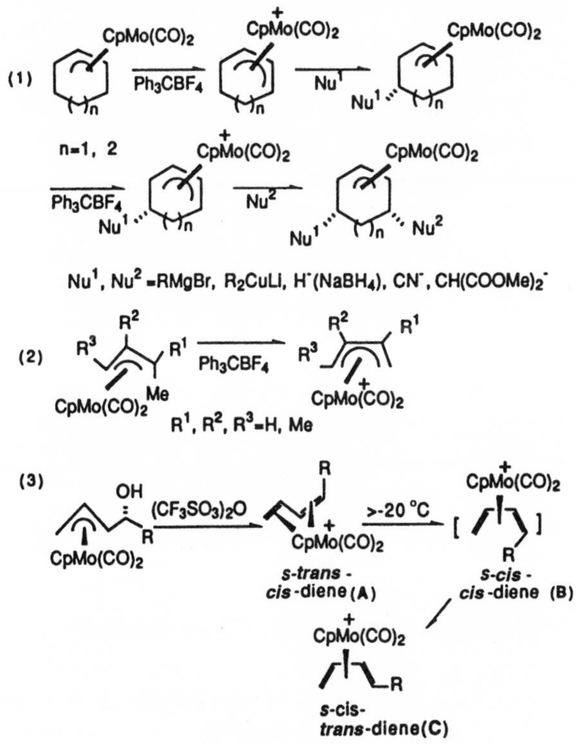

(1)
CpMo(CO)2
Ph3CBF4
CpMo(CO)2
Nu1
CpMo(CO)2
n=1, 2
Ph3CBF4
Nu2
Nu1, Nu2 =RMgBr, R2CuLi, H⁻(NaBH4), CN⁻, CH(COOMe)2⁻
(2)
R1, R2, R3=H, Me
(3)
OH
(CF3SO3)2O
>-20 °C
s-trans - cis-diene (A)
s-cis - cis-diene (B)
s-cis- trans-diene(C)

Scheme 15.

organometallic alkylating reagents to regenerate the allyl compounds. When this reaction is performed on these two cyclic systems, it enables a double nucleophilic addition by means of repeated hydride abstraction and nucleophilic addition.

In an acyclic system, the η^4-s-trans-diene cation is generated from protonation of a trans-η^3-pentadienyl[61] or an α-alkoxylallyl compound.[29] In the latter case, the leaving ROH group is displaced by the CpMo(CO)$_2$ fragment in an intramolecular S$_N$2 fashion to give the η^4-s-trans-cis -diene cation **A**. All reported s-trans-diene cations are stable only at low temperatures (< –40 °C, CD$_2$Cl$_2$). In this case the s-trans-diene cation undergoes irreversible transformation to the more stable s-cis-trans-diene form **C** via an s-cis-cis-diene intermediate **B** as the temperature is raised. The isomerization of intermediate **B** to **C** operates by a ring-flipping mechanism.[17]

Intramolecular nucleophilic attack at the π-diene cation proceeds relatively smoothly providing that the products have a five- or six-membered ring. Green reported[62] representative cases shown in Scheme 16. (η^5-Indenyl)Mo(CO)$_2$(CH$_3$CN)$_2$ **55** proved a useful reagent for intramolecular cyclization of the molybdenum-η^4-dienes **56** and **57** to the η^3-tetrahydrofuran **58** and -pyran **59** compounds. A molybdenum-π-complex of bicyclic spiroether **61** is likewise generated from diene **60** by desilylation. In the presence of DBU, the branched CR$_2^-$(R = COOMe,

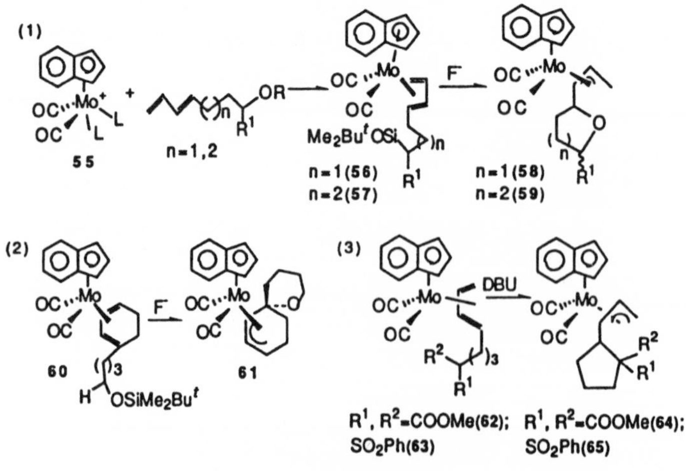

Scheme 16.

SO_2Ph) terminus of **62** and **63** undergo intramolecular carbon–carbon formation to yield the π-allyl complexes **64, 65** of cyclopentane products.

In cyclic Mo-π-diene cations, the methylene protons adjacent to the π-diene group, **66** and **67**, are highly acidic and readily deprotonated by tertiary amine to give the cyclic pentadienyl compounds **68** and **69**.[63] Deprotonation of the corresponding protons on the acyclic *s-cis*-diene cation is much less common. Only one case is reported[64] (Eq. 2, Scheme 17). An unexpected result was obtained in deprotonating *s-cis*-butadiene cation **72** by $LiN(SiMe_3)_2$ yielded a η^3-buta-2,3-dien-1-yl compound **1**.[21,64] Reprotonation of this η^3-allyl complex fails to regenerate the original diene cation, but instead yields η^4-vinylketene species **2** by incorporating a carbonyl group into the Mo=C bond of the η^3-vinyl carbenium intermediate **A**.

III. MULTIPLE FUNCTIONALIZATION AND SYNTHETIC APPLICATIONS

A. Cyclic π-Allyl Systems

With various versatile demetallation methods for compounds of $CpMo(CO)_2$(π-allyl) and their facile transformation to π-diene cations, implementing stereocontrolled multiple functionalization of these compounds appears ultimately important for application to organic synthesis.

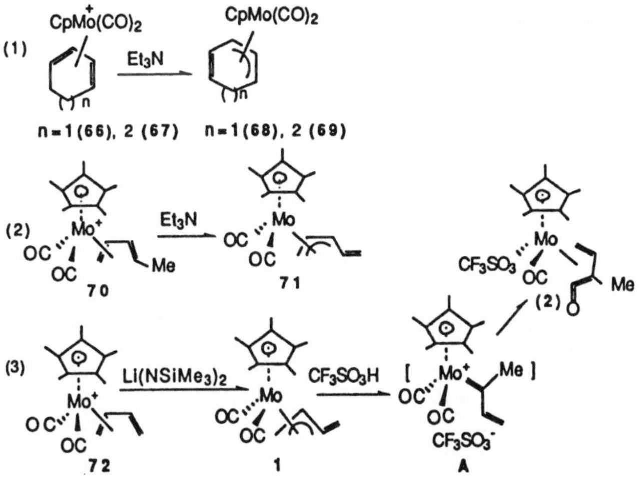

Scheme 17.

A cyclic molecule is generally conformationally more rigid than its acyclic counterpart, hence stereocontrolled functionalization of the former proceeds more readily. As mentioned previously (Scheme 15), the facile interconversion between allyl and diene in a cyclic system makes it a prominent target for implementing stereocontrolled functionalization. According to this method, it is easy to establish two *anti* substituents R^1 and R^2 adjacent to a π-allyl fragment on cyclohexane and -heptane rings[48-52] (Scheme 18, Eq. 1). Pearson used the dialkylation cyclohexane products[52] to generate pericyclic compounds. Treatment of **73** with excess iodine gives the tetrahydrobenzofuran derivative **74**, and the reaction proceeds via internal nucleophilic attack on the allyl cation by the enol form of the keto ester group. MCPBA-oxidation of **75** induces internal cyclization to yield the tricyclic compound **76**. I_2-promoted lactonization of **78** gives **79** that is further converted to acyclic molecule **81** to construct a C-4 to C-9 subunit of tylosin or carbomycin. Finally,

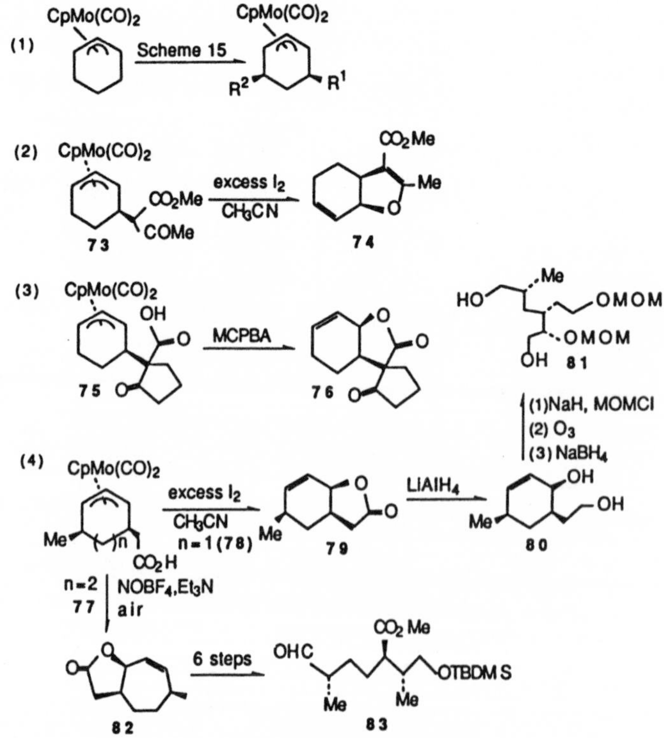

Scheme 18.

even-membered ring compound **77** is converted to acyclic derivative **83** that is promising as a building unit of diverse macrolides and ionophores.

Because cyclopentadienone is unstable at ambient temperature, its organic utilization is difficult to develop. A molybdenum η^4-cyclopentadienone cation[65] **84** was reported in which nucleophilic attack on organolithium, -magnesium, and -copper reagents occurs at the C-α carbon to give η^3-allyl ketone compound **85**, rather than the common Michael reaction. Demetallation of **85** is achieved with ICO_2CF_3. Its pathway is similar to Pearson's I_2 oxidation method[51,52] in which an electrophilic-Mo(IV)(**A**) intermediate is created. The similarity of the two methods is best represented by Eqs. 2 and 3. When a ketone group is present, as in complexes **87** and **89**, intramolecular cyclization occurs via enol attack at the Mo(iv)-allyl fragment to give *cis*-type bicyclic and tricyclic enol compounds **88** and **90**.

Because Ph_3CBF_4 undergoes hydride abstraction from cyclic π-allyl compounds, $CpMo(CO)_2$ may be regarded as an electron-releasing group to stabilize formation of the adjacent carbocation. It is also possible to generate a carbanion when a cyano group is present, such as in **91** and **92**[66,67] that shows potent nucleophilicity to react with aldehydes and

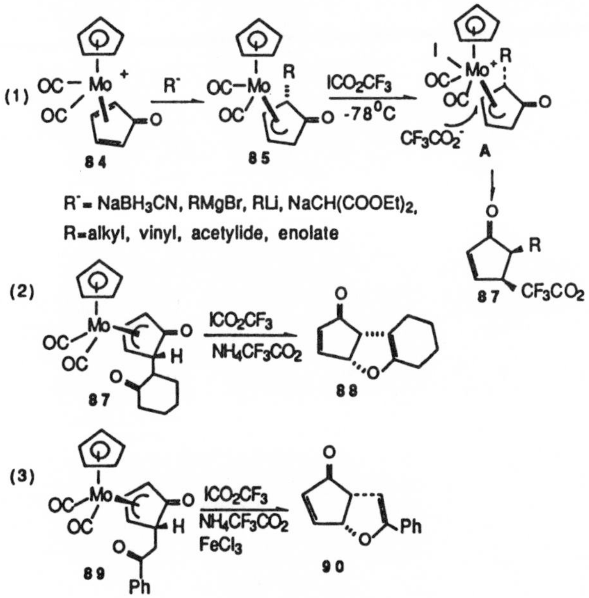

Scheme 19.

enone to construct a quaternary carbon center. When t-BuLi is used as a deprotonating reagent, a lithiated $C\alpha$-carbanion is generated through decyanation via a radical mechanism, and this *in situ* newly generated carbonanion **A** undergoes an alkylation reaction with electrophiles like protons, aldehydes, and α,β-unsaturated esters.

Because the enolate of a six-membered ring compound **93** is easily generated by LDA,[63] this anion has been utilized to synthesize double-alkylation product **95** by alkylation with electrophiles, for example, alkyl halides, acyl halides, aldehydes, and vinyl sulfones, shown in Scheme 21. Hydrolysis of dialkylated product **95** provides the acid **96** which is further converted to bicyclic lactone **97** in excellent yields. It is difficult to generate the enolate of the eight-membered ring compound **98**[68] by a conventional base such as LDA. Its alkyation reaction is accomplished by conversion to its enol silanes **99**. The trimethylsilane enol is function-alized either by conversion to the enolate by MeLi (step ii), followed by alkylation with MeI, by direct conversion to the ketol following the Ru bottom protocol (step iii), or by treatment with phenylselenenyl chloride (step iv). The newly generated hydroxyl, methyl and phenylseleneny group of **100–102** lie on the metal face of the ring. This unexpected stereochemistry is attributed to a particular boat conformation that makes the metal face of the olefin group less sterically hindered. Repetition of these reactions on **102** affords *cis*-disubstituted products **103 a–c**. A 3,

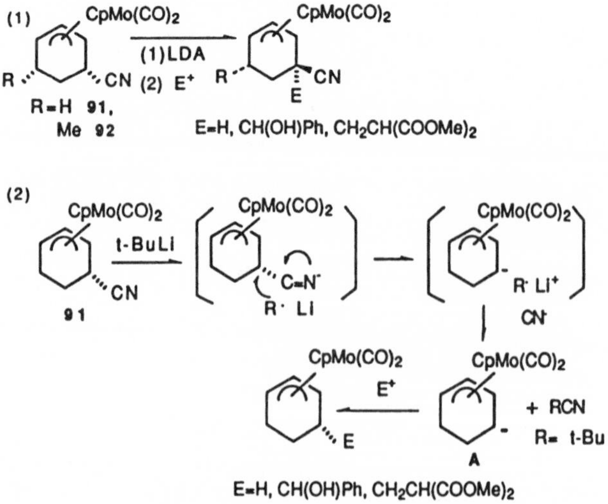

Scheme 20.

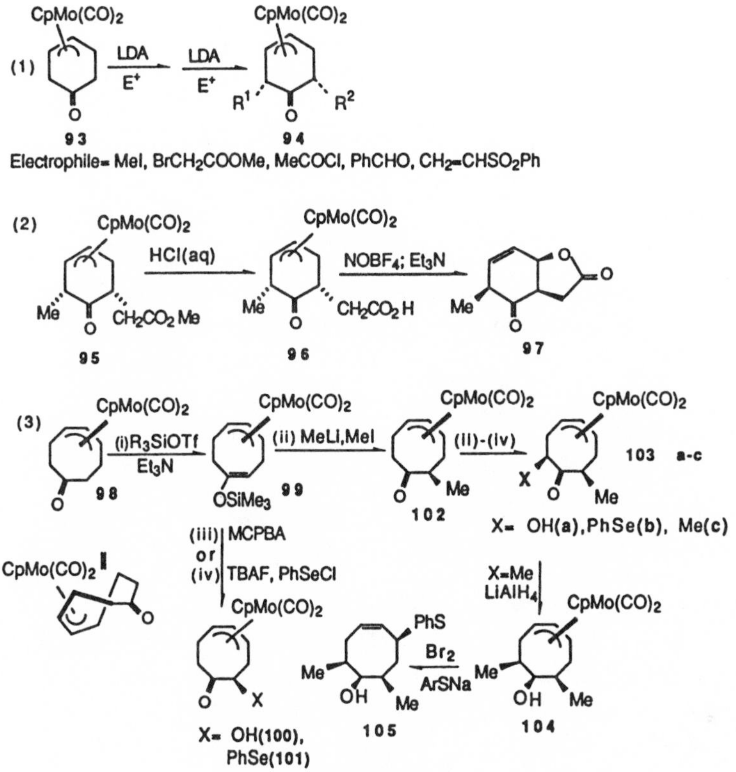

Scheme 21.

5, 6, 8-tetrasubstituted octene **105** of all *cis* configuration is generated from **103c** through constructions of further functionalization and demetallation.

Because it is similar to R_3Si^{53} as an electron-donating group, the $CpMo(CO)_2$ fragment is capable of stabilizing a β-carbocation center. By analogy with the allyl compounds of silane, borane, and tin,[53-55] $CpMo(CO)_2(\eta^3$-cyclohexadienyl)[69] and its 6-substituted derivatives **106** a–c undergo BF_3-promoted electrophilic addition to aldehydes and α,β-unsaturated enones by generating cyclohexadiene cationic precipitates **107**, as shown in Scheme 22.

Demetallation of these salts with anhydrous Me_3NO in CH_2Cl_2 liberates free cyclohexadiene compounds **108** and **109**. For the aldehyde reactions, the diastereomeric selectivities **109a** and **109b** improve with increasing sizes of aldehydes and only one isomer **109a** was observed for dienes derived from **106b** and **106c**. The stereochemical outcome of

Scheme 22. M=CpMo(CO)$_2$

this reaction is conceivable based on an open transitional structure (Eq. 2) in which the aldehyde carbonyl group lies *trans* to the dienyl C=C bond and the R^1 and R^2 groups are mutually staggered to minimize steric hindrance. This conformation is particularly favored with increasing sizes of R^1 and R^2, which are expected to contribute to greater stereoselection of the major isomer **109a**, consistent with observations.

Further elaboration of these molybdenum-π-cyclohexadiene cations for stereoselective synthesis of cyclohexene compounds was achieved. Addition of NaBH$_4$, RMgBr, and LiCH(COOMe)$_2$ to diene **107** gives the η^3-4,6-*cis*-disubstituted cyclohexadienyl compounds **110 a–d**. After conversion of **110b** to its acid form **111**, a NOBF$_4$-promoted intramolecular cyclization of this monoacid compound delivers bicyclic lactone **112** with a 50% yield.

Addition of a chiral sulfoximinyl ester enolate to π-molybdenum-diene cations was reported.[70] Enantiomeric excess was determined for the molybdenum cyclohexadiene and -heptadiene cations with diverse (+)**113a** and (–)**113b** sulfoximinyl ester enolates. The nature of the N-substituent (i.e., tosyl, alkyl or silyl) and the enolate countercation have a pronounced influence on the stereochemical outcome. Relative to chiral recognition, the use of a sulfoximine having (+)-(*S*) stereochemistry at sulfur, such as **113a**, preferentially leads to the formation of (+)-(*S*)-monoester derivatives **114a**, whereas the (–)-(*R*) sulfoximine enolates **113b** produce the (–)-(*R*) monoester compound **114b**. Good

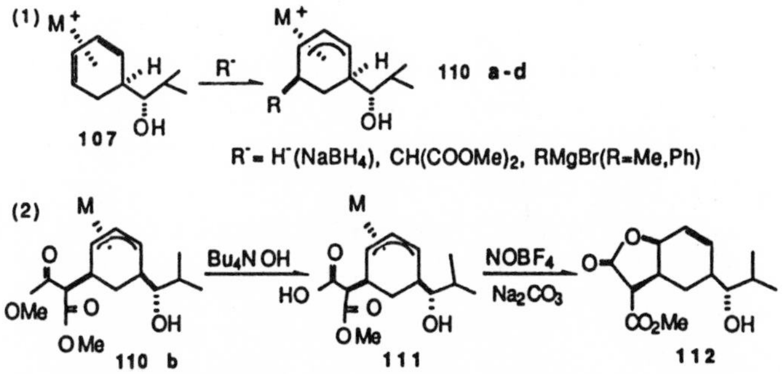

Scheme 23. $(M=CpMo(CO)_2)$

enantiomeric excess values are obtained for R = TBDMS(t-butyldimethylsilyl-) and DMTS(dimethylthexylsilyl-) with the countercations M = Na and K.

Addition of chiral oxazolidinone enolates[71] to two molybdenum-diene cations **66** and **67** was investigated, and the results were more satisfactory for the cyclohexadiene system than for its seven-membered counterpart,

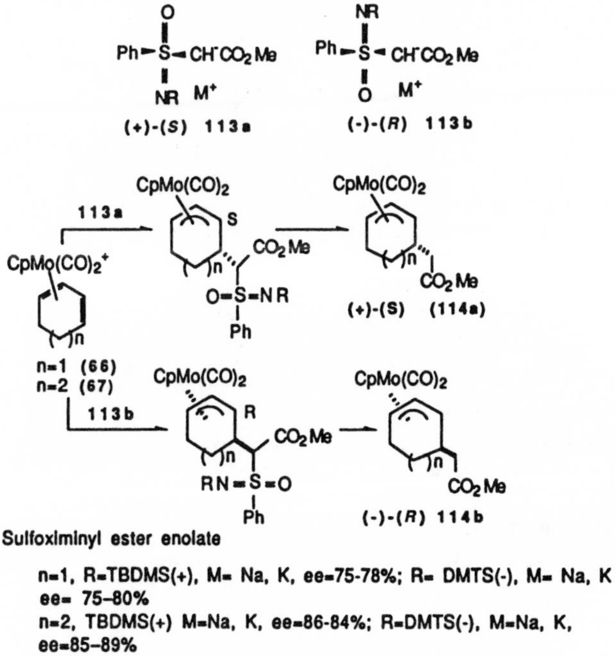

Sulfoximinyl ester enolate

n=1, R=TBDMS(+), M= Na, K, ee=75-78%; R= DMTS(-), M= Na, K
ee= 75–80%
n=2, TBDMS(+) M=Na, K, ee=86-84%; R=DMTS(-), M=Na, K,
ee=85–89%

Scheme 24.

shown in Scheme 25. The enolates of chiral oxazolidinones **115a** and **115b** add preferentially to the pro-*S* terminus of the diene moiety to give (*S*)-**116**, whereas the enolates of chiral oxazolidinones **115 c–e** add preferentially to the pro-*R* terminus to give (*R*)-**117**. The stereochemical outcome of utilizing enolates of chiral oxazolidinone (Scheme 25) or sulfoximinyl ester (Scheme 24) to add to molybdenum-diene cations is elucidated based on an open transitional structure with a synclinal arrangement of the C=C bonds of an enolate and a diene-Mo fragment, as shown in Eq. 4.

Inexpensive and commercially available D- and L-arabinose are useful starting materials for large-scale preparation of the enantiomerically pure chiral molybdenum diene[36] of 2H-pyrans **118** (1*S*) and **119** (1*R*), respectively. Being potent electrophiles, these air-stable cations react with various nucleophiles, for example, H^-, LiR(R=alkyl, enolate, alkynyl, vinyl) and R_1MgBr (R = vinyl, aryl), at the coordinated enolic diene

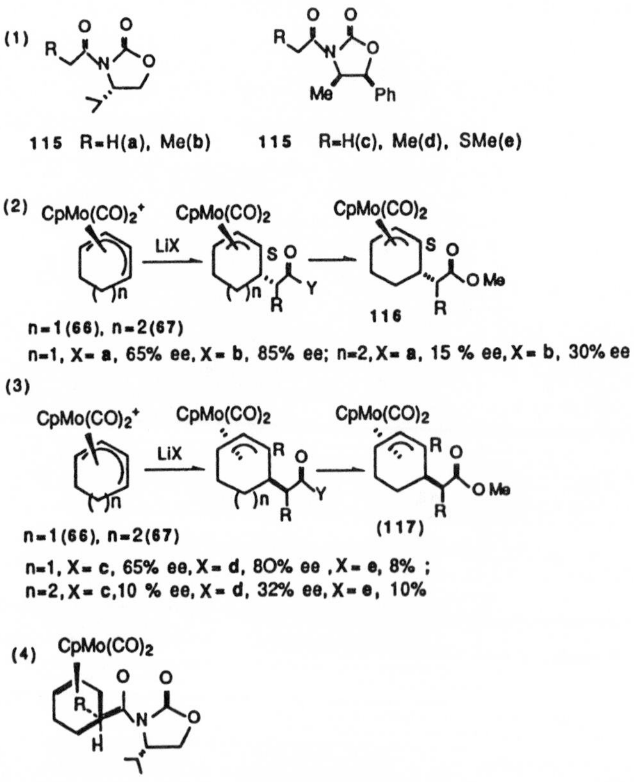

(1)
115 R=H(a), Me(b)
115 R=H(c), Me(d), SMe(e)
(2) CpMo(CO)2+ CpMo(CO)2 CpMo(CO)2
LiX
116
n=1(66), n=2(67)
n=1, X= a, 65% ee, X= b, 85% ee; n=2, X= a, 15 % ee, X= b, 30% ee
(3)
CpMo(CO)2+ CpMo(CO)2 CpMo(CO)2
LiX
(117)
n=1(66), n=2(67)
n=1, X= c, 65% ee, X= d, 80% ee, X= e, 8% ;
n=2, X= c, 10 % ee, X= d, 32% ee, X= e, 10%
(4) CpMo(CO)2

Scheme 25.

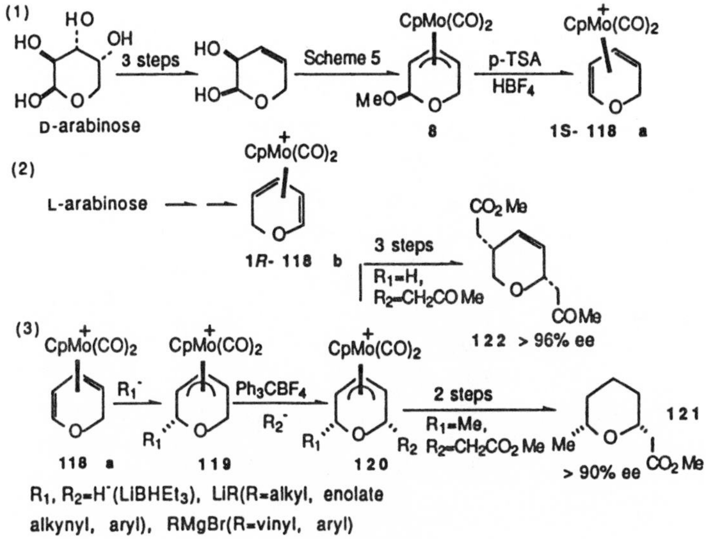

(1)

HO
HO OH
HO O
D-arabinose
3 steps →
HO
HO O
MeO O
Scheme 5 →
8
$\overset{+}{CpMo(CO)_2}$
p-TSA
HBF$_4$ →
$\overset{+}{CpMo(CO)_2}$
1S- 118 a

(2)

L-arabinose — — →
$\overset{+}{CpMo(CO)_2}$
1R- 118 b
3 steps
R$_1$=H,
R$_2$=CH$_2$CO Me →
CO$_2$Me
O
COMe
122 > 96% ee

(3)

$\overset{+}{CpMo(CO)_2}$
O
118 a
→ R$_1$⁻
$\overset{+}{CpMo(CO)_2}$
R$_1$ O
119
Ph$_3$CBF$_4$
R$_2$⁻ →
$\overset{+}{CpMo(CO)_2}$
R$_1$ O
120
R$_2$
2 steps
R$_1$=Me,
R$_2$=CH$_2$CO$_2$Me →
Me O
CO$_2$Me
121
> 90% ee

R$_1$, R$_2$=H⁻(LIBHEt$_3$), LiR(R=alkyl, enolate
alkynyl, aryl), RMgBr(R=vinyl, aryl)

Scheme 26.

carbon terminus to give π-allyl compounds **119**. Further treatment of **119** with Ph$_3$CBF$_4$ regenerates the chiral diene cations that were subject to second nucleophilic attack R$_2^-$ to give the π-allyl compound **120** of cis-2,5-disubstituted pyran. This reaction provides an enantiospecific route to cis-disubstituted 5,6-dihydro-2H-pyran and cis-2,6-disubstituted tetrahydropyran, represented by **121** and **122**. In these cases the enantiomeric excess values exceeded 90%.

Liebeskind et al. prepared optically pure π-allyl compounds of δ-lactone **123** from the chiral acetate,[32] and the latter was directly synthesized from tri-o-acetyl-D-glucal in 70% yield. Alkylation of chiral compound **123** at the carbonyl group with Et$_3$OBF$_4$ delivers the methoxyl cation **124** that is further subject to nucleophilic attack of R$_1^-$ at the enolic η^4-diene carbon to give syn-ethoxyl π-allyl compounds **125**. The syn-methoxyl group of **125** is removed with Ph$_3$CBF$_4$ to give **126** although the cis CpMo(CO)$_2$ fragment feasibly imposes a strong steric hindrance. Further attack on the chiral π-diene cation **126** by nucleophiles R$_2^-$, such as sodium borohydride, lithium carbanion, and Grignard reagent, delivers 2,2,6-trisubstituted pyrans compounds **127**. This work provides a strategy for converting a lactonyl carbonyl group to a stereocontrolled quaternary carbon center having any desired substituents R^1 and R^2. By this method, stereoselective synthesis of cis- and trans-2,6-disubstituted

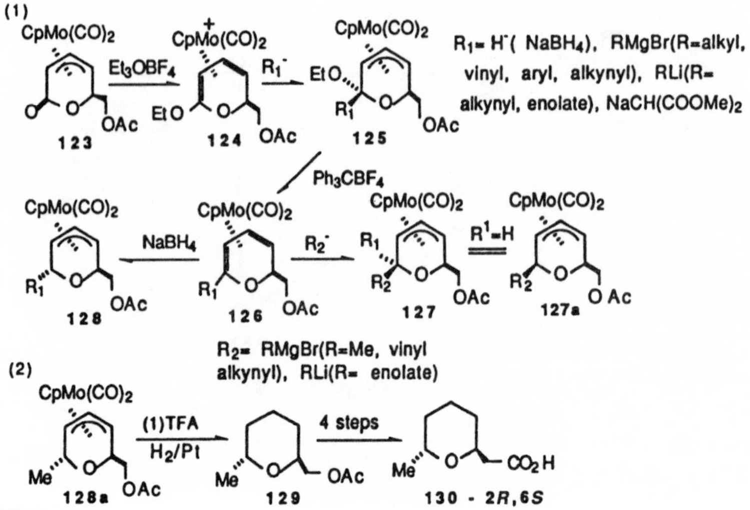

Scheme 27.

pyranyl compounds, such as **127a** and **128** is achieved by selecting R^2 or R^1 = H. To show the application of this reaction, enantiomerically pure *cis*-2,6-disubstituted pyran **130** was obtained from **128a** in yields >85%.

B. Acyclic Systems

Implementing stereocontrolled functionalization of molybdenum π-allyl compounds for organic synthesis of acyclic systems has attracted little attention. We first investigated the enolate chemistry of $CpMo(CO)_2(\eta^3\text{-}1\text{-}COCH_3C_3H_4)$ **131**.[29] Deprotonation of **131** by LDA generates the enolate **I** that reacts with aldehydes to give two diastereomeric products **132a,b**, easily separable on a column. The diastereoselectivity **132a/132b** is reasonable except for bulky Me_3CCHO. For all molybdenum α-ketone allyl compounds reported, the allyl and ketone groups retain a sickled-shape conformation, and reaction of $NaBH_4$ with these compounds proceeds with high stereoselectivity via hydride attack *trans* to the metal fragment. We utilized major aldol products **132a** and their allyllic α, γ -1,3-diol derivatives **133** for stereoselective syntheses of 3-oxo-5-*R*-2-vinyl-tetrahydrofuran **135** and 3-hydroxy-5-*R*-2-vinyl-tetrahydro-furans **134** based on $NOBF_4$-promoted demetallation.[50] The yields were fair ranging from 50–60%. The stereochemistry of these products indicates that the ether rings are formed from

Scheme 28. (M=CpMo(CO)₂)

intramolecular attack of CHRO⁻ at the allyl carbon opposite the CpMo(CO)₂ fragment.

We converted allylic α-secondary alcohol **136** and α, γ-1,3-diols **133** to the corresponding CpMo(NO)Cl derivatives **137** and **140** according to Faller's method. The reactions are given in Scheme 29. Reaction of these π-allyl compounds with aldehydes in CH₂Cl₂ in the presence of MeOH give acetal derivatives of 1,3-diol **138** and 1,3,5-triol **141** in overall yields of ca. 30%. The specific proton positions on the acetal rings were determined by proton NOE spectra. The relatively small overall yields are attributed to the difficult preparation of CpMo(NO)Cl(π-allyl) compounds **137** and **140** of which the α-hydroxy group is prone to ionization in an acidic medium (Scheme 15, Eq. 3). To account for the stereochemical outcome, we suggest a chairlike transition-state structure according to Faller's proposal. Scheme 29 shows four possible transition states **A–D** for all possible diastereomers. States **A** and **C** suffer more steric hindrance with the bulky pseudoaxial cyclopentadienyl group than **B** and **D**. Structure **D** is less favored than **B** because of a stronger 1,3-diaxial interaction involving the phenyl group. Thus structure **B** controls the product stereochemistry.

The α-hydroxyl allyl compounds presented in Scheme 30 are easily ionized selectively by (CF₃SO₂)₂O at −78 °C in anhydrous ether to generate *s-trans-cis*-diene cationic precipitates via an intramolecular S_N2 mechanism. The resulting *s-trans-cis*-1,3-diene cationic precipitates are thermally unstable as temperatures increase to 25 °C and decompose liberating free dienes. We utilized this method for chemoselectively synthesizing *cis*-1,3-dienes and *cis*-1,3,5-hexatrienes[50] with 50–70%

Scheme 29. $M=CpMo(CO)_2$

isolated yields and isomeric purity > 93%. Na_2CO_3 was used to prevent any acid-promoted isomerization.

Addition of organocopper reagents to *s-trans*-diene cation **A** in Scheme 31 gives π-1-*anti*-3-*syn*-allyl compounds **143** with ca. 50%

R^1-H, R^2=CH₂CH(OH)Ph, CH₂Ph, CH₂CH-CHPh; yields 58–61%; isomeric
purity >96:4 ; R^1-Me, R^2=(CH₂)₂Ph, CH₂CH-CHPh; yields 50–51%;
isomeric purity >95:5 ; R^1-H, R^2= CH-CHPh, CH-CHC₄H₃O; yields 61–68%;
isomeric purity >93:7

Scheme 30. $(M=CpMo(CO)_2)$

Scheme 31. $(M=CpMo(CO)_2)$

yield. Initially we attempted to prepare their nitrosyl cations, but discovered later that additions of $NOBF_4$ in excess proportion to these dicarbonyl compounds produce organic isoxazole compounds **144** in appreciable amounts[50,72] (yields 40–50%). A mechanism to form isoxazole is proposed in Scheme 31. The role of nitrosonium ion may be twofold: (1) to oxidize secondary alcohols to ketones,[73] and (2) to promote insertion[74] of a nitrosyl group into the π-*anti*-allylic compounds. Evidence for the former role comes from isolation of dieneone **145** after quenching the reaction with Na_2CO_3 during the course of isoxazole production. The nature of the second step is unclear. According to our proposal, isoxazole arises from the attack of transient $NO^-(HNO)$[75] on the electrophilic allyl cation **B** at the *anti*-carbon terminus. NO insertion at the allylic *syn*-carbon of **B** gives a *cis*-olefin which is not a favorable pathway. Further abstraction of hydrogen from the resulting allylic nitroso compound **C** produces an oximine **D** which is expected to give isoxazole **144** after intramolecular cyclization.

Conjugate addition of organocopper reagents to a molybdenum π-allyl enone compounds **146** was investigated.[76] Similar to $CpFe(CO)PPh_3(\eta^1$-enone)[77,78] and $Cr(CO)_3$(1-enone-2-OR-benzene),[79] *s-cis*(CO/C=C) enone conformer **146a** occurs in the crystal structure according to X-ray measurements and is also the major dissolved species in equilibrium with minor *s-trans* conformer **146b** on the basis of ^{1}H-NMR spectra. Steric interaction between H^3 and H^6 hydrogens destabilizes the *s-trans*-enone species **146b**. In contrast with the reported iron and chromium enone compounds,[77-79] minor *s-trans*-enone conformer **146b** apparently reacts more rapidly with organocopper reagent than the *s-cis*-enone conformer to control product stereochemistry. We assume here that organocopper reagents approach the enone group of the two conformers **146 a,b** opposite the metal fragment. The results appear in Scheme 32. At –40 °C, the reactions proceed with reasonable diastereoselectivity, **147a/147b** = 3/1–6/1, for various substituents of R^1 and R^2. This ratio improves in the presence of BF$_3$·Et$_2$O (equimolar) and only **147a** is produced. The coordinated BF$_3$·O=C fragment of *s-cis*-enone conformer **146a** is expected to exert steric hindrance with its C=C bond, which thus decreases its concentration. Michael addition to the *s-cis*-enone conformer **146a** is less sterically hindered because its vinyl group is farther from the metal center than in the *s-trans*-form **146b**. The electronic effect must be favorable for the *s-trans* isomer **146b** to account for its greater reactivities.

We employed this reaction to stereoselectively synthesize 2,3,4,5-tetrasubstituted tetrahydrofuran compounds. Treating one major product (**147a** R′ = Ph, R″ = Me) with LiN(SiMe$_3$)$_2$ selectively generates cis-enolate anion **A** that condenses with aldehydes, presumably via a cyclic

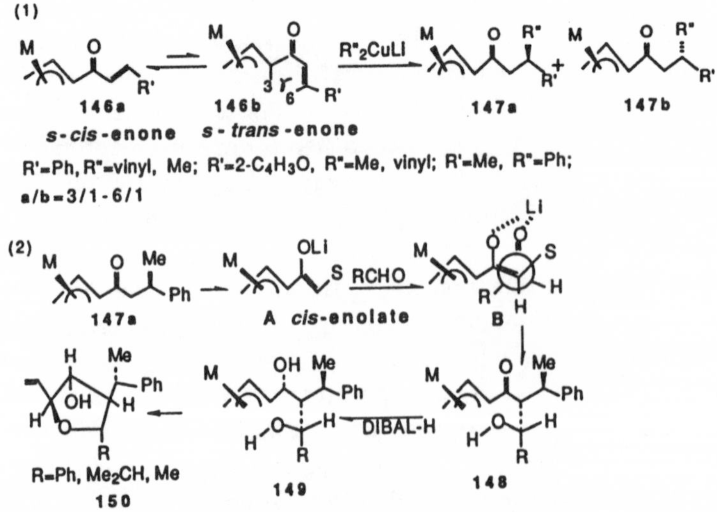

Scheme 32.

transitional structure, to provide the aldol product with excellent yields and selectivities. Evidence for the proposed cyclic transitional structure is inferred from the X-ray structure of **148** (R = Me, Ph). According to the Fourier difference map, an intramolecular hydrogen bond exists within OH$^{\bullet\bullet}$O=C that locks the remaining three-carbon chain into a boatlike conformation. With Li$^+$ replacing H$^+$, the transition state **B** is generated, which has two mutually *trans* R and CHMePh substituents to minimize steric hindrance.[80] Although *cis*-enolate **A** and *s-cis* enone conformers **146a** have identical structural skeletons, their roles in the corresponding reactions are distinct. Further reduction of **148** with DIBAL-H, gives the allylic α-hydroxyl alcohol **149** as a single diastereomer which undergoes a subsequent NOBF$_4$-promoted cyclization to deliver tetrasubstituted furan compounds **150**. The specific proton positions of the ring**150** were confirmed by proton NOE-difference spectra.

Diels–Alder reactions of π-allyl enone compounds **146a, b** with cyclopentadiene in refluxing THF were investigated. Four diastereomeric products **151 a–d** were isolated and separated in pure forms on a SiO$_2$ column and by fractional crystallization. Each product representative of **151 a–d** was structurally characterized by X-ray crystallography.[82] The product ratios are specified in Scheme 33. The combined yields are ca. 40–50%. If CpMo(CO)$_2$ is regarded as a stereodirecting fragment, products **151 a–d** are envisaged as deriving from the addition of cyclopentadiene to *s-trans*-enone conformer **146b** and *s-cis* conformer **146a** in *endo* and *exo* fashions, respectively,[81] as illustrated in Scheme 33. In refluxing THF, the two major products are **151b** and **151c** which indicates that *cis*-enone conformer **146a** dominates the reaction pathway. In contrast with their Michael reactions, the role of *s-trans*-enone conformer **146b** is less pronounced here. In presence of BF$_3$·Et$_2$O, *s-trans*-isomer **146b** controls the diastereofacial selectivity and gives the *endo* addition products exclusively in cold THF. These enone compounds fail to react with other electron-rich olefins including *trans*-1-methoxy-3-(trimethylsilyoxy)-1,3-butadiene, 2,3-dimethybutadiene, and 1-methoxylcyclohexadiene, even in the presence of Lewis acid. The weak dienophilic behavior of these enones is attributed to their intrinsic stability due to electron delocalization over the enone and metal-allyl groups.

Addition of CF$_3$SO$_3$H to the η^3-2-carbomethoxypentadienyl compounds of molybdenum and tungsten compounds **153** in diethyl ether at −40 °C generates *s-trans*-diene cation **154a**[43] that is stable at this tem-

THF(reflux, 24h)
R=Me, 151a : 151b : 151c : 151d = 14.6 : 43.9 : 28.1 : 13.4 ; 45% yields
R=3-NO$_2$C$_6$H$_4$ 151a : 151b : 151c : 151d = 12.1 : 32.9 : 37.8 : 17.2; 41% yields
R=COOMe 151a : 151b : 151c : 151d = 28.9 : 36.8 : 32.2 : 2.1 ; 50% yields

BF$_3$· Et$_2$O(Et$_2$O, 0 °C, 2h)
R=Me, 151a : 151b : 151c : 151d = 90 : 10 : 0 : 0 ; 83% yields
R=3-NO$_2$C$_6$H$_4$ 151a : 151b : 151c : 151d = 86 :14 : 0 : 0 ; 78% yields
R=COOMe 151a : 151b : 151c : 151d = 87 : 13 : 0 :0 ; 80% yields

Scheme 33. (M=CpMo(CO)$_2$)

perature. When the temperature raised above −10 °C, the cation undergoes irreversible conformational isomerization to the more stable *s-cis*-
diene cation **154b**. Nucleophilic attack on these two cations is an
alternative to the Michael reaction that, however, is not operable on the
free 2-carboxylated-1,3-diene because of its facile dimerization. For
s-trans-diene cation **154a**, hard bases, such as water, alcohols, and bulky
amines, add highly regio- and stereoselectively at the δ-carbon to give
155a as a single diastereomer. Softer bases, such as thiols, tend slightly
to a non-Michael reaction pathway with δ/α = 1.5. Alkylation with
organocuprate reagents tends toward the Michael reaction product (α-
addition, **155b**) with α/δ >2.

Nucleophilic addition to *s-cis*-diene cation **154b** is less facile than its
s-trans-diene isomer **154a**. Water and methanol fail to give any addition
products, and the yields are generally low. All nucleophiles including
hydroxide, methoxide, thiols, amines, and organocuprates unambigu-
ously add to the α-carbon of the diene to afford **156a**.

The tendency for addition at the δ-carbon of the *s-trans*-diene carbon
of **154a** is attributed to the relative instability of the α-addition product
155b that experiences steric hindrance between the planar *anti*-car-
bomethoxyl and the CHCMe allylic π-bond. The tendency toward the
α-carbon of organocuprates reflects its intrinsic preference for the con-

Scheme 34. $(M=CpMo(CO)_2, CpW(CO)_2)$

jugated enone group,[84] and its carbon–carbon forming reaction is irreversible. Molybdenum π-allyl compounds bearing α-hydroxyl and α-alkoxyl groups are easily displaced by metal fragments to undergo ionization in a weakly acidic medium which is generally a reaction condition for nucleophiles like water and alcohol. The formation process of these products is considered reversible. Therefore the δ-addition products of **155a** given by hard-base nucleophiles, such as H_2O and ROH, are believed to be thermodynamically favorable because of their facile reversibility.

To rationalize the α-addition regiochemistry for *s-cis*-diene cation **154b** is more straightforward. Nucleophilic additions at α- and δ-carbons of this cation, both give the *anti*-substituted CH_2Nu (α-addition) and CHMeNu (δ-addition) π-allyl compounds of **156a** and **156b**, respectively. The former is less sterically hindered than the latter according to comparison of the steric interaction of their *anti*-substituents with the *cis*-allylic π-bond, that is, CH_2Nu versus CHMeNu. Moreover, nucleophilic attack at the α-carbon is also electronically favorable because of the electronic influence of the neighboring carbomethoxyl group. According to both kinetic and thermodynamical considerations, α-addition at the *s-cis*-diene cation is ultimately important.

Although a metal-*s-trans*-diene cation is more chemically reactive and more thermodynamically unstable than its η^4-*s-cis*-diene cation,[85,86] it is still possible to achieve formation of carbon–carbon bonds involving the former as reaction intermediates. A representative case is given in Scheme 35 that shows BF_3-promoted addition of $CpMo(CO)_2(\eta^3$-pentadienyl) **157** to aldehydes and methyl vinyl ketone in cold toluene to afford η^4-*s-trans*-diene cationic precipitates **A**.[87] Although this cation is isolable and retains exceptionally high electrophilicity, the process of forming the carbon–carbon bond is not stereoselective. Furthermore, nucleophiles, such as water and methanol, add equally to the α- and δ-carbons (Scheme 35) so that four diastereomers **159 a,b** are produced. We utilized this method to synthesize α-functionalized *trans*-dienes **158** through Me_3NO-promoted demetallation reactions. In the presence of $BF_3{\cdot}Et_2O$, the reactions of molybdenum-η^3-heptatrienyl **160** with benzaldehyde, methyl vinyl ketone, and cyclohexenone likewise give isolable *s-trans*-diene cationic precipitates **B**, which, after Me_3NO

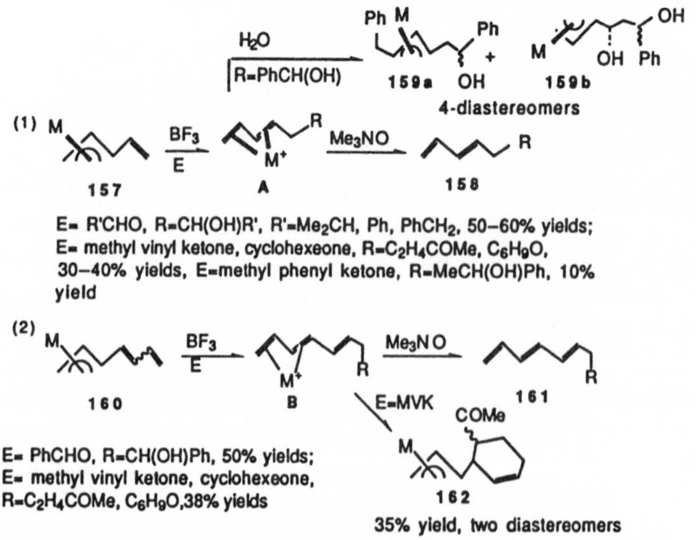
H2O
R=PhCH(OH)
Ph M
Ph
OH
159a
M
OH
OH Ph
159b
4-diastereomers
(1) M
BF3
E
R
M+
Me3NO
R
157
A
158
E= R'CHO, R=CH(OH)R', R'=Me2CH, Ph, PhCH2, 50–60% yields;
E= methyl vinyl ketone, cyclohexeone, R=C2H4COMe, C6H9O,
30–40% yields, E=methyl phenyl ketone, R=MeCH(OH)Ph, 10%
yield
(2) M
BF3
E
M+
R
Me3NO
R
160
B
161
E=MVK
COMe
M
162
35% yield, two diastereomers
E= PhCHO, R=CH(OH)Ph, 50% yields;
E= methyl vinyl ketone, cyclohexeone,
R=C2H4COMe, C6H9O,38% yields

Scheme 35.

demetallation, afford functionalized trienyl compounds **161** with moderate yields. In the case of methyl vinyl ketone, hydrolysis of the salt produces a six-membered-ring π-allyl compound **162** as a 1:1 mixture of two diastereomers with a 35% yield.

The problem that the reaction in Eq. (1) lacks stereo- and regio-selectivity was solved by introducing a substituent at the allylic C(2)-carbon, such as for **153**,[88] that is condensed with aldehydes by $BF_3 \cdot Et_2O$, stereospecifically, in cold toluene to give *s-trans*-diene precipitate **A**. Further hydrolysis of these salts with H_2O/THF at -40 °C produces tungsten-allyllic-α, γ-diol **163 a,b** with good diastereoselectivity. The selectivities **163a/163b** increase with increasingly sterically demanding aldehydes. The regio-controlled attack of H_2O at the δ-diene carbon is rationalized by the presence of a sterically more hindered *anti*-COOR group in the α-addition product, as mentioned before (Scheme 34).

These BF_3-catalyzed reactions are synthetically useful because they simultaneously create two asymmetrical secondary hydroxylic carbons in a one-pot synthesis. To demonstrate this synthetic utility, we converted tungsten-allyllic α, γ-diol **163a** to lactone **164** that subsequently undergoes $NOBF_4$-promoted intramolecular cyclization, after Ce(IV) oxidation, to afford bicyclic α-methylene-γ-butyrolactone **165** with good yields. It is required to convert **164** to acetate derivatives **166** if external nucleophilic attack on their allyl-NO cations is carried out. Otherwise intramolecular cyclization reaction to give **165** occurs. Treating the NO salts of acetate compounds with PhSNa delivers the furanone as a single diastereomer **167** with good yields. $NaBH_4$ reduction of these salts and then Ce (iv) oxidation yields α-methylene-γ-butyrolactones **168 a,b** as the major product with **168a/168b** >4.5. Apparently, H^- here adds preferentially to the more substituted terminus of the allyl moiety. This

R	a/b
CH_2Ph	a only
Ph	89:11
$CHMe_2$	91: 9
Me	75: 25

Scheme 36. $M=CpW(CO)_2$

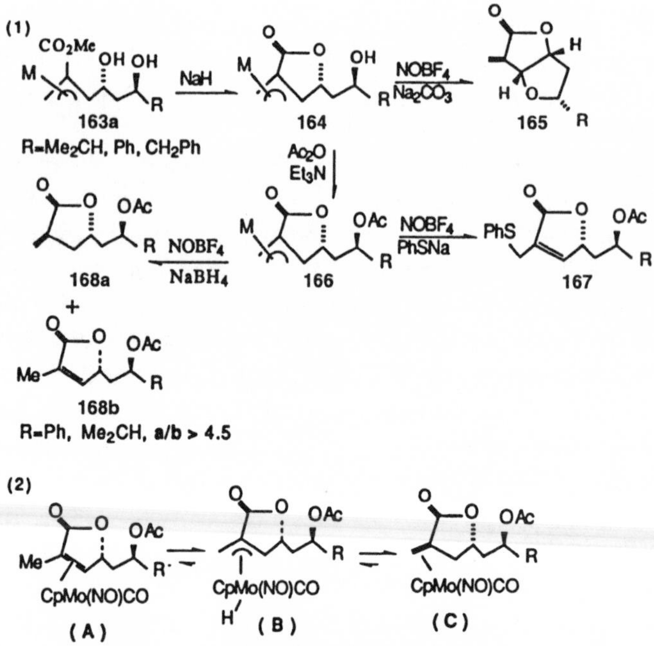

Scheme 37.

regiochemistry differs somewhat from that of **167** which is more consistent with Faller's model (Scheme 8). This discrepancy is attributed to further isomerization of the kinetic product **A** to the more stable form **C** via a Mo(II)-hydride allyl intermediate **B**.

We also examined the reactions of **153** with α,β-unsaturated enones in doubly molar proportions in the presence of $BF_3 \cdot Et_2O$ that produced two diastereomeric metal allyl compounds **169** having six-membered rings.[89] These products are envisaged as an exceptional [2+2+2] cycloaddition of a vinyl group and two enone molecules. This special feature is distinct from the expected Michael reaction pathway. To test the possibility of the Michael reaction, we slowly added methyl vinyl ketone in an equimolar proportion to the η^3-pentadienyl compound and methyl vinyl ketone which afforded only the same composition of **169a** and **169b**. Attempts to quench the reaction by H_2O during its course gave no sign of a 1:1 product.

We also studied the reaction of $CpW(CO)_2$(1-allenyl-2-COOMe-allyl) **170** and methyl vinyl ketone which likewise gives two diastereomeric [2+2+2] cycloadditive structures **171a,b**. The stereochemistries of the two products are identical to those of the reactions derived from **153** according to X-ray structural analysis.[90]

To account for the stereochemical outcome, we propose a mechanism in Scheme 38 that involves a metal-*s-trans*-diene cationic intermediate **I**. A near chairlike transition state structure[91] is proposed to account for the stereochemistry. The states **A** and **B** represent the most likely transition states that have the most bulky complexed RCO(BF$_3$) group situated at the equatorial site to avoid 1,3-diaxial steric hindrance. A further comparison between the two structures indicates that **A** is the preferred structure because it avoids steric hindrance between two neighboring equatorial substituents, as in **B**. The presence of an electron-withdrawing group, such as carbomethoxyl, is required for this cyclization. No cyclization occurs for other η^3-pentadienyl compounds, including that having the substituent CH=CH$_2$ at the C(2)-allylic carbon position.

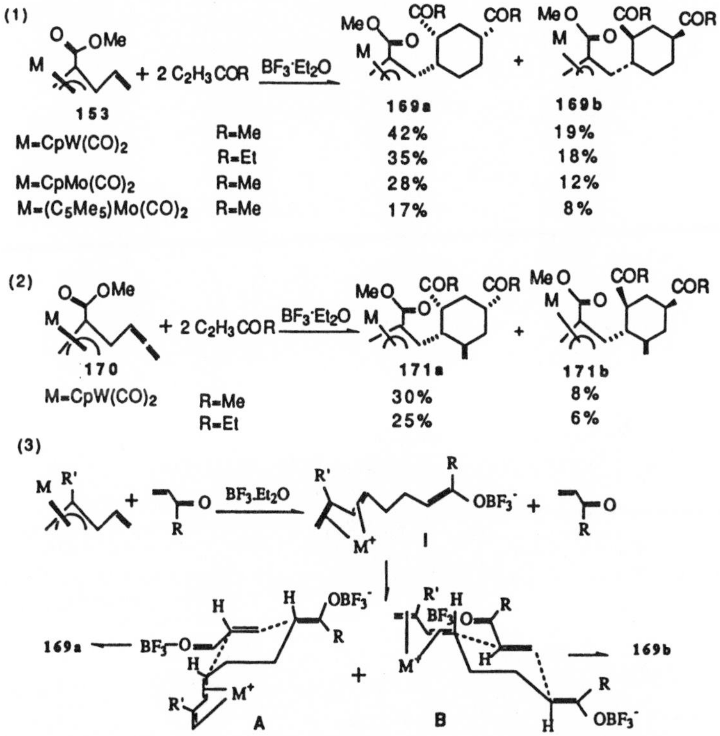

153	**169a**	**169b**
M=CpW(CO)$_2$ R=Me	42%	19%
R=Et	35%	18%
M=CpMo(CO)$_2$ R=Me	28%	12%
M=(C$_5$Me$_5$)Mo(CO)$_2$ R=Me	17%	8%

170	**171a**	**171b**
M=CpW(CO)$_2$ R=Me	30%	8%
R=Et	25%	6%

M=CpM(CO)$_2$, R'=COOMe, R=Me, Et ; the R group of A was omited for clarity)

Scheme 38.

A new cyclization [$2\pi+2\pi+2$-allyl-π] was observed for CpMo(CO)$_2$(2-vinylallyl) **172** and electrophiles like tetracyanoethylene, phenyl isocynate, and benzaldehyde.[92,93] In regard to the potency of this reaction, the vinyl group of the π-allyl enone compound **178** (Scheme 39, Eq. 2) also undergoes cyclization with TCNE to give the bicyclic compound **179**. It was proved that these cyclizations proceed via trimethylenemethane zwitterionic intermediates **A**. These trimethylene-methane cationic precipitates **A** and **B** are generated and isolated from the BF$_3$-promoted addition of **172** with aldehydes and α,β-unsaturated enones in cold toluene. In the presence of Na$_2$CO$_3$, the trimethylene-methane cation **B**, derived from benzaldehyde and **172**, undergoes sub-sequent intramolecular cyclization to give the η^3-allyl pyranyl compound **176** as a mixture of two diastereomers, and this result provides direct evidence for the proposed mechanism. To pursue further utilization of

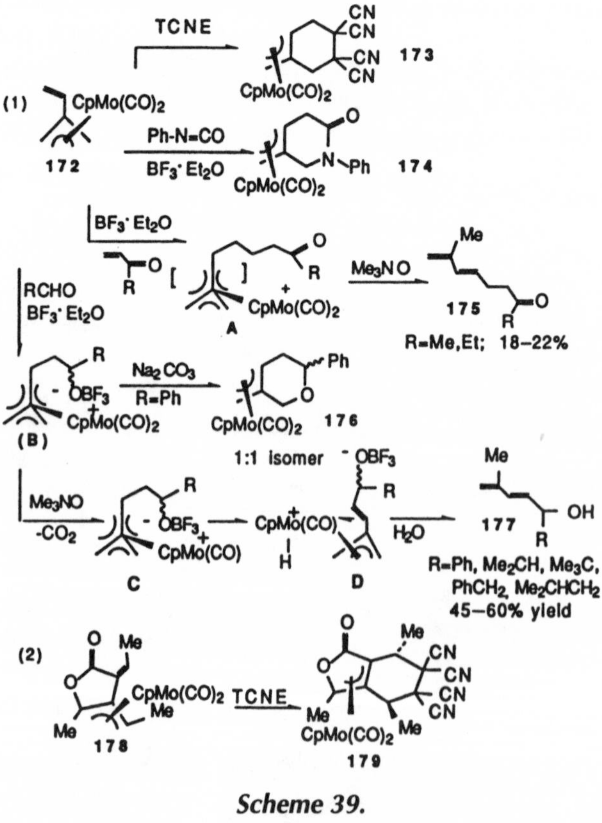

Scheme 39.

this new reaction, we demetallated these trimethylenemethane cations by Me_3NO in CH_2Cl_2 to give 5-functionalized 2-methyl-1,3-dienes[177,175] with reasonable yields (> 50%) for aldehydes and ca. 20% yields for unsaturated enones. We labeled with deuterium by starting with 50% deuterium content on the $C=CH_2$ end of **172**. ^{2}H-NMR spectra of the diene, derived from deuterated **172** and PhCHO, revealed that both CH_3 and $=CH–CH(OH)$ of **177** (R = Ph), were deuterated at a ratio of ca. 3 : 2. To explain the mechanism of formation of the organic products, we propose that, after losing CO by Me_3NO-promoted decarbonylation, the newly generated 16e-species **C** eliminates the neighboring acidic methylene protons to form 18e- η^5-isopentadienyl hydride intermediates **D**, which after reductive elimination are expected to liberate the observed dienes, as shown in Eq. 3.

The enolate of $CpMo(CO)2(\eta^3$-2-acetylallyl) **180** is readily generated by lithium diisopropylamide at –78 °C that condenses with aldehyde to give the aldol products in yields > 85%.[94] The reaction of ethyl enolate $CpMo(CO)_2(\eta^3$-2-COCHCMeLi-allyl) **A** with aldehydes shows no diastereoselectivities to give a 1:1 *syn/anti* mixture of products **186a,b**. The diastereoselectivity is improved with $(n$-butyl$)_2$BOTf (in equimolar proportions) that preferably gives the *syn*-aldol product **186a**. The *syn/anti* ratios improve with increasing sizes of aldehydes. The $(n$-butyl$)_2$BOTf is believed to provide a cyclized transition state to control the selectivity better.[80]

The crystal structures of these aldol products reveal that a strong intramolecular hydrogen bond exists between hydroxyl and α-ketone groups. It is easy to envisage the existence of such a hydrogen bond if one considers the effect of the electron-donating power of $CpMo(CO)_2$ that increases the basicity of the α-ketone group with a resonance represented by **B**.

We utilized this hydrogen-locked chairlike conformation to induce stereoselective generation of a 1,3-diol by reducing the compounds with Bu_4NBH_4 in methanol/benzene (1:1 ratio). The *syn* isomer is the major product. That the *syn/anti* selectivities increase with sterically demanding aldehydes is attributed to the importance of the pseudo-chairlike conformation of **181** that has the bulky aldehyde R group in the equatorial position and that hydride adds from the axial position. PhSNa attack of the nitroso-allyl cations of **182a** affords the *syn*-diol compounds **183** with good yields. Treating molybdenum *syn*-1,3-diol **182a** with 1.5 equiv. of I_2 in CH_3CN gives 2-*R*-4-hydroxyl-5-methylene-tetrahydropyran compounds **184**. We also attempted to reduce *syn*-aldo products **186a** with

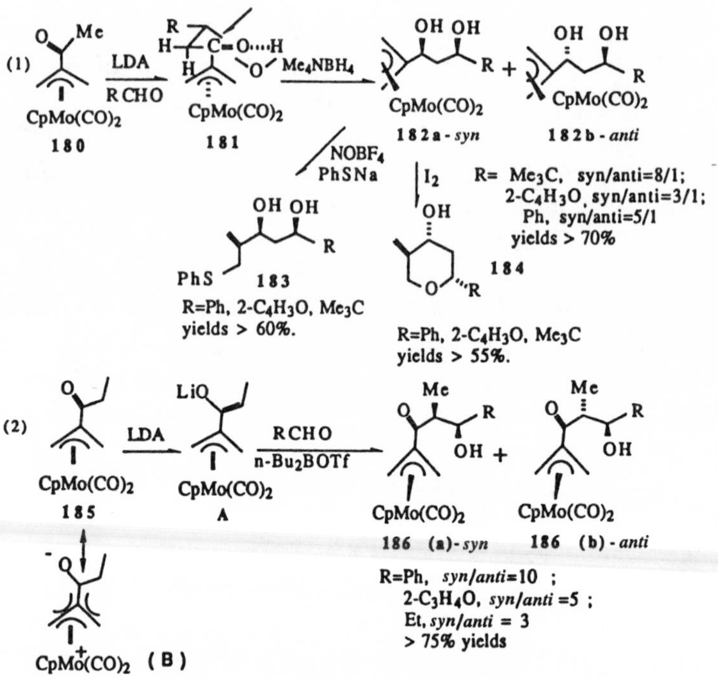
(1)
LDA
RCHO
Me_4NBH_4
$CpMo(CO)_2$
180
181
182a-*syn*
182b-*anti*
$NOBF_4$
PhSNa
I_2
R= Me_3C, syn/anti=8/1;
$2-C_4H_3O$, syn/anti=3/1;
Ph, syn/anti=5/1
yields > 70%
183
184
R=Ph, $2-C_4H_3O$, Me_3C
yields > 60%.
R=Ph, $2-C_4H_3O$, Me_3C
yields > 55%.
(2)
LDA
RCHO
$n-Bu_2BOTf$
185
A
186 (a)-*syn*
186 (b)-*anti*
(B)
R=Ph, *syn/anti*=10 ;
$2-C_3H_4O$, *syn/anti* =5 ;
Et, *syn/anti* = 3
> 75% yields

Scheme 40.

DIBAL-H or Bu_4NBH_4. The resulting products were too unstable during column chromatography for further utilization.

ACKNOWLEDGMENTS

Our research in this field has been supported by the National Science Council, Republic of China. The author is grateful to a number of students whose names appear in the papers cited. The assistance of Mr. and Ms. Lin-Hwun Shieu, Kuei-Wen Liang, Ching-Hsiu Lee, and Shwu-Ju Shieh for typing and editing this article is acknowledged.

REFERENCES

1. Trofimenko, S. *Chem. Rev.* **1993**, *93*, 943.
2. (a) Etienne, M.; White, P. S.; Templeton, J. L. *J. Am. Chem. Soc.* **1991**, *113*, 2324; (b) Feng, S. G.; White, P.; Templeton, J. L. *J. Am Chem. Soc.* **1992**, *114*, 2951; (c) Feng, S. G.; Templeton, J. L. *J. Am. Chem. Soc.* **1989**, *111*, 6477; (d) Luan, L.; Brookhart M.; Templeton, J. L. *Organometallics* **1992**, *11*, 1433; (e) Feng, S. G.; Templeton, J. L. *Organometallics* **1992**, *11*, 2168.

3. (a) Blosch, L. L.; Abboud, K.; Boncella, J. M. *J. Am. Chem. Soc.* **1991**, *113*, 7066; (b) Blosch, L. L.; Gamble, A. S.; Abboud, K.; Boncella, J. M. *Organometallics* **1992**, *11*, 2342; (c) Kim, H. P.; Kim, S.; Jacobson, R. A.; Angelici, R. J. *Organometallics* **1986**, *5*, 2481; (d) Doyle, R. A.; Angelici, R. J. *J. Am. Chem. Soc.* **1990**, *112*, 94.

4. (a) Curtis, M. D.; Shiu, K. B.; Butler, W. M.; Huffmann, J. C. *J. Am. Chem. Soc.* **1986**, *108*, 3335; (b) Eagle, A. A.; Tiekink, E. R. T.; Young, C. G. *J. Chem. Soc., Chem. Commun.* **1991**, 1746; (c) Eagle, A. A.; Tiekink, E. R. T.; Young, C. G. *Organometallics* **1992**, *11*, 2934.

5. Drew, S. A.; Jeffery, J. C.; Pilotti, M. U.; Stone, F. G. A. *J. Am. Chem. Soc.* **1990**, *112*, 6148 and references therein.

6. (a) Pearson, A. J. In *Advances in Metal-Organic Chemistry*; Liebeskind, L. S., Ed.; JAI Press: London, 1989, Vol. 1, p. 1; (b) Pearson, A. J. *Synlett.* **1990**, 10.

7. Backvall, J.-E. In *Advances in Metal-Organic Chemistry*; Liebeskind, L. S., Ed.; JAI Press: London, 1989, Vol. 1, p. 135.

8. Blyston, S. L. *Chem. Rev.* **1989**, *89*, 1663.

9. (a) Pearson, A. J. In *Comprehensive Organometallic Chemistry*; Wilkenson, G.; Stone, F. G. A.; Abel, E. W., Eds. Pergamon Press: Oxford, 1982; Vol. 8, p. 939; (b) Pearson, A. J. *Metallo-Organic Chemistry*; Wiley: Chichester, 1985; p. 257.

10. Davies, S. G. In *Organotransition Metal Chemistry: Applications to Organic Synthesis*; Pergamon Press: New York, 1982; Chap. 5.

11. Astruc, D. *Topics in Current Chemistry*, Herrmann, W. A., Ed.; Springer-Verlag: Berlin, 1992; Vol. 160, p. 47.

12. Gree, R. *Synthesis* **1989**, 341.

13. Uemura, M. In *Advances in Metal-Organic Chemistry*; Liebeskind, L. S., Ed.; JAI Press: London, 1991; Vol. II, p. 195.

14. Wulff, W. D. In *Advances in Metal-Organic Chemistry*; Liebeskind, L. S., Ed.; JAI Press: London, 1989; Vol. I, p. 209.

15. Sollddie-Cavallo, A. In *Advances in Metal-Organic Chemistry*; Liebeskind L. S., Ed.; JAI Press: London, 1989; Vol. I, p. 99.

16. Faller, J. W.; Rosan, A. M. *Ann. N.Y. Acad. Sci.* **1977**, *295*, 186.

17. Faller, J. W.; Rosan, A. M. *J. Am. Chem. Soc.* **1976**, *98*, 3388.

18. Faller, J. W.; Rosan, A. M. *J. Am. Chem. Soc.* **1977**, *99*, 4858.

19. Adams, R. D.; Chodosh, D. F.; Faller, J. W.; Rosan, A. M. *J. Am Chem. Soc.* **1979**, *101*, 2570.

20. Faller, J. W.; Murray, H. H.; White, D. L.; Chao, K. H. *Organometallics* **1983**, *2*, 400.

21. Faller, J. W.; John, J. A.; Mazziert, M. R. *Tetrahedron Lett.* **1989**, *30*, 1769.

22. Benyunes, S. A.; Deeth, R. J.; Fries, A.; Green, M.; McPartlin M.; Nation, C. B. *J. Chem. Soc., Dalton Trans.* **1992**, 3543.

23. (a) Huffman, M. A.; Liebeskind, L. S.; Pennington, W. T. *Organometallics* **1992**, *11*, 255; (b) Huffman, M. A.; Liebeskind, L. S. *J. Am. Chem. Soc.* **1990**, *112*, 8617.

24. Wulff, W. D.; Gilberson, S. R.; Springer, J. P. *J. Am. Chem. Soc.* **1986**, *108*, 520.

25. Anderson, B. A.; Wulff, W. D.; Rheingold, A. L. *J. Am. Chem. Soc.* **1990**, *112*, 8615.

26. Morris, K. G.; Saberi, S. P.; Thomas, S. E. *J. Chem. Soc., Chem. Commun.* **1993**, 209.

27. Eisch, J. J.; King, R. B., Eds.; *Organometallic Synthesis*; Academic Press: New York, 1965; Vol. 1, p. 114.

28. (a) Abel, E. W.; Moorehouse, S. *J. Chem. Soc., Dalton Trans.* **1973**, 1706; (b) Gibson, D. H.; Hsu, W. L.; Lin, D. S. *J. Organomet. Chem.* **1979**, *172*.

29. (a) Vong, W. J.; Peng, S. M.; Liu, R. S. *Organometallics* **1990**, *9*, 2187; (b) Lee, T. W.; Liu, R. S. *Organometallics* **1988**, *7*, 878; (c) Vong, W. J.; Peng, S. M.; Lin, S. H.; Lin, W. J.; Liu, R. S. *J. Am. Chem. Soc.* **1991**, *113*, 573.

30. Hayter, R. G. *J. Organomet. Chem.* **1968**, *13*, p. 1.

31. (a) Faller, J. W.; Lambert, C. *Tetrahedron* **1985**, *41*, 5755; (b) Faller, J. W.; Linnebarrier, D. *Organometallics* **1988**, *7*, 1670.

32. Rubio, A.; Liebeskind, L. S. *J. Am. Chem. Soc.* **1993**, *115*, 891.

33. McCallum, J. S.; Sterbenz, J. T.; Liebeskind, L. S. *Organometallics* **1993**, *12*, 927.

34. Faller, J. W. unpublished results (personally communicated).

35. (a) Trost, B. M.; Weber, L.; Strege, P. E.; Fullerton, T. J.; Dietsche, T. J. *J. Am. Chem. Soc.* **1978**, *100*, 3416; (b) Backvall, J. E. *Acc. Chem. Res.* **1983**, *16*, 335.

36. Hansson, S.; Miller, J. F.; Liebeskind, L. S. *J. Am. Chem. Soc.* **1990**, *112*, 9660; (b) Liebeskind, L. S. In *International Conferences on Organic and Inorganic Chemistry on the Occasion of the 60th Chinese Chemical Society*, Taipei, Taiwan, 1992.

37. Watson, P. L.; Bergman, R. G. *J. Am. Chem. Soc.* **1979**, *101*, 2055.

38. (a) Roustan, J. L.; Merour, J. Y.; Carrier, C.; Benaim, J.; Cadiot, P. *J. Organomet. Chem.* **1979**, *169*, 39; (b) Yang, G. M.; Lee, G. H.; Peng, S. M.; Liu, R. S. *J. Chem. Soc., Chem. Commun.* **1991**, 478.

39. Green, M.; Naythi, J. Z.; Scott, C.; Stone, F. G. A.; Welch, A. J.; Woodward, P. *J. Chem. Soc., Dalton Trans.* **1978**, 1067.

40. Bottrill, M.; Green, M. *J. Chem. Soc. Dalton Trans.* **1979**, 821.

41. Charrier, C.; Collin, J.; Merour, J. E.; Roustan, J. L. *J. Organomet. Chem.* **1978**, *162*, 57.

42. Tseng, T.-W.; Wu, I. Y.; Lin, Y. C.; Cheng, C. T.; Cheng, M. C.; Tsai, Y. J.; Chen, M. C.; Wang, Y. *Organometallics* **1991**, *10*, 43.

43. Cheng, M. H.; Ho, Y. H.; Lee, G. H.; Peng, S. M.; Liu, R. S. *J. Chem. Soc., Chem. Commun.* **1991**, 697.

44. Benaim, J.; Giulieri, F. *J. Organomet. Chem.* **1979**, *165*, 28.

45. Faller, J. W.; Chao, K. H. *J. Am. Chem. Soc.* **1983**, *105*, 3893.

46. Faller, J. W.; Chen, C. C.; Marttina, M. J.; Jakubowski, A. *J. Organomet. Chem.* **1973**, *52*, 361.

47. Faller, J. W.; Lambert, C.; Mazzier, M. R. *J. Organomet. Chem.* **1990**, *383*, 161.

48. Faller, J. W.; Chao, K. H. *Organometallics* **1984**, *3*, 400.

49. Pearson, A. J.; Khan, M. N. I. *J. Org. Chem.* **1985**, *50*, 5276.

50. Lin, S. H.; Lee, G. H.; Peng, S. M.; Liu, R. S. *Organometallics* **1993**, *12*, 2591.

51. Pearson, A. J.; Khan, M. N. I. *J. Am. Chem. Soc.* **1984**, *106*, 1872.

52. Pearson, A. J.; Khan, M. N. I.; Clardy, J. C.; Ho, C. H. *J. Am. Chem. Soc.* **1985**, *107*, 2748.

53. (a) Hosomi, A. *Acc. Chem. Res.* **1988**, *21*, 200; (b) Sukurai, H. *Synlett.* **1989**, 1.

54. (a) Hoffman, R. W. *Angew. Chem., Int. Ed. Engl.* **1982**, *21*, 555; (b) Brown, H.C.; Ramachandran, P. V. *Pure Appl. Chem.* **1991**, *63*, 307.

55. (a) Rousch, W. R.; Walts, A. E.; Hoong, L. K. *J. Am. Chem. Soc.* **1985**, *107*, 8186; (b) Rousch, W. R.; Halterman, R. L. *J. Am. Chem. Soc.* **1986**, *108*, 294.

56. Faller, J. W.; Shvo, Y. *J. Am. Chem. Soc.* **1980**, *102*, 5398.

57. Faller, J. W.; Shvo, Y.; Chao, K. H.; Murray, H. H. *J. Organomet. Chem.* **1982**, *226*, 251.

58. Faller, J. W.; Linnebarrier, D. L. *J. Am. Chem. Soc.* **1989**, *111*, 1937.

59. Faller, J. W.; Diverdi, M. J.; John, J. A. *Tetrahedron Lett.* **1991**, *32*, 1271.

60. Faller, J. W.; Nguyen, J. T.; Ellis, W.; Mazzieri, M. R. *Organometallics* **1993**, *12*, 1434.

61. Benyunes, S. A.; Green, M.; Grimshire, M. J. *Organometallics* **1989**, *8*, 2268.

62. Baxter, J. S.; Green, M.; Lee, T. V. *J. Chem. Soc., Chem. Commun.* **1989**, 1595.

63. Pearson, A. J.; Mallik, S.; Mortezaei, R.; Perry, W. D.; Shively, R. J.; Youngs, W. J. *J. Am. Chem. Soc.* **1990**, *112*, 8034.

64. Benyunes, S. A.; Green, M.; McPartlin, M.; Nation, B. M. *J. Chem. Soc., Chem. Commun.* **1989**, 1887.

65. Liebeskind, L. S.; Bombrum, A. *J. Am. Chem. Soc.* **1991**, *113*, 8736.

66. Pearson, A. J.; Holden, M. S.; Simpson, R. B. *Tetrahedron Lett.* **1986**, 4121.

67. Simpson, A. J.; Khetani, V. *J. Am. Chem. Soc.* **1989**, *111*, 6778.

68. Pearson, A. J.; Mallik, S.; Pinkerton, A.; Adams, J. P.; Zheng, S. *J. Org. Chem.* **1992**, *57*, 2910.

69. Wang, S. H.; Cheng, Y. C.; Lee, G. H.; Peng, S. M.; Liu, R. S. *Organometallics* **1993**, *12*, 3282.

70. Pearson, A. J.; Blystone, S. L.; Nar, H.; Pinkerton, A. A.; Roden, B. A.; Yoon, J. *J. Am. Chem. Soc.* **1989**, *111*, 134.

71. Pearson, A. J.; Khetani, V. D.; Roden, B. A. *J. Org. Chem.* **1989**, *54*, 5141.

72. Lin, S. H.; Peng, S. M.; Liu, R. S. *J. Chem. Soc., Chem. Commun.* **1992**, 615.

73. (a) Olah, G. A. *Aldrichim. Acta* **1979**, *12*, 189; (b) Olah, G. A.; Salem, G.; Staral, S. J.; Ho, T. L. *J. Org. Chem.* **1978**, *43*, 173.

74. Schoonover, M. W.; Baker, E. C.; Eisenberger, R. *J. Am. Chem. Soc.* **1979**, *101*, 1880.

75. Ichinose, N.; Mizuno, K.; Tamai, T.; Otsuji, Y. *Chem. Lett.* **1988**, 233.

76. Lin, S. H.; Cheng, W. J.; Liao, Y. L.; Wang, S. L.; Lee, G. H.; Peng, S. M.; Liu, R. S. *J. Chem. Soc., Chem. Commun.*, in press, **1993**.

77. Davies, S. G.; Dordor-Hedgecock, I. M.; Sutton, K. H.; Davies, S. G.; Jones, R. H.; Prout, K. *Tetrahedron* **1986**, *42*, 209.

78. Davies, S. G.; Walker, J. C. *J. Chem. Soc., Chem. Commun.* **1985**, 209.

79. (a) Uemura, M.; Oda, H.; Minami, T.; Shiro, M.; Hayashi, Y. *Organometallics* **1992**, *11*, 3782; (b) Uemura, M.; Oda, H.; Minami, T.; Hayashi, Y. *Organometallics* **1992**, *11*, 3782.

80. Heathcock, C. H. In *Asymmetric Synthesis, Vol. 3. Stereodifferentiating Addition Reactions Part B.* Morrison, J. D., Ed.; Academic Press: London, 1984; Chap. 2, p. 111.

81. Carey, F. A.; Sunderberg, R. J. *Advanced Organic Chemistry, Part B., Reactions and Synthesis*, 3rd ed.; Plenum Press: New York, 1990; Chap. 6.

82. Cheng, W. J.; Liu, R. S. unpublished results.

83. (a) McIntosh, J. M.; Sieler, R. A. *J. Org. Chem.* **1978**, *43*, 4431; (b) Goldberg, O.; Dreiding, A. J. *Helv. Chim. Acta* **1976**, *59*, 220.

84. Peel, M. R.; Johnson, C. R. In *The Chemistry of Enones*, Part II; Patai, S.; Rappoport, Z., Eds.; John Wiley: New York, 1989; Chap. 21, p. 1090.

85. Erker, G.; Wicker, J.; Engel, K.; Rosenfeldt, F.; Dietrich, W.; Kruger, C. *J. Am. Chem. Soc.* **1980**, *102*, 6344.

86. Nakamura, A.; Yasuda, H. *Angew. Chem., Int. Ed. Engl.* **1987**, *26*, 723.

87. Lin, S. H.; Yang, Y. J.; Liu, R. S. *J. Chem. Soc., Chem. Commun.* **1991**, 1004.

88. Cheng, M. H.; Ho, Y. H.; Wang, S. L.; Cheng, C. Y.; Peng, S. M.; Lee, G. H.; Liu, R. S. *J. Chem. Soc., Chem. Commun.* **1992**, 45.

89. Cheng, M. H.; Lin, S. H.; Chow, J. F.; Lee, G. H.; Peng, S. M.; Liu, R. S. *J. Chem. Soc., Chem. Commun.* **1993**, 757.

90. Lin, Y. H.; Liu, R. S. unpublished results.

91. Mikami, K.; Shimizu, M. *Chem. Rev.* **1992**, 1028.

92. (a) Yang, G. M.; Lee, G. H.; Peng, S. M.; Liu, R. S. *Organometallics* **1991**, *10*, 3600; (b) Yang, G. M.; Lee, G. H.; Peng, S. M.; Liu, R. S. *Organometallics* **1992**, *11*, 3444.

93. Su, G. M.; Lee, G. H.; Peng, S. M.; Liu, R. S. *J. Chem. Soc., Chem. Commun.* **1992**, 215.

94. Liao, M. Y.; Lee, G. H.; Liu, R. S.; Liu, R. S. unpublished results.

SYNTHESIS OF BIARYLS VIA THE CROSS-COUPLING REACTION OF ARYLBORONIC ACIDS

Norio Miyaura

Advances in Metal-Organic Chemistry
Volume 6, pages 187–243.
Copyright © 1998 by JAI Press Inc.
All rights of reproduction in any form reserved.
ISBN: 0-7623-0206-2

I. INTRODUCTION

The mixed Ullman reaction in which two haloarenes couple in the presence of copper powder is a general route for preparing symmetrical and unsymmetrical biaryls.[1] Homocoupling haloarenes with a Ni catalyst and zinc powder is a convenient alternative for improved yields of symmetrical biaryls.[2] Although those procedures are still commonplace in the literature, the Ni- or Pd-catalyzed cross-coupling reactions of arylmetallics with aryl electrophiles are productive and provide reliable results in synthesizing unsymmetrical biaryls.[3-5] In 1972, Kumada and Tamao first reported the reaction of Grignard reagents with alkenyl or aryl halides significantly catalyzed by Ni(II) complexes. After this discovery, a number of other Ni- or Pd-catalyzed reactions including aryl-aryl coupling were developed. The methods presently available for aryl-aryl coupling are the reactions of ArLi,[6] ArMgX,[7] ArZnX,[8] ArBX$_2$, ArSiX$_3$,[9] and ArSnX$_3$.[10] These reagents are derived from the same materials, starting from metallation of arenes or a Grignard-type reaction of haloarenes, and are expected to undergo similar coupling reactions with haloarenes. But some limitations are encountered in obtaining unsymmetrical biaryls without homocoupling, highly functionalized or sterically hindered biaryls, and the stoichiometric conditions relative to metal reagents and halides. Thus, much attention has been recently focused on arylboronic acids because they are thermally stable, nontoxic, relatively inert to water and oxygen, and their coupling reaction is highly selective (Eq. 1).[11-14]

$$
\text{Ph—B(OH)}_2 \;+\; \text{Br—Ar} \quad \xrightarrow[\substack{\text{aq. Na}_2\text{CO}_3 \\ \text{benzene, reflux}}]{\text{Pd(PPh}_3)_4} \quad \text{Ph—Ar} \tag{1}
$$

The first palladium-catalyzed biaryl coupling of arylboronic acids was carried out in a heterogeneous solution using benzene and aqueous sodium carbonate.[15] However, the reaction is feasible with various catalysts, bases, and solvents which affect the rate of coupling and the selectivity of products. This review discusses the reaction mechanisms, the reaction conditions, various approaches for improving biaryl synthesis, and the synthetic applications for fine chemical and functional materials. The following abbreviations are used for palladium ligands: Ph$_2$P(CH$_2$)$_n$PPh$_2$: dppe (n = 2), dppp (n = 3), dppb (n = 4); dppf is 1,1'-bis(diphenylphosphino)ferrocene.

II. SYNTHESIS OF ARYLBORONIC ACIDS

Arylboronic acids are prepared by transmetallation between an organometallic reagent and an appropriate boron reagent.[16] In laboratory-scale preparation, the use of Grignard reagents or lithium reagents and trialkyl borates is the most common from the standpoint of availability and easy preparation (Eqs. 2 and 3).

$$ArLi \xrightarrow{B(O^iPr)_3} \xrightarrow{H_2O} ArB(OH)_2 \qquad (2)$$

$$ArMgX \xrightarrow{B(OMe)_3} \xrightarrow{H_2O} ArB(OH)_2 \qquad (3)$$

Other organometallics of Al, Si, Zn, Sn, and Hg are used, depending on the availability of the reagents. The transmetallation of $B(OR)_3$ with Ar-M (M = Li, MgX) at low temperature (usually at -78 °C) first proceeds through the formation of a relatively unstable $[ArB(OR)_3]M$ in equilibrium with $ArB(OR)_2$ and ROM (Eq. 4).[17] If $[ArB(OR)_3]M$ is

$$B(OR)_3 + ArM \longrightarrow [ArB(OR)_3]M \longrightarrow ArB(OR)_2 + MOMe$$

$$\xrightarrow{RM} [Ar_2B(OR)_2]M \longrightarrow Ar_2BOR + MOMe$$

$$\longrightarrow Ar_3B \xrightarrow{RM} [Ar_4B]M \qquad (4)$$

cleanly formed and if equilibrium favors this complex, the boronic ester is formed selectively. Otherwise, successive steps give rise to the di-, tri-, or tetraarylborates. Triisopropyl borate is the best of the available alkyl borates to prevent such side reactions for organolithiums, thus allowing the preparation of a number of arylboronic acids with high yields, often over 90%.

Ortho lithiation of arenes directed by $CONR_2$,[18] $OCONR_2$,[19] OMOM,[20] or $NHCOR$[21,22] (Eq. 5), or the halogen-lithium exchange reaction (Eq. 6)[23] provides various aryllithiums regioselectively. *In situ* treatment of these lithium intermediates with trimethyl- or tri(isopropyl) borate gives variously functionalized arylboronic acids. Trapping aryllithium or aryl-magnesium intermediates with trimethylchlorosilane, followed by transmetallation to BCl_3 or BBr_3[24,25] is a convenient alternative when a pure aryllithium or arylmagnesium intermediate is not directly available (Eqs. 5 and 6).

$$(5)$$

$$(6)$$

2-(BocHN)-4-methoxyphenylboronic acid (Eq. 5).[22] A solution of *t*-BuLi in pentane (1.7 M, 340 mmol) is added to a solution of *t*-butyl 4-methoxy-carbanilate (136 mmol) in ether (500 mL) at −20 °C. After stirring for 5 h at −20 °C, (MeO)$_3$B (408 mmol) is added. The resulting viscous solution is swirled manually for 5 min, then allowed to warm to 23 °C and to stand for 12 h. The solution is partitioned between saturated aqueous NH$_4$Cl (500 mL) and ethyl acetate (500 mL). The aqueous layer is extracted further with ethyl acetate (2 × 500 mL) and the combined organic layers are dried over MgSO$_4$. The product is purified by flash chromatography (2.5% MeOH in CH$_2$Cl$_2$ → 10% MeOH in CH$_2$Cl$_2$) to provide boronic acid (19.9 g, 55%). R_f = 0.43 (10% MeOH in CH$_2$Cl$_2$).

4-Bromo-2,5-dihexylphenylboronic Acid (Eq. 6).[23] A solution of BBr$_3$ (3.77 mmol) in CH$_2$Cl$_2$ (12 mL), is added to a solution of 1-bromo-2,5-dihexyl-4-trimethylsilylbenzene (2.52 mmol) in CH$_2$Cl$_2$ (20 mL), and then the mixture is stirred for 3 h. Water (10 mL) is added and the layers are separated. The aqueous layer is washed twice with CH$_2$Cl$_2$, and the combined organic phase is dried over MgSO$_4$. Chromatographic separation on silica gel with (1) hexane and (2) CH$_2$Cl$_2$ gives the boronic acid (94%), mp 85.5 °C, R_f = 0.27 (hexane/ethyl acetate=3/1).

The mercuration of arenes followed by transmetallation to BH$_3$[26] or BCl$_3$[27] is advantageous over the lithiation route in synthesizing indole-

3-boronic acid (Eq. 7).[28] Iodonolysis of the ate-complex obtained from

$$\text{(7)}$$

triethylborane and 3-lithiopyridine[29(a)] or 3-lithioquinoline[29(b)] gives exceptionally air-stable diethylborane derivatives (Eq. 8). Direct borylation

$$\text{(8)}$$

of benzene with boron trichloride is an economical process for phenylboronic acid, but the procedure results in a mixture of *o*- and *p*-isomers with toluene (Eq. 9).[30]

$$\text{(9)}$$

Boronic acids generally present a host of difficulties in analysis and isolation. The principal difficulty is their spontaneous condensation to the boroxines [(ArBO)$_3$] in varying degrees. Thus, NMR spectroscopy in CDCl$_3$ exhibits two pairs of signals corresponding to free boronic acid and boroxine. However, they give a simple signal of [ArB(OH)$_3$]Na in a solution of NaOD in D$_2$O. Boronic acid itself cannot be chromatographed, but dehydration to the corresponding boroxine allows separation through silica gel (see the experimental procedures for Eqs. 5 and 6). Reaction of arylboronic acids with 1,2- or 1,3-alkanediols provides thermally stable cyclic esters which are isolable by distillation and analyzable by GC. Some of the pinacol esters of arylboronic acids are chromatographed on silica gel with the solvent containing a few percent of MeOH. Alternatively, the synthesis of arylboronic acids can be directly subjected to cross-coupling without isolating boronic acids, because the reaction shown in Eq. 2 quantitatively produces arylboronic acids and the palladium-catalyzed biaryl coupling tolerates alcohol, water, and inorganic salts. The method obviates the need for isolating boronic acids, often difficult because of their high water solubility.[31]

Arylboronic acids are stable to air and water, and the C–B bond is inert to many electrophiles, thus allowing functionalization of parent arylboronic acids. Radical bromination at the benzylic carbon,[32] nitration with fuming nitric acid,[33] and oxidation of the methyl group to the acid[34] are shown in Eq. 10. The weakly electron-withdrawing property of the

$$(10)$$

$B(OH)_2$ group ($\sigma_p = 0.12$) directs electrophilic substitution at the *meta* position. Amino derivatives are prepared by a sequence of nitration and catalytic hydrogenation of arylboronic acids.[33]

The cross-coupling protocol provides a valuable method for homologation of arylboronic acids or direct borylation of haloarenes. The coupling takes place at the Sn–C bond in preference to the boronic ester because the B–C bond is inert to biaryl coupling in the absence of a base (Eq. 11).[35] Bulky 1,3-diphenylpropanediol protects boronic acid during

$$(11)$$

coupling with an organozinc reagent (Eq. 12).[36]

$$(12)$$

The palladium-catalyzed cross-coupling reaction of tetra(alkoxo)diborons provides a method for direct borylation of aryl iodides, bromides,[37] or triflates[38] (Eq. 13). Because bis(pinacolato)diboron is

See, the experimental procedure in the text: p-Me$_2$NC$_6$H$_4$I (90%), (13)
p-MeOC$_6$H$_4$I (82%), p-BrC$_6$H$_4$I (86%), p-MeO$_2$CC$_6$H$_4$Br (86%),
p-NO$_2$C$_6$H$_4$Br (86%)
p-MeSC$_6$H$_4$OTf (81%),o-MeOC$_6$H$_4$OTf (80%), o-O$_2$NC$_6$H$_4$OTf (60%)
p-ClC$_6$H$_4$OTf (91%), p-OHCC$_6$H$_4$OTf (91%).

hermally stable and easily handled in air, the reagent is useful as the oron nucleophile for the cross-coupling reaction with organic electrophiles. The reaction is accelerated in polar solvents, for example, DMSO > DMF > dioxane > toluene, in the presence of PdCl$_2$(dppf) and KOAc. Arylboronic acids having an electron-donating group, such as p-NMe$_2$ and p-OMe, are synthesized from the corresponding iodides, but aryl bromides or triflates give good results for p-CO$_2$Me, p-COMe, p-NO$_2$, and p-CN derivatives because the electron-withdrawing substituents enhance the rate of coupling.

Pinacolborane is a unique boron nucleophile for a similar coupling eaction with aryl iodides or bromides (Eq. 14).[39] It is interesting that

p-MeOC$_6$H$_4$I (77%), p-MeC$_6$H$_4$I (79%), p-EtO$_2$CC$_6$H$_4$I (79%),
p-NO$_2$C$_6$H$_4$I (84%), p-Me$_2$NC$_6$H$_4$Br (79% at 100 °C)

ester, ketone, and nitro groups remain intact during the coupling reaction of pinacolborane at 80 °C. The reaction is accelerated in the presence of an electron-donating group whereas a withdrawing group slows down he rate of coupling, the electronic effect of which is the reverse of the cross-coupling reactions of diborons and other organoboronic acids. A mechanism involving σ-bond metathesis between ArX and HPdB(OR)$_2$ s proposed for providing ArPdB(OR)$_2$ and HX.

Typical Procedure for Eq. 13.[37] A flask charged with $PdCl_2$(dppf) (0.03 mmol), KOAc (3 mmol), bis(pinacolato)diboron (1.1 mmol) is flushed with nitrogen. DMSO (6 mL) and haloarene (1 mmol) are added, and the mixture is stirred for 1–24 h at 80 °C. The product is extracted with benzene, washed with water, and dried over $MgSO_4$. Kugelrohr distillation gives the arylboronate. Coupling with triflates is carried out under the following reaction conditions.[38] The flask is charged with $PdCl_2$(dppf) (0.03 mmol), dppf (0.03 mmol), KOAc (3 mmol), and bis(pinacolato)diboron (1.1 mmol), and flushed with nitrogen. Dioxane (6 mL) and aryl triflate (1.0 mmol) are added, and then the resulting mixture is stirred at 80 °C for 6–24 h.

Typical Procedure for Eq. 14.[39] A mixture of $PdCl_2(PPh_3)_2$ (0.03 mmol), dioxane (4 mL), ArI (1 mmol), Et_3N (3 mmol), and pinacolborane (1.5 mmol) is stirred for 1–5 h at 80 °C under nitrogen. The reaction mixture is diluted with benzene, washed with water, and dried over $MgSO_4$. Kugelrohr distillation gives a pure arylboronate.

III. CATALYTIC CYCLE OF CROSS-COUPLING

A general catalytic cycle for the cross-coupling reaction of organometallics, which involves an oxidative addition-transmetallation-reductive elimination sequence, is depicted in Fig. 1.[11,40] Although each step involves further processes including ligand exchange and *cis-trans* isomerization, it is significant that the great majority of cross-coupling reactions catalyzed by Ni(0), Pd(0), and Fe(I) are all rationalized in terms of this common and powerful catalytic cycle.[3–5]

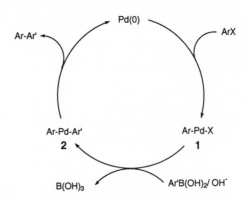

Figure 1. Catalytic Cycle

Oxidative addition is often the rate-determining step in a catalytic cycle. The relative reactivity decreases in the order of I > Br > OTf >> Cl. Although the palladium complexes exhibit excellent catalytic activity for various aryl iodides, bromides, and triflates, the biaryl-coupling reactions of less reactive chloroarenes and aryl mesylates are best carried out by nickel(0) complexes.[41-44] For example, the palladium-catalyzed reaction of chloroarenes is limitedly used for substituents with σ > 0.45, whereas all substituents in a range of -0.83 to 0.66 accommodate the nickel-catalyzed reaction well (Table 1).[45]

The Hammett correlation for the addition of chlorobenzenes to a Pd(0) or Ni(0) complex is shown in Figs. 2[46] and 3.[47]

$Ni(PPh_3)_4$ exhibits a unique Hammett correlation in that reactivity is increased linearly by electron-withdrawing groups with σ > 0.23 and it is insensitive to electron-donating substituents with σ < 0.23. This is in sharp contrast to the oxidative addition to the palladium(0) complex that reveals a linear correlation for both donating and withdrawing groups. Thus, the presence of electron-donating groups retards palladium-catalyzed reactions, but the nickel-catalyzed reaction of chloroarenes is insensitive to such donating groups. Oxidative additions to both Pd(0) and Ni(0) are rationalized by the aromatic nucleophilic substitution mechanism reported by Milstein (Eq. 15).[46] However, the mechanism of oxidative addition of electron-rich chloroarenes to Ni(0) still remains unclarified.

Table 1. Pd- versus Ni-Catalyzed Cross-Coupling Reaction of Phenylboronic Acid with Chloroarenes

			Yield %	
Entry	X=	σ	$Pd(PPh_3)_4{}^a$	$NiCl_2(dppf)^b$
1	4-C≡N	0.66	80	74
2	4-COCH$_3$	0.50	64	96
3	4-CO$_2$Me	0.45	40	87
4	3-OMe	0.12	< 1	72
5	3-CH$_3$	−0.07	< 1	89
6	4-NMe$_2$	−0.83	0	85

Notes: ^{a}PhB(OH)$_2$ (1.1 equiv), ArCl, Na$_2$CO$_3$ (1.5 equiv) and Pd(PPh$_3$)$_4$ (3 mol%) in aqueous DME at 80 °C. ^{b}PhB(OH)$_2$ (1.1 equiv), ArCl, K$_3$PO$_4$ (3 equivs) and NiCl$_2$(dppf) (3 mol%) in dioxane at 80 °C. The nickel catalyst was reduced with BuLi (4 equivs) prior to the coupling.

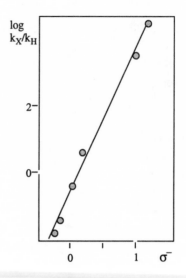

Figure 2. Oxidative Addition of ArCl to Palladium(0)-dppb

$$PhCl \xrightarrow{M(0)} \left[\text{◁} \overset{M^+}{\underset{Cl}{\diagup}} \longrightarrow \text{◁} \overset{M}{\underset{Cl}{+}} \right] \longrightarrow \text{◯}-M\text{-}Cl \qquad (15)$$

M = Pd, Ni

The cross-coupling reaction of organoboron compounds takes place in the presence of base which is quite different from the related coupling reactions of other organometallics. Available information indicates that the aryl group on the boron atom transfers to Ar-Pd- X **1** by the following two processes (Eqs. 16 and 17).[11] The nucleophilicity of the aryl group on boron is enhanced by quarternization with a negatively charged base (Eq. 16). Although there is no direct evidence that boronate anions, such

$$(HO)_2B\text{-}Ar' \underset{}{\overset{HO^{\ominus}}{\rightleftharpoons}} (HO)_3\overset{\ominus}{B}\text{—}Ar' \overset{}{\underset{Ar-Pd-X}{\diagdown}} \longrightarrow Ar\text{-}Pd\text{-}Ar' \qquad (16)$$
$$\mathbf{2}$$

$$Ar\text{—}Pd\text{—}X$$
$$\downarrow RO^{\ominus}$$
$$Ar\text{—}Pd\text{—}OR \xrightarrow{Ar'B(OH)_2} \left[\begin{array}{c} \overset{\oplus}{Ar\text{—}Pd\text{-----}OR} \\ Ar'\text{——}\overset{\ominus}{\underset{OH}{B}}\text{—}OH \end{array} \right] S_E \text{(coord)} \longrightarrow Ar\text{-}Pd\text{-}Ar' \qquad (17)$$

RO = MeO, HO, AcO   **2**

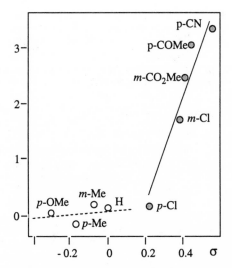

Figure 3. Oxidative Addition of ArCl to Ni(PPh₃)₄

as ArB(OH)₃⁻, are responsible for transmetallation, a related complex [ArBBu₃]Li readily undergoes cross-coupling with organic halides.[45] Thus, it is quite reasonable to assume that boronate anions participate in transmetallation. Biaryl coupling of arylboronic acids with aryl halides at pH = 7–8.5 is retarded relative to the reaction at pH = 9.5–11.75. The pK_a of phenylboronic acid is 8.8, suggesting the formation of the hydroxyboronate anion [ArB(OH)₃⁻] at pH > pK_a and its transmetallation to palladium(II) halides.

Biaryl coupling of ate-complexes obtained from organolithiums and boronic esters may involve a similar transmetallation process (Eq. 18).[48]

$$\text{(18)}$$

R = p-MeO₂C (95%), o-MeO₂C (83%), p-MeCO (95%), p-NC (80%)

Fluoride salts are effective for the cross-coupling reactions of 1-alkenyl- and arylboronic acids (Eq. 19).[49] The species that undergoes transmetallation is assumedly an organo(trifluoro)borate ion.

An alternative transmetallation process is that whereby organoboron compounds readily transfer their organic groups to (alkoxo)palladium(II) complexes under neutral conditions (Eq. 17).

$$PhB(OH)_2 \xrightarrow{F^-} Ph-\overset{\overset{F}{|}}{\underset{\underset{F}{|}}{B}}\overset{\ominus}{-}F \xrightarrow[Pd(PPh_3)_4 \,/\, DME \,/\, 80\ °C]{ArX} Ph\text{-}Ar$$

$$\text{(19)}$$

ArX = p-CH$_2$CN (92%), p-NHCOCF$_3$ (92%), p-C$_6$H$_4$Br (45%),
p-CH$_2$CH$_2$OTs (63%), p-CH$_2$CH$_2$OAc (78%)

It is known that the halogen or OTf ligand on Ar-Pd-X is readily displaced by an alkoxy, hydroxy, or acetoxy anion to provide a basic Pd-OR complex.[50–52] Indeed, the (hydroxo)palladium complex reported by Alper[51] gives a cross-coupled product (70%) together with biphenyl (15%) (Eq. 20). The electronic effects of substituted arylboronic acids

$$\overset{PPh_3}{\underset{PPh_3}{Ph\text{-}Pd\text{-}OH}}{}_{/2} + (HO)_2B\text{-}\langle\!\!\rangle\text{-}OMe \xrightarrow{THF,\ rt} \langle\!\!\rangle\text{-}\langle\!\!\rangle\text{-}OMe \quad \text{(20)}$$

$$70\%$$

are consistent with the S$_E$2 (coord) mechanism involving a rate-determining coordination of the RO ligand to the boron atom.[53] As a result of complex formation, transfer of an activated organic group from boron to palladium takes place. Such complexation before migration is one of the crucial steps essential in all ionic reactions of organoboron compounds, namely, the well-known intramolecular 1,2-migration from the organoborane/electrophile complex. High reactivity of oxo-palladium complexes for transmetallating organoboron compounds is attributed to the high reactivity of the Pd–O bond, which consists of a soft acid and a hard base combination, and the high oxophilicity of the boron center. The basicity of PdOH is not known, but the related complexes, such as PtH(OH)[P(^{i}Pr)$_2$] or *trans*-Pt(OH)(Ph)(PPh$_3$)$_2$, are reported to be more basic than NaOH.[54]

It is not yet obvious in many reactions which process shown in Eqs. 16 and 17 predominates; However, the formation of a hydroxo- or

$$Pd(PPh_3)_4 \xrightarrow{PhBr} \overset{PPh_3}{\underset{PPh_3}{Ph\text{-}Pd\text{-}Br}}$$

$$\Big\downarrow AcO^{\ominus} \qquad \text{(21)}$$

$$\overset{PPh_3}{\underset{PPh_3}{Ph\text{-}B(OR)_2}} \longleftarrow \overset{PPh_3}{\underset{PPh_3}{Ph\text{-}Pd\text{-}B(OR)_2}} \overset{diboron}{\longleftarrow} \overset{PPh_3}{\underset{PPh_3}{Ph\text{-}Pd\text{-}OAc}}$$

acetatopalladium(II) intermediate should be considered one of the crucial transmetallation processes in base/palladium-induced cross-coupling reactions.[11] Borylation of aryl halides and triflates with the diboron (Eq. 13) involves such transmetallation between $ArPd(OAc) \cdot L_2$ and diboron.[37,38] Treating *trans*-[PhPdBr(PPh$_3$)$_2$] with KOAc in DMSO at room temperature quantitatively gives *trans*-[PhPd(OAc)(PPh$_3$)$_2$] which undergoes transmetallation with diboron giving a 67% yield of the pinacol ester of phenylboronic acid at room temperature. Arylboronic acids couple with Ph_2IX (X = BF_4, OTs, OTf), PhI(OH)OTs,[55] or ArN_2BF_4[56] without the assistance of a base (Eq. 22). ArPdOH or [ArPd-

$$\text{(22)}$$

a) Pd(PPh$_3$)$_4$, aq. Na$_2$CO$_3$ in DME at r.t.   98%
b) Pd(PPh$_3$)$_4$ in DME at r.t.   96%

$(OH_2)]^+$ derived from $ArPd^+$ and water is proposed as the reactive species involved in transmetallation. $PhB(OH)_2$, Ph_2BOMe, and Ph_3B readily transfer the phenyl group to *cis*-[Pt(S)$_2$L$_2$]$^{2+}$ (L = PMe$_3$, PPh$_3$; S = MeOH, H$_2$O) at 0 °C to give *trans*-[Pt(Ph)(S)L$_2$]$^+$, whereas the corresponding Ph_4Si and Ph_4Sn do not react with the complex.[57] The mechanism is consistent with transmetallation to the oxo-platinum(II) species of *cis*-[Pt(OR)(S)L$_2$]$^+$ (R = H or Me) produced by deprotonating *cis*-[Pt(ROH)L$_2$]$^{2+}$.

IV. REACTION CONDITIONS FOR CROSS-COUPLING

A. Effect of Catalysts

A wide variety of palladium(0) catalysts or precursors are used for cross-coupling reactions. Pd(PPh$_3$)$_4$ has been most common, but PdCl$_2$(PPh$_3$)$_2$ and Pd(OAc)$_2$ itself or plus a phosphine ligand are convenient because they are stable to air and are reduced *in situ* to active Pd(0) complexes (Eqs. 23, 24,[58,59] and 25[51,60]).

PdCl$_2$(PPh$_3$)$_2$ + 2 ArB(OH)$_2$ + 2 OH⁻ $\longrightarrow$ Pd(PPh$_3$)$_2$ + Ar-Ar + 2 Cl⁻   (23)

Pd(OAc)$_2$ + n PPh$_3$ + H$_2$O $\longrightarrow$ Pd(PPh$_3$)$_{n-1}$ + O=PPh$_3$ + 2 AcOH   (24)

PdCl$_2$(PPh$_3$)$_2$ + 2 OH⁻ $\longrightarrow$ Pd(PPh$_3$) + O=PPh$_3$ + 2 Cl⁻ + H$_2$O   (25)

A combination of $Pd_2(dba)_3$ or $Pd(dba)_2$ and a phosphine ligand is also used for preparing various palladium(0)-phosphine complexes.[61] $NiCl_2(dppf)$ or $NiCl_2(PPh_3)_2$ are reduced *in situ* by heating the reaction mixture of arylboronic acid, catalyst, and anhydrous K_3PO_4 in dioxane at 95 °C.[43] However, the reduction of $NiCl_2 \cdot L_2$ with BuLi or DIBAL in the presence or absence of an additional phosphine ligand is recommended because reducing the NiX_2 complex to an active Ni(0) species is much slower than that of the corresponding PdX_2 complexes.[42]

Various phosphine ligands are effective in stabilizing the Pd(0) species during the cross-coupling reaction of arylboronic acids. Palladium complexes that contain fewer than four triphenylphosphines or bulky phosphines, such as tris(o-tolyl)phosphine, tris(2,4,6-trimethoxy-phenyl)phosphine, and triphenylarsine, provide highly reactive catalysts for oxidative addition because of the ready formation of the coordinate unsaturated palladium species.[62] For bidentate ligands, oxidative addition is generally favored by basic ligands (electronic effects) and ligands possessing a small bite angle (steric effects).[61] The superiority of $PdCl_2(dppb)$ or $PdCl_2(dppp)$ over $Pd(PPh_3)_4$ is demonstrated in the cross-coupling reaction of chloroarenes[63] and heteroary chlorides.[64] $PdCl_2(dppf)$ and $NiCl_2(dppf)$ are efficient catalysts in the reaction of aryl mesylates[44] and chlorides.[42,43] Tris(o-tolyl)phosphine has been used as a bulky ligand with low coordination ability, but it is recently reported that its addition to $Pd(OAc)_2$ yields a palladacycle which exhibits high catalytic efficiency for biaryl coupling, often exceeding a 75,000 turnover number (Eq. 26).[65]

$$ \text{(26)} $$

L = DMSO, DMF, MeCN, PPh₃

The order of coupling rates with representative palladium sources is shown in Eq. 27.[66] Coupling with 4-iodonitrobenzene is greatly acceler-

$$ \text{(27)} $$

$Pd(PPh_3)_4$	8 h	23%
$Pd_2(dba)_3 + 2\,PPh_3$	8 h	98%
$Pd_2(dba)_3$	20 min	98%
$Pd(OAc)_2$	45 min	98%

ated by palladium catalysts without a phosphine ligand. The inhibitory effect of the phosphine ligand can be attributed to its large steric hindrance during oxidative addition and transmetallation.

A fine metallic palladium obtained *in situ* from Pd(OAc)$_2$ is indeed an excellent catalyst for biaryl coupling at room temperature in the aqueous phase (Eq. 28).[67] The reaction is further accelerated in the

$$
\begin{array}{c}
\text{HOOC} \\
\text{HO} \overset{}{\underset{}{\bigcirc}} \text{Br}
\end{array}
\xrightarrow[\substack{\text{Pd(OAc)}_2,\ \text{Na}_2\text{CO}_3\ \text{in H}_2\text{O} \\ 20\ ^\circ\text{C for 2 h}}]{\text{PhB(OH)}_2}
\begin{array}{c}
\text{HOOC} \\
\text{HO} \overset{}{\underset{}{\bigcirc}} \overset{}{\underset{}{\bigcirc}} \\
99\%
\end{array}
\qquad (28)
$$

presence of 1 equivalent of Bu$_4$NBr for bromoarenes having an electron-donating or electron-withdrawing group.[68] Pd/C is studied as an economical catalyst for synthesizing liquid crystals (Eq. 29).[69]

$$
\begin{array}{c}
\overset{O}{\underset{O}{\bigcirc}}\text{---}\bigcirc\text{---}\bigcirc\text{---Br} + (\text{HO})_2\text{B}\overset{F\quad F}{\underset{}{\bigcirc}}
\end{array}
\xrightarrow[\substack{5\%\ \text{Pd/C} \\ \text{Na}_2\text{CO}_3\ \text{in 95\% EtOH} \\ 80\ ^\circ\text{C for 19h}}]{}
\ 94\%
\qquad (29)
$$

Such catalysts without a phosphine ligand have an advantage in avoiding contamination by both phosphine and elemental palladium in the final products. However, the use of more than two phosphine ligands is generally recommended to avoid precipitating metallic palladium because complete conversion is not always possible under "ligandless" conditions, especially for electron-rich or sterically hindered haloarenes. Pd(PPh$_3$)$_4$ is widely used and exhibits excellent catalytic activity and selectivity in various biaryl couplings, though a side reaction gives a coupling product of phosphine-bound aryls, which is discussed in Section IV D.

Biaryl coupling in aqueous solvents offers advantages in large-scale industrial processes because of the simplicity of catalyst-product separation and the economy and safety of using water as the solvent. Reactions are carried out in a single, basic aqueous phase or in a two-phase, basic aqueous-organic medium by using a water soluble phosphine ligand (Eq. 30).[70] The industrial process for *o*-tolylbenzonitrile is carried out in a mixed solvent of DMSO and water (Eq. 31).[71]

$$\text{NaO-}\overset{\overset{\displaystyle O}{\|}}{\underset{\overset{\displaystyle \|}{O}}{S}}\!\!-\!\!\langle\rangle\!\!-\!\!Br \quad \xrightarrow[\substack{Pd[PPh_2(m\text{-}C_6H_4SO_3Na)]_3 \\ Na_2CO_3 \text{ in EtOH/H}_2O \text{ (6/4)} \\ \text{at } 80\ ^\circ C}]{4\text{-MeC}_6H_4B(OH)_2} \quad \text{NaO-}\overset{\overset{\displaystyle O}{\|}}{\underset{\overset{\displaystyle \|}{O}}{S}}\!\!-\!\!\langle\rangle\!\!-\!\!\langle\rangle\!\!-\!\!CH_3 \tag{30}$$

78%

$$\underset{Cl}{\overset{C\equiv N}{\langle\rangle}} \quad \xrightarrow[\substack{Pd(OAc)_2 \,/\, P(m\text{-}C_6H_4SO_3Na)_3 \\ Na_2CO_3 \text{ in DMSO/H}_2O \text{ at } 80\ ^\circ C}]{4\text{-MeC}_6H_4B(OH)_2} \quad \underset{}{\overset{C\equiv N}{\langle\rangle}}\!\!-\!\!\langle\rangle\!\!-\!\!CH_3 \tag{31}$$

91%

Typical Procedure for Eq. 30.[70] A solution of $PPh_2(m\text{-}C_6H_4SO_3Na)$ (0.863 mmol) in ethanol (5 mL) is added to an aqueous solution of Na_2PdCl_4 (2.42 mL, 0.242 mmol). After stirring for 5 min, the mixture is treated with activated zinc powder (0.09 g) for 1 h. The mixture is filtered through Celite, concentrated in vacuo, and layered with *t*-BuOH to give air-sensitive yellow crystals of $Pd[PPh_2(m\text{-}C_6H_4SO_3Na]_3(H_2O)_4$ with a yield of 94%. A solution of *p*-tolylboronic acid (1 mmol) in EtOH (5 mL) is added to a solution of $4\text{-BrC}_6H_4SO_3Na$ (1.25 mmol), Na_2CO_3 (2 mmol), and the above catalyst (0.06 mmol) in water (5 mL). The mixture is stirred for 7 h at 80 °C and then cooled to room temperature. The precipitated solid is filtered, washed with water, benzene, and ether and gives the analytically pure biaryl. The filtrate from the initial filtration is concentrated in vacuo and refiltered to give a second crop of the biaryl with a yield of 0.21 g (78%).

A palladium-phosphine complex supported on polystyrene-PPh_2 resin catalyzes the cross-coupling reaction of various organoboron compounds with organic halides or triflates under conditions comparable to that of the homogeneous catalysts, such as $Pd(PPh_3)_4$ (Eq. 32).[72] The catalyst is

$$\underset{PPh_2}{\overset{}{\overline{P}\!\!-\!\!\langle\rangle\!\!-\!\!}} \quad \xrightarrow[\substack{2.\ NH_2NH_2}]{1.\ PdCl_2} \quad \substack{\text{Polymer bound} \\ \text{palladium(0)}} \tag{32}$$

easily separated from the reaction mixture and reused more than 10 times with no decrease in activity.

Preparation of Polymer-Bound Palladium(0) Complex (Eq. 32).[72] A 100%-chloromethylated polystyrene resin is treated with $LiPPh_2$ at 25 °C for 48 h. The reaction of the resin thus obtained (5 g) with $PdCl_2$ (0.75 g) yields a yellow polymeric Pd(II) complex. A mixture of the polymer-palladium(II) complex (5 g, 3.71 mmol of Pd) and PPh_3 (7.42 mmol) in ethanol (50 mL) is stirred for 3 h. Then the hydrazine hydrate (3.71 mmol) is added to the mixture. After stirring for 2 h, the resulting polymer is filtered, washed with ethanol and ether, and dried in vacuo to give a dark green polymer-bound palladium(0) complex (P/Pd = 3.9).

B. Effect of Bases

The cross-coupling reaction of organoboronic acids proceeds in the presence of a negatively charged base, such as sodium or potassium carbonate, phosphate, or hydroxide, which is used as an aqueous solution or as suspension in toluene, dioxane, DME, or DMF. A combination of base and phase-transfer catalyst, such as Bu$_4$NX, is also used successfully.[68,73]

Aqueous Na$_2$CO$_3$ is a mild base effective for the biaryl coupling of various arylboronic acids. However, the reaction of mesitylboronic acid with iodobenzene reveals the following order of reactivity: Ba(OH)$_2$ > NaOH > K$_3$PO$_4$ > Na$_2$CO$_3$ > NaHCO$_3$ (Table 2).[74]

Aqueous Ba(OH)$_2$ exhibits an exceptionally high accelerating effect allowing sterically hindered coupling for various 2,6,2′-trisubstituted biaryls (Eq. 33).[74] Similarly, the barium base gives excellent results[20,75]

$$(33)$$

ArX: 2-MeOC$_6$H$_4$I (80%), 2-ClC$_6$H$_4$I (94%), 2-bromonaphthalene (86%)

Table 2. Effects of Base on the Coupling of Mesitylboronic Acid with Iodobenzene[a]

		Yield (%)[b]		
Solvent	Base	8 h	24 h	48 h
Benzene/H$_2$O	Na$_2$CO$_3$	25 (6)	77 (12)	85 (26)
	Ba(OH)$_2$	92 (13)		
DME/H$_2$O	Na$_2$CO$_3$	50 (1)	66 (2)	83 (7)
	Cs$_2$CO$_3$	93 (0)		
	K$_3$PO$_4$	70 (0)	83 (3)	
	NaOH	95 (2)		
	Ba(OH)$_2$	99 (2)[c]		
	TlOH	74 (20)[d]		

Notes: [a]PhI (1 equiv), mesityl B(OH)$_2$ (1.1 equiv), base (1.5 equiv) at 80 °C.

[b]Yield of mesitylene obtained by protodeboronation is shown in the parentheses.

[c]After 4 h at 80 °C.

[d]At room temperature for 10 min.

in the reverse combination of *ortho*-disubstituted haloarenes and *ortho*-monosubstituted arylboronic acids (Eq. 34).[75(a)] Because transmetalla-

$$
\begin{array}{c}
\text{Me} \quad \text{Br} \\
\text{Me} \quad \text{—OCH}_3 \\
\text{Me} \quad \text{Br}
\end{array}
\quad
\begin{array}{c}
\text{Me} \\
(HO)_2B\text{—} \\
\end{array}
\quad
\xrightarrow[\substack{\text{Pd(OAc)}_2 + 2P(o\text{-tolyl})_3 \\ \text{aq. Ba(OH)}_2 \text{ in DME,} \\ \text{reflux for 4 h}}]{}
\quad
\begin{array}{c}
\text{Me} \quad \text{Me} \\
\text{—OCH}_3 \\
\text{Me} \quad \text{Me}
\end{array}
\quad 95\% \quad (34)
$$

tion involves nucleophilic substitution of Pd-X (Eq. 16 or 17), a strong base is recommended when several *ortho* substituents are in the halides and the boronic acids or a strong donating group is in the halides because these substituents retard the transmetallation step.

Aqueous TlOH solution, which produces highly insoluble salts similar to the barium base, completes the coupling reaction of mesitylboronic acid with iodobenzene within 30 min at room temperature (Table 2 and the procedure in the text).[45] However, this excellent base for mesityl-boronic acid, unfortunately, does not work well for other less substituted arylboronic acids. Aryl- and 1-alkenylboronic acids transmetallate to thallium salts.[16(b)]

Typical Procedure for Eq. 34.[75(a)] A flask is charged with (*o*-tolyl)$_3$P (6 mmol), Pd(OAc)$_2$ (3 mmol), *o*-tolylboronic acid (12.2 mmol), Ba(OH)$_2$·8H$_2$O (90 mmol), 2,6-dibromo-3,5-dimethylanisole (29.6 mmol), DME (120 mL), and H$_2$O (24 mL). The mixture is degassed in vacuo at −20 °C and flushed with argon. After heating at reflux for 3 h, the mixture is poured into aqueous NH$_4$Cl, extracted with ether, and dried over Na$_2$SO$_4$. Chromatography on silica gel (Et$_2$O/hexane = 1/200 to 1/10) gives a mixture of the diastereoisomers (94%). Recrystallization from MeOH gives *meso*- (45%) and *dl*-3,5-dimethyl-2,6-bis(2-methyl-phenyl)anisole (38%).

Biaryl Coupling with TlOH (Table 2).[45] Under argon, a mixture of Tl$_2$SO$_4$ (20 mmol) and Ba(OH)$_2$·8H$_2$O (20 mmol) in water (18 mL) is stirred for 5 h at room temperature. Filtration through a Celite pad on a sintered-glass filter gives a colorless solution of TlOH (ca. 1 M).

Pd(PPh$_3$)$_4$ (0.02 mmol) and mesitylboronic acid (1.6 mmol) are dissolved in DME (6 mL) under argon. Iodobenzene (1 mmol) and aqueous TlOH (1M, 2 mmol) are added successively, and then the mixture is stirred for

30 min at room temperature. GC analysis reveals the formation of 2,4,6-trimethylbiphenyl (94%) and mesitylene (46%).

The protodeborylation of arylboronic acids and the protection of functional groups sensitive to bases are difficulties observed during the biaryl-coupling reaction. Even if there is no great steric hindrance, reaction under aqueous conditions gives undesirable results due to competitive hydrolytic deboronation which is accelerated by the presence of adjacent heteroatoms and *ortho*-substituents. 2-Pyridyl-, 2,6-dimethoxy-, and 2-formyl benzeneboronic acids are highly sensitive to hydrolytic B–C bond cleavage. For example, 3-formyl-2-thienylboronic acid fails to couple with 2-bromothiophene because of its fast protodeboronation.[76] Although the reaction of Eq. 35 finally achieved an 80% yield, a large

$$(35)$$

excess of the boronic acid was added repeatedly until the aryl iodide partner was completely consumed.[77]

The following are the rates for the cleavage of $XC_6H_4B(OH)_2$ with water at pH 6.7 relative to phenylboronic acid.[78] 2,6-dimethoxy (125), 2-F (77), 2-Cl (59), 2-MeO (11), 4-MeO (4.2), 2-Me (2.5), 3-F (2.3), 3-Me (2), 4-F (1.7). For arylboronic acids sensitive to water, the reaction under nonaqueous conditions is desirable. The coupling of 2-formylphenylboronic acid with 2-iodotoluene at 80 °C in an aqueous solution of Na_2CO_3 and DME results in a 54% yield of biaryl accompanied by benzaldehyde (39%). The yield, however, improves to 89% with the 1,3-propanediol ester of boronic acid and anhydrous K_3PO_4 suspended in DMF at 80 °C (Eq. 36).[74] 2-Pyridylboronic acid is highly susceptible to

$$(36)$$

ArX: 2-MeC$_6$H$_4$I (89%), iodomesitylene (73%), 2-MOMOC$_6$H$_4$I (85%), 2-MeO$_2$CC$_6$H$_4$Br (63%), 2-AcHNC$_6$H$_4$I (79%)

hydrolytic protodeboronation, but anhydrous conditions using solid KOH and Bu_4NI in benzene yield 2,2'-bipyridine (Eq. 37).[79] However,

$$\begin{array}{c}\text{2-Br-3,5-dimethylpyridine} \\ \xrightarrow{\hspace{2cm}} \\ Pd(PPh_3)_4 \text{ anhyd. KOH} \\ Et_4NI, \text{ benzene at 80 °C}\end{array}$$

60%   (37)

Negishi's coupling using arylzincs[8] or Stille's coupling with arylstannanes[10] is perhaps a more general alternative in such cases.

Typical Procedure for Eq. 36.[74] BuLi (1.5 M in hexane, 6 mmol) is added to a solution of 2-bromobenzaldehyde dimethyl acetal (6 mmol) in ether (10 mL) at –78 °C. After stirring for 1 h, $B(OMe)_3$ (6 mmol) in ether (3 mL) is added dropwise. The mixture is stirred for 30 min at –78°C, allowed to warm to 20 °C, to stand for 2 h, and then is treated with aqueous 1 M-HCl (10 mL) for 1 h. The boronic acid is extracted with ether and washed with water. 1,3-Propanediol (6 mmol) and excess $MgSO_4$ are added and the mixture stands for 16 h. Kugelrohr distillation gives 2-formylphenylboronic acid 1,3-propanediol ester (57%, bp 100–110 °C/0.1 mmHg). A flask is flushed with nitrogen and charged with $Pd(PPh_3)_4$ (0.03 mmol), anhydrous K_3PO_4 (1.5 mmol, a commercially available $K_3PO_4\cdot nH_2O$ is dried in vacuo at 150 °C), the above boronic ester (1.1 mmol), and 2-iodotoluene (1 mmol) in DMF (6 mL). The mixture is heated at 80 °C for 4 h to yield the biaryl (96%) and benzaldehyde (10%).

Other difficulties encountered during biaryl coupling under basic conditions are saponification of esters, racemization of optically active compounds, or Aldol condensation of carbonyl compounds. These difficulties associated with bases are overcome by using bases heterogeneously. Esters remain intact during cross-coupling in two-phase systems using aqueous Na_2CO_3 and benzene[15,80] or solid $K_3PO_4\cdot nH_2O$ suspended in DMF, dioxane, or toluene.[81] The synthesis of arylalanine suffers from racemization induced by the base, but the optically pure compound is finally obtained when using anhydrous K_2CO_3 suspended in toluene (Eq. 38).[82]

$$\begin{array}{c}\text{ArB(OH)}_2 \\ \xrightarrow{\hspace{2cm}} \\ Pd(PPh_3)_4 \\ K_2CO_3 \\ \text{toluene} \\ 90 °C, 2 h\end{array}$$

75-94%   (38)

Fluoride salts, such as CsF and Bu_4NF (2–3 equivs) are mild bases that accelerate the coupling reaction of substrates sensitive to bases (Eq. 19).[49,63,83] Triethylamine induces biaryl coupling at 100 °C in DMF or EtOH, though it is limitedly used for electron-deficient haloarenes having a withdrawing group (Eq. 39).[84–86]

$$\begin{array}{c}\text{2-CH}_3\text{OC}_6\text{H}_4\text{B(OH)}_2\\\hline \text{Pd(OAc)}_2/2\text{P(o-tolyl)}_3\\\text{Et}_3\text{N (3 equivs) in DMF}\\\text{100 °C for 2-3 h}\end{array}$$

70%

(39)

C. Electrophiles

Triflates are valuable as partners in cross-coupling reactions[87] because they are easy to access from phenols. The reaction, assisted by $K_3PO_4 \cdot nH_2O$ or K_2CO_3 suspended in dioxane or toluene, gives biaryls with high yields (Eqs. 38 and 40).[81] Aqueous bases also have been

90%

(40)

a) $4\text{-MeC}_6\text{H}_4\text{B(OH)}_2$, $Pd(PPh_3)_4$, $CsCO_3$ (88%). b) NaH/Tf_2O (89%)
c) $4\text{-FC}_6\text{H}_4\text{B(OH)}_2$, $Pd(PPh_3)_4$, solid K_3PO_4 in dioxane at 80 °C

successfully used to synthesize electron-rich or sterically hindered biaryls (Eq. 41).[88,89] The order of reactivity of halides and triflates is I >

$$\begin{array}{c}\text{C}_6\text{Cl}_2\text{H}_3\text{B(OH)}_2\\\hline \text{Pd(PPh}_3)_4, \text{ aq. Na}_2\text{CO}_3,\\\text{LiCl (2 equivs)}\\\text{toluene at 95 °C}\end{array}$$

53%

(41)

Br > OTf >> Cl. Thus, sequential cross-coupling of halotriflates provides unsymmetrical terphenyls when two leaving groups with different reactivity are selected.[90] An alternative method is a stepwise, double cross-coupling of halophenol derivatives (Eq. 40).[81,91] However, care must be taken because the phosphine ligand changes the order of reactivity between the triflate and the bromide. Bulky monodentate phosphine ligands react in the order Br > OTf, and bidentate ligands, such as dppp and dppb, reverse the order to OTf > Br.[92]

Biaryl coupling of triflates often fails to proceed because the catalysts decompose, precipitating metallic palladium at an early stage of the reaction, caused presumably by side reaction between phosphine and triflates leading to phosphonium salts $[(ArPPh_3)OTf]$.[93] Adding ~1–2 equivalents of lithium or potassium bromide or chloride is often effective in preventing catalyst decomposition (Eq. 41),[26,81,88] by converting the labile cationic palladium(II) species to organopalladium(II) halides (Eq. 42).[10]

$$ArOTf \xrightarrow{\text{Pd(0)}} [Ar\text{-}Pd(II)][OTf]^- \xrightarrow{\text{LiX}} Ar\text{-}Pd\text{-}X + LiOTf \qquad (42)$$

One of the challenges is to extend the reaction from triflates to less reactive arylsulfonates. The palladium-catalyzed reaction of aryl mesylates results in low yields because of their slow oxidative addition to a palladium(0) complex, but they readily participate in the nickel-catalyzed cross-coupling reaction at 80°–100 °C. A nickel(0) species incorporating dppf ligand, obtained by $in\ situ$ reduction of 10 mol% of $NiCl_2(dppf)$ with zinc dust, is recognized as the most effective catalyst (Eq. 43).[44] However, again the reaction needs to be optimized because a

$$\text{(Ph)}\text{-}OSO_2Me \xrightarrow[\substack{NiCl_2(dppf)/\ Zn \\ K_3PO_4/\ \text{dioxane at 100 °C for 24 h}}]{(HO)_2B\text{-}\text{(Ar)}\text{-}OMe} \text{(Ph)}\text{-}\text{(Ar)}\text{-}OMe \qquad (43)$$

$$81\%$$

similar reaction of the ate-complexes of arylboronates with $NiCl_2(PPh_3)_2$, recently reported, occurs at room temperature (Eq. 18).[48]

Typical Procedure (Eq. 43).[44] A 125 mL-Schlenk tube is charged with phenyl mesylate (0.5 mmol), 4-methoxyphenylboronic acid (0.55 mmol), $NiCl_2(dppf)$ (0.05 mmol), zinc powder (0.86 mmol), K_3PO_4 (1.5 mmol), and a magnetic stirring bar. The contents are dried at 25 °C under reduced pressure (1×10^{-3} mmHg) for 3 h. Then, the tube is filled with N_2, followed by three evacuation-filling cycles. Freshly distilled dioxane (1 mL) is added, and the mixture is stirred for 24 h at 100 °C. Chromatography on silica gel gives 4-methoxybiphenyl (81%) and unreacted phenyl mesylate (18%).

Chloroarenes are an economical and easily available substrate, but they are rarely used for the palladium-catalyzed cross-coupling reaction of arylboronic acids because of their slow oxidative addition to palladium(0). Thus, palladium catalysts have been limitedly used for activated chloroarenes, such as chloropyridines (Eq. 44)[64] and electron-deficient chloroarenes.[63]

$$R-\underset{N}{\overset{\displaystyle \fbox{}}{}}-Cl \quad \xrightarrow[\substack{\text{PdCl}_2\text{(dppp)/ aq. Na}_2\text{CO}_3 \\ \text{benzene/EtOH, reflux for 48 h}}]{\text{PhB(OH)}_2} \quad R-\underset{N}{\overset{\displaystyle \fbox{}}{}}-\fbox{} \qquad (44)$$

3-chloropyridine (71%), 3-hydroxy-2-chloropyridine (15%),
2,5-dichloropyridine (83%)

Biaryl coupling of various chloroarenes having an electron-withdrawing and an electron-donating group takes place in the presence of a nickel catalyst (Eq. 45).[41–43] The nickel(0) complex, prepared *in situ* from

$$\underset{R}{\overset{\displaystyle \fbox{}}{}}-Cl \quad \xrightarrow[\substack{\text{NiCl}_2\text{(dppf)/BuLi} \\ \text{K}_3\text{PO}_4\text{/ dioxane at 80 °C for 16 h}}]{\text{PhB(OH)}_2} \quad \underset{R}{\overset{\displaystyle \fbox{}}{}}-\fbox{} \qquad (45)$$

4-CH$_3$COC$_6$H$_4$Cl (96%), 4-HOCC$_6$H$_4$Cl (89%), 4-MeO$_2$CC$_6$H$_4$Cl (87%),
4-NO$_2$C$_6$H$_4$Cl (0%), 4-AcHNC$_6$H$_4$Cl (81%), 4-MeOC$_6$H$_4$Cl (82%),
4-H$_2$NC$_6$H$_4$Cl (90%), 4-Me$_2$NC$_6$H$_4$Cl (85%)

NiCl$_2$·(dppf) or NiCl$_2$·2PPh$_3$/2PPh$_3$ and *n*-BuLi (4 equivs), is a very efficient catalyst. Nickel(0) complexes are highly reactive, but they are a very labile species that slowly decomposes during prolonged reaction time. Thus, the use of an additional phosphine ligand to NiCl$_2$(dppf) or NiCl$_2$(PPh$_3$)$_2$ is often advantageous to achieve high yields. The palladium-catalyzed reaction of arylboronic acid is best conducted with an aqueous base, but water should be avoided in the nickel-catalyzed reaction. The rate-determining role of the oxidative addition step in the coupling reaction of chloroarenes is suggested by a plot of the relative reactivity of substituted chloroarenes versus the σ constants. A small change in the rate constants ($\rho = 0.74$) in the range of -0.27 to 0.06 and a high electronic effect ($\rho = 5.26$) from 0.37 to 0.66 are identical to that observed in the oxidative addition of chloroarenes to Ni(PPh$_3$)$_4$ (see Fig. 3).

Procedure for Eq. 45.[42] A flask is charged with NiCl$_2$(dppf) (0.03 mmol) and dppf (0.03 mmol) and flushed with argon. The catalyst is dissolved in dioxane (6 mL) and then treated with *n*-BuLi (0.12 mmol) at room temperature for 10 min to give a solution of a nickel(0) complex. 2-Methylphenylboronic acid (1.1 mmol), K$_3$PO$_4$·nH$_2$O (3 mmol), and 4-amino-2,4-dimethoxychlorobenzene (1 mmol) are added through the neck of the flask while maintaining a slow stream of argon. The mixture is stirred at 80 °C for 16 h. Chromatography over silica gel (hexane/ethyl acetate = 2/1) gives the desired biaryl (91%).

Arenediazonium salts ArN$_2$BF$_4$ are an excellent substrate available from aromatic amines for carrying out oxidative addition under mild conditions.[56] The desired cross-coupling with aryl(trifluoro)borates (ArBF$_3$)K is achieved at room temperature with 5 mol% of Pd(OAc)$_2$ or with the paradacycle shown in Eq. 26 (Eq. 46).[94] Biaryl coupling of arene

$$ \text{(46)} $$

iodonium salts, which readily proceeds at room temperature under neutral conditions, is shown in Eq. 22.[55]

Typical Procedure for Eq. 46.[94] A concentrated aqueous solution of KHF$_2$ (3.3 equivs) is added to a methanol solution of ArB(OH)$_2$ at room temperature. After stirring for 10 min, the solvent is evaporated in vacuo. The resulting solid is extracted with several portions of acetone. Evaporation of acetone gives crude ArBF$_3$K which is recrystallized or reprecipitated from acetone/ether. Yields are more than 95%.

A suspension of ArN$_2$BF$_4$ (1 mmol), ArBF$_3$K (1.2 mmol), and Pd(OAc)$_2$ (5 mol%) in dioxane (4 mL) is stirred at 20 °C in a foil-covered flask. The reaction is followed by measuring nitrogen gas evolution.

D. Side Reactions and By-Products

Cross-coupling reactions catalyzed by palladium or nickel complexes often afford unexpected by-products including homocoupling products of electrophiles and organometallics and coupling products of phosphine-bound aryls. Reaction of Grignard reagent has suffered from the homocoupling resulting from the metal-halogen exchange (Eq. 47),

$$\text{ArX} + \text{Ar'MgX} \rightleftharpoons \text{ArMgX} + \text{Ar'X} \qquad (47)$$

but this type of by-product is not reported in the reaction of arylboronic acids. Metathesis of RMX to R_2M and MX_2 (M = Ni, Pd) produces dimers of electrophiles and organometallics (Eq. 48).[95] Oxidative addition of

$$2\ \text{Ar-Pd-X} \longrightarrow \begin{array}{c} \text{X—Pd—Ar} \\ \diagup\!\diagdown \\ \text{X—Pd—Ar} \end{array} \longrightarrow \left[\begin{array}{l} \text{Ar-Pd-Ar} \xrightarrow{\hspace{2cm}} \text{Ar-Ar} \\ + \\ \text{PdX}_2 \xrightarrow{2\ \text{Ar'B(OH)}_2} \text{Ar'-Ar'} \end{array} \right. \qquad (48)$$

metal–carbon bonds to low-valent transition metals is another route leading to dimers of organometallics (Eq. 49).[96] Homocoupling of elec-

$$\text{Ar-M-X} \xrightarrow{\text{Pd(0)}} \text{Ar-Pd-MX} \xrightleftharpoons \text{Ar-Ar} \qquad (49)$$
$$\text{M = Hg, Tl, Sn, Pb}$$

trophiles is observed in the oxidative addition of organic halides to nickel(0) complexes via an electron-transfer mechanism (Eq. 50).[97]

$$\text{ArX} \xrightarrow[\text{electron transfer}]{\text{Ni(0)}} \text{Ni(I)X} + \text{Ar}\bullet \xrightarrow{\hspace{2cm}} \text{Ar-Ar} \qquad (50)$$

Although there are several probable processes leading to such by-products, arylboronic acids undergo a clean cross-coupling reaction in the presence of a palladium or nickel catalyst.

The by-products observed are the coupling products of phosphine-bound aryls and dimers of arylboronic acids. The oxidative addition of ArX to palladium(0) to afford **3**, followed by a sequence of transmetallation-reductive elimination yields a cross-coupling product **4**. Triarylphosphines are an excellent ligand for stabilizing the palladium(0) species. However, an undesirable side reaction of the aryl-exchange between palladium- and phosphine-bound aryls $(3 \rightarrow 5)$[98,99] leads to the coupling product **6** of phosphine-bound aryls (Eq. 51).

$$\text{Ar-X} \xrightarrow[k_1]{\text{Pd(PPh}_3)_4} \underset{\underset{\text{PPh}_3}{|}}{\overset{\overset{\text{PPh}_3}{|}}{\text{Ar}-\text{Pd}-\text{X}}} \xrightarrow[k_2]{\text{Ar'B(OH)}_2} \text{Ar-Ar'}$$

3   **4**

$$k_3 \updownarrow \quad \text{Intramolecular aryl exchange} \tag{51}$$

$$\underset{\underset{\text{PPh}_3}{|}}{\overset{\overset{\text{Ar}-\text{PPh}_2}{|}}{\text{Ph}-\text{Pd}-\text{X}}} \xrightarrow[k_4]{\text{Ar'B(OH)}_2} \text{Ph-Ar'}$$

5   **6**

Biaryl coupling of Eq. 52[100] indeed suffers from rapid incorporation of the phenyl group of Ph_3P. A bulky phosphine ligand of $(o\text{-MeOC}_6\text{H}_4)_3P$ is effective in retarding such a side reaction while maintaining a high yield of the desired product.

$$\tag{52}$$

$$\text{Ar} = \text{—} \langle \bigcirc \rangle \qquad 81\%, \; 7/8 = 67 / 33$$

$$\text{MeO} \text{—} \langle \bigcirc \rangle \qquad 82\%, \; 7/8 = 96 / 4$$

As shown in Table 3,[45] coupling with a phosphine ligand (Ph-Ar') is high with electron-rich haloarenes (Entries 5–7), whereas it is very low with electron-deficient haloarenes (Entries 1–4). The presence of an *ortho*-substituent reduces such coupling of the phosphine ligand (Entries 8 and 9). It is also interesting to note that bromoarenes always afford better selectivity than the corresponding iodides (entries 3, 4, 6, 7, 8 and 9), whereas iodoarenes are widely used because of their high reactivity in oxidative addition to palladium(0). Thus, the phosphine-bound aryls participate in the cross-coupling reactions of electron-rich haloarenes without the steric hindrance of *ortho* substitution.

The cross-coupling reaction of *p*-tolylboronic acid with *p*-iodo- or *p*-bromoanisole is accompanied by the formation of undesired biaryls

Table 3. Effects of substituents on the Selectivity of Biaryl Coupling between *p*-Tolylboronic Acid [Ar'B(OH)$_2$] and Haloarenes (ArX)^a

Entry	Haloarene ArX	Hammett σ	Distribution of Biaryls (%)		
			Ph-Ar'	Ar'-Ar'	Ar-Ar'
1	*p*-MeCOC$_6$H$_4$Br	+0.847	<1	1	98
2	*p*-ClC$_6$H$_4$I	+0.227	0	3	97
3	*m*-MeOC$_6$H$_4$I	+ 0.115	6	2	92
4	*m*-MeOC$_6$H$_4$Br	+0.115	4	1	95
5	*p*-MeC$_6$H$_4$I	−0.170	29	–	71
6	*p*-MeOC$_6$H$_4$I	−0.268	49	3	48
7	*p*-MeOC$_6$H$_4$Br	−0.268	33	1	66
8	*o*-MeOC$_6$H$_4$I		3	2	95
9	*o*-MeOC$_6$H$_4$Br		0	2	98

Note: ^{a}All reactions were carried out at 80 °C for 3 h using Pd(PPh$_3$)$_4$ (3 mol%), haloarene (1 mmol), *p*-tolylboronic acid (1.1 mmol) and Na$_2$CO$_3$ (2 mmol) in DME/H$_2$O (4/1, 5 ml).

derived from triphenylphosphine **9** and boronic acid **10** (Eq. 53 and Table 4).[45]

(53)

Coupling with a phosphine-bound phenyl is high with triphenyl-phosphine complexes, though the palladium-phosphine complexes retard the formation of **10** (Entries 1 and 9). The reaction is more selective when using a bulky ligand of *o*-tolylphosphine (Entry 5) or a bidentate ligand, such as dppf (Entries 6, 7, and 11), the results of which are similar to that of tris(*o*-methoxyphenyl)phosphine. A phosphine-free palladium catalyst, prepared *in situ* from Pd(OAc)$_2$, completely avoids the formation of **9**. However, unfortunately it increases homocoupling of arylboronic acid **10** (Entries 8 and 12–14) and is not effective for bromoarenes (Entry 8). In such cases, a strong base, such as K$_3$PO$_4$ or NaOH, is highly effective in minimizing the formation of **9**. Selectivity

Table 4. Effects of Catalysts, Bases, and Solvents on the Selectivity of Biaryl Coupling of p-Tolylboronic Acid and 4-Iodo- or 4-Bromoanisole (Eq. 53)[a]

					Conversion[b]	Distribution (%)		
Entry	X=	Catalyst	Solvent	Base	(%)	9	10	11
1	Br	Pd(PPh$_3$)$_4$	DME	Na$_2$CO$_3$	100	33	1	66
2	Br	Pd(PPh$_3$)$_4$	DME	K$_3$PO$_4$	88	4	2	94
3	Br	Pd(PPh$_3$)$_4$	DME	NaOH	82	1	7	92
4	Br	Pd(PPh$_3$)$_4$	DME	NaOH[c]	84	0	6	94
5	Br	Pd(OAc)$_2$ + 5(o-tolyl)$_3$P	DME	Na$_2$CO$_3$	100		10	90
6	Br	PdCl$_2$(dppf)	DME	Na$_2$CO$_3$	89	12	3	85
7	Br	PdCl$_2$(dppf)	DME	K$_3$PO$_4$	100	8	1	91
8	Br	Pd(OAc)$_2$	DME	Na$_2$CO$_3$	38	0	20	80
9	I	Pd(PPh$_3$)$_4$	DME	K$_3$PO$_4$	100	41	<1	59
10	I	Pd(PPh$_3$)$_4$	DMSO	K$_3$PO$_4$	98	15	2	83
11	I	PdCl$_2$(dppf)	DME	K$_3$PO$_4$	99	11	3	86
12	I	Pd(OAc)$_2$	DME	K$_3$PO$_4$	81[d]	0	11	89
13	I	Pd(OAc)$_2$	DMSO	K$_3$PO$_4$	100[d]	0	11	89
14	I	Pd(OAc)$_2$	EtOH	K$_3$PO$_4$	89[d]	0	6	94

Notes: [a] A solution of p-tolylboronic acid (1.1 mmol), 4-haloanisole (1 mmol), a catalyst (0.03 mmol), base (2 mmol) in solvent (5 mL, organic solvent/H$_2$O = 4/1) was stirred for 3 h at 80 °C.
[b] The reaction times to achieve a 100% conversion were not optimized.
[c] NaOH (3 mmol) was used.
[d] At 50 °C for 3 h.

of 94% is achieved when using Pd(PPh$_3$)$_4$ and aqueous K$_3$PO$_4$ or NaOH (Entries 2–4).

Aryl exchange occurs before transmetallation. Thus, the transmetallation rate constant k_2 which is low for steric and electronic reasons results in increasing the coupling product of phosphine-bound aryls (Eq. 51). Transmetallation is slowed down when electron-rich haloarenes and weak bases are used and accelerated with electron-deficient haloarenes because the transmetallation shown in Eqs. 16 and 17 involves nucleophilic substitution of Pd-X. The reported equilibrium ratio of **3/5** is 4/96 at 60 °C when Ar is p-methoxyphenyl.[98] Thus, it is quite reasonable that the reaction accompanies a large amount of **6** when the rate constant of transmetallation k_2 is lower than k_3. A strong base, polar solvent, and a sterically less hindered bidentate ligand, such as dppf, increase k_2. The formation of yields of **9** in p-iodoanisole higher than the bromo derivative is similarly rationalized by the relative values of k_2 and k_3 because transmetallation to Pd-Br is faster than that of Pd-I, which is reverse to the order in oxidative addition.[101,68] The steric hindrance of *ortho* substitution in both phosphines and haloarenes retards the aryl-exchange equilibration ($3\rightleftharpoons5$).

A very minor amount of homocoupling biaryl is derived during the reduction of a palladium(II) or nickel(II) halide complex with arylboronic acid (Eq. 23) or by the metathetic reaction shown in Eq. 48. However, a large number of homocoupling products of arylboronic acids are reported in literature. The mechanism proceeding through oxidative addition of the C–B bond to palladium(0) is recently proposed as the route to homocoupling (Eq. 54).[102] The oxidative addition of the C–B

$$\begin{array}{ccc}
& \text{ArB(OH)}_2 & \\
\text{Pd(0)} & \longrightarrow & \text{Ar-Pd-B(OH)}_2 \\
\uparrow & & \downarrow \\
& & \\
\text{Ar-Ar} & \text{B(OH)}_2 & \text{ArB(OH)}_2 \\
+ & | & \longleftarrow \\
\text{(HO)}_2\text{BB(OH)}_2 & \text{Ar-Pd-B(OH)}_2 & \\
& | & \\
& \text{Ar} &
\end{array}$$

(54)

bond to palladium(0) has been postulated in the C–B bond protonolysis[103] and the Heck-type reaction of aryl- or 1-alkenylboronic acids,[104] and all these reactions are reported to be catalyzed by phosphine-free palladium. The Pd(OAc)$_2$-catalyzed reactions in Entries 12–14 in Table 4 indeed result in higher yields (~10%) of homocoupling biaryl **10** than that of the palladium-phosphine complexes, but it seems quite unlikely that Eq. 54 will afford high yields of homocoupling products.

Careful consideration must be given to oxygen during experiments involving the cross-coupling reaction. When the reaction mixture is exposed to air, arylboronic acid readily produces a homocoupling biaryl **10** in the presence of a palladium(0) catalyst and base (Eq. 55).[45] The

$$2 \text{ } \underset{}{\bigcirc}\text{-B(OH)}_2 + \text{O}_2 \xrightarrow[\substack{\text{aq. Na}_2\text{CO}_3 \\ \text{benzene at r.t.}}]{\text{Pd(PPh}_3)_4} \underset{}{\bigcirc}\text{-}\underset{}{\bigcirc} + \underset{}{\bigcirc}\text{-OH}$$

(55)

30 min	31%	-
60 min	54%	-
120 min	64%	-
300 min	70%	30%

reaction is slow in the absence of a base, but it is very fast in the presence of an aqueous base. Thus, oxygen contamination in the solvent causes the homocoupling of arylboronic acid. It is also probable that such dimerization takes place during the workup operation in air when there is unreacted arylboronic acid.

Although the mechanism has not yet been elucidated, the catalytic cycle shown in Eq. 56 can be reasonably assumed. Transmetallation to

$$(56)$$

$Pd(OH)(OOH)(PPh_3)_2{}^{105}$ yields a homocoupling biaryl and hydrogen peroxide. Thus, this type of dimerization competes with the formation of phenols[106] (Eq. 55). However, even if the experimental operation strictly controls oxygen, another slow process leading to such a dimer is unavoidable when "naked palladium" is used as the catalyst.

V. SYNTHESIS OF SYMMETRICAL BIARYLS-HOMOCOUPLING

The oxidative homocoupling reaction of organometallic reagents provides a simple method for synthesizing symmetrical biaryls. However, the reaction has not been extensively studied because the dimerization of aryl halides, rather than arylmetal reagents, is synthetically a straightforward route to such biaryls.

Unlike the cross-coupling reaction discussed previously, homocoupling proceeds through a stepwise, double transmetallation of two organic groups on boron to PdX_2, followed by reductive elimination of a dimer. Thus, the use of a suitable reoxidant is essential for recycling the palladium catalyst, similar to the Wacker oxidation process. The selective oxidation of a palladium(0) species in the presence of arylboronic acid can be required for the oxidant. Oxidation with $Cu(OAc)_2{}^{107}$ or molecular oxygen[106] has been reported as a method for palladium-catalyzed dimerization of arylboronic acids (Eq. 57).

$$2 \quad \text{Ar-B(OH)}_2 \quad \xrightarrow{\text{a or b}} \quad \text{biaryl} \qquad (57)$$

a) $Cu(OAc)_2$, $Pd(OAc)_2/2PPh_3/NaOAc$ in MeOH at reflux
b) O_2, $Pd(OAc)_2$, aq. Na_2CO_3 in EtOH at r. t. (see, Eqs. 55 and 56)

VI. SYNTHESIS OF UNSYMMETRICAL BIARYLS

A. Directed *Ortho*-Metallation Cross-Coupling Sequence

The ready availability of *ortho*-functionalized arylboronic acids by a metallation-boronation sequence provides a synthetic link to the cross-coupling protocol, which allows syntheses of various polycyclic heteroaromatics via cyclization between two *ortho* functionalities. The synthesis of arylboronic acids having an CON^iPr_2,[18,108] $OCONEt_2$,[19,109] NH^tBoc,[21,22,110–113] or *ortho*-CHO,[76,114,115] and their coupling with various halides have been extensively studied by Gronowitz, Snieckus, and Rocca.[13,14]

Cyclization between 2-NHBoc and 2'-CHO spontaneously furnishes a pyridine ring of perlolidine (Eq. 58).[116] The synthesis can be achieved

$$(58)$$

by a combinations of either 2-BocHN arylboronic acid and 2-formyl aryl halides (Eq. 58) or 2-formyl arylboronic acids and 2-amino- or 2-acylamino aryl halides (Eq. 59).[114] However, the former combination or the use of 2-acethylamino haloarenes is recommended because the protodeboronation of 2-formyl arylboronic acids is often high during slow coupling with electron-rich amino haloarenes (Eq. 59).[14] Steric hin-

$$\text{(59)}$$

R = H (38%), R = MeCO (77%)

drance of *ortho* functionalities slows down the rate of coupling, but it is not a major factor in forming 2,2'-disubstituted biaryls unless one of the *ortho* groups is exceedingly bulky.

The sequence has considerable scope for synthesizing various biaryls and heterocycles, which have been previously reviewed[13,14] and are also shown Sections VI C and D. 2-Amido arylboronic acid is used for synthesizing 2,10-diazaphenanthrenes, the structure of which is found in numerous alkaloids of the "Marine sponge" family (Eq. 60).[116]

$$\text{(60)}$$

91-99%   X = F, Cl

Procedure for Eq. 60.[18,116] To a solution of s-BuLi (39.2 mmol) and TMEDA (39.2 mmol) in THF (200 mL) at −78 °C under argon is added a solution of N,N-diisopropylbenzamide (35.7 mmol) in THF (20 mL). The mixture is stirred for 45 min and treated with $(MeO)_3B$ (107 mmol). The mixture is allowed to warm to ambient temperature over 12 h, cooled to 0 °C, and acidified to pH 6.5 with 5% aqueous HCl. Removal of THF in vacuo followed by standard workup affords the crude 2-(diisopropyl-carbamoyl)-phenylboronic acid (95%). A mixture of $Pd(PPh_3)_4$ (0.03 mmol), the above boronic acid (1 mmol), 3-amino-2-fluoro-4-iodopyrid-ine (1 mmol) in toluene (10 mL), and aqueous K_2CO_3 (2M, 1 mL) is refluxed for 24 h. Filtration, extraction with toluene, drying over $MgSO_4$, and flash chromatography on silica gel (Et_2O/hexane = 8/2) give N,N-di-isopropyl-2-(3-amino-2-fluoro-4-pyridyl)benzamide (99%).

B.   Double Cross-Couplings in a One-Pot Procedure

Stepwise, double cross-coupling with dihaloarenes provides a method for synthesizing teraryls, quateraryls, and other higher order polyaryls.[117] Iodo-, bromo- and chloroarenes are well differentiated under standard conditions, thus allowing stepwise, double cross-coupling of two aryl-boronic acids (Eq. 61).[118]

$$(61)$$

This approach has been successfully applied to various polyaryls. For example, the synthesis of oligophenylene rods is shown in Eq. 62.[23,119]

$$(62)$$

The trimethylsilyl (TMS) or bromo group in the quaterphenyl is converted into the corresponding boronic acid by either of the procedures shown in Eq. 6, which allows further cross-coupling to synthesize oligophenylene rods.

Ready availability of arylboronic esters from aryl halides or triflates (Eqs. 13 and 14) now offers a one-pot, two-step procedure for synthesizing unsymmetrical biaryls. The cross-coupling reaction of bis(pinacolato)diboron with triflate in dioxane is followed by a subsequent coupling with another triflate in the presence of K_3PO_4 to furnish an unsymmetrical biaryl from two triflates (Eq. 63).[38] The synthesis from two different

$$(63)$$

a) bis(pinacolato)diboron (1.1 equiv), PdCl$_2$(dppf) + dppf, KOAc in dioxane at 80 °C,
b) 4-NCC$_6$H$_4$OTf (1 equiv), PdCl$_2$(dppf), K_3PO_4 in dioxane at 80 °C

aryl halides is similarly achieved with high yields by carrying out both reactions at 80 °C in DMF (Eq. 64).[120] Solid $K_3PO_4 \cdot nH_2O$ suspended in

(64)

83%

DMF or dioxane, an aqueous solution of K_3PO_4, or aqueous KOH are recommended as the base for pinacol esters because their coupling reaction is very slow with Na_2CO_3.

Typical Procedure for Eq. 63.[38] A flask is charged with PdCl$_2$(dppf) (0.03 mmol), dppf (0.03 mmol), KOAc (3 mmol), and bis(pinacolato)diboron (1.1 mmol) and flushed with nitrogen. Dioxane (6 mL) and aryl 4-acyl-phenyl triflate (1.1 mmol) are added, and then the resulting mixture is stirred at 80 °C for 6 h. After cooling to room temperature, K_3PO_4 (3 mmol), PdCl$_2$(dppf) (0.03 mmol), and 4-cyanophenyl triflate (1.0 mmol) are added. Then the mixture is heated to 80 °C for 16 h and yields 4-acetyl-4'-cyanobiphenyl (93%).

The diboron offers a direct and efficient method for synthesizing the boronic ester in the solid phase which hitherto has met with little success using classical methodology. A solid-phase boronate is quantitatively obtained by treating a polymer-bound iodobenzamide with the diboron at 80 °C for 20 h in the presence of PdCl$_2$(dppf) and KOAc (Eq. 65).[121]

(65)

90%

Subsequent coupling with iodo- or bromoarenes, followed by hydrolysis with trifluoroacetic acid, furnishes various biaryls. The robot synthesis or parallel synthesis on the surface of resin will be the topic of further accounts of this research.

C. Fine Chemicals and Functional Materials

Considerable attention has recently focused on synthesizing biaryls and other polyaryls because they are useful in designing functional molecules and functional materials. The semirigid structure of biaryls allows rational design of various molecular recognition compounds including drugs and super molecules, whose active sites must be fixed in a proper direction and distance. The large steric hindrance and restricted rotation of biaryls have provided various new metal ligands for asymmetric synthesis. The coupling protocol is a versatile tool for functionalizing the polymer surface which is applicable to combinatorial synthesis and polyphenylene synthesis.

Synthesis of Drugs

The isomeric michellamines A, B, and C are active against human immunodeficiency virus (HIV)-1 and HIV-2 strains. Michellamines A and B, both having identical tetrahydroisoquinoline structures and different configurations at the biaryl axes, are synthesized by a simultaneous double-coupling of the di-triflate (Eq. 66).[89,122–124] A highly hindered

$$(66)$$

biaryl coupling leading to a mixture of two atropodiastereoisomers arising from restricted rotation is achieved by using the barium base, as discussed in Section IV B.

Renin-angiotensin plays a critical role in regulating blood pressure. A sequence of direct *ortho*-metallation to give the key boronic acid and its cross-coupling reaction was reported by Merck in USA as a two-step procedure for the angiotensin II receptor antagonist (losartan) (Eq. 67).[125]

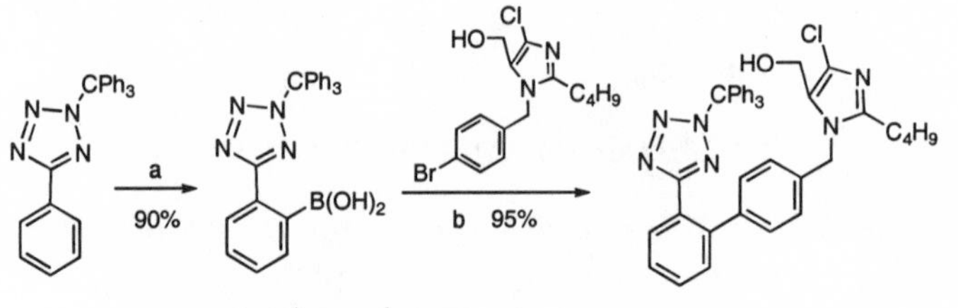

(67)

a) 1. BuLi, 2. B(O^{i}Pr)$_3$, 3. ^{i}PrOH/NH$_4$Cl/H$_2$O.
b) Pd(OAc)$_2$•4 PPh$_3$, aq. K$_2$CO$_3$ in THF/DME at reflux

A highly efficient, convergent approach overcomes many of the draw-backs associated with previously reported syntheses, thus providing a method for large-scale production using inexpensive and nonhazardous reagents. Tributylphosphine (10 mol%) is added before isolation which avoids the precipitation of palladium and minimizes contamination with elemental palladium to less than 18 ppm.

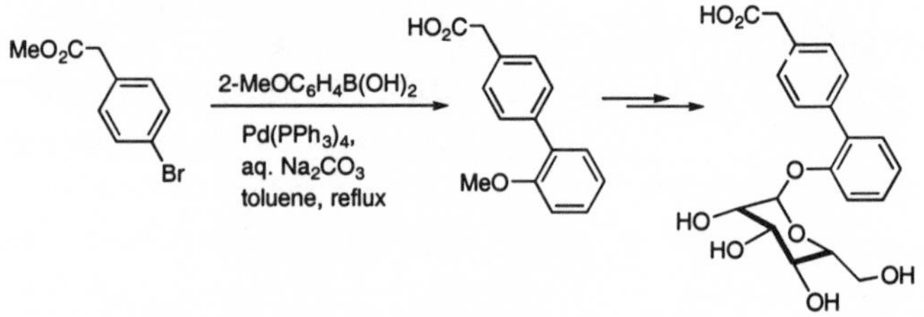

(68)

Palladium-mediated biaryl coupling was used as the key step to prepare a novel class of non-oligosaccharide selectin inhibitors, which were designed by Texas Biotechnology on the basis of the structural information about the sLeX tetrasaccharide.[126] The replacement of three sugar molecules of the native tetrasaccharide, while retaining a calcium recognition site (fucose) and a carboxylic acid, led to a new design of non-oligosaccharide selectin inhibitors which have greater *in vitro* potency (Eq. 68).

Arylboronic acids undergo cross-coupling with vinyl halides or triflates under conditions similar to biaryl coupling. The introduction of aryl moieties directly at the 2-position of carbapenum was first demonstrated by the Stille coupling of arylstannanes, but the procedure is being reinvestigated by Merck in USA because of the toxic by-products which are difficult to remove, especially in large-scale production (Eq. 69).[127]

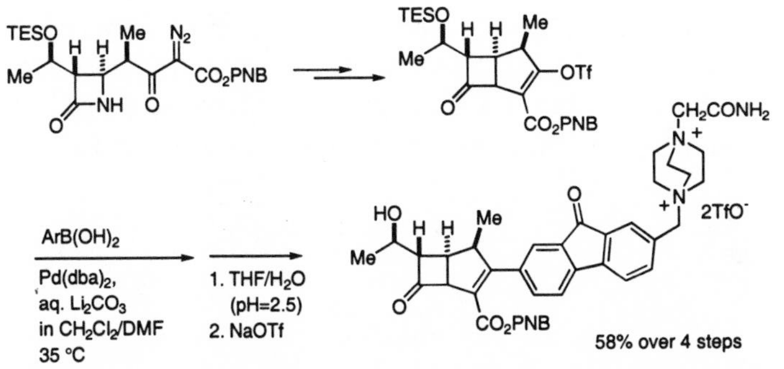

(69)

The reaction suffered from low yields due to the instability of the starting enol triflate. However, phosphine-free palladium was recognized as an excellent catalyst for carrying out the reaction at low temperature (see the discussion in Section IV A). The overall yield of four steps from TES diazo ester is 58%.

A homochiral boronate derived from the chiral imidazolidinone template yields various biaryls with little or no loss of stereochemical integrity. Although the amino ester shown in Eq. 38 undergoes partial racemization during biaryl coupling, a recent method of Novartis in USA using Seebach's imidazolidinone made the assessment of the diastereoselectivity exceedingly simple (Eq. 70).[128] The imidazolidinones

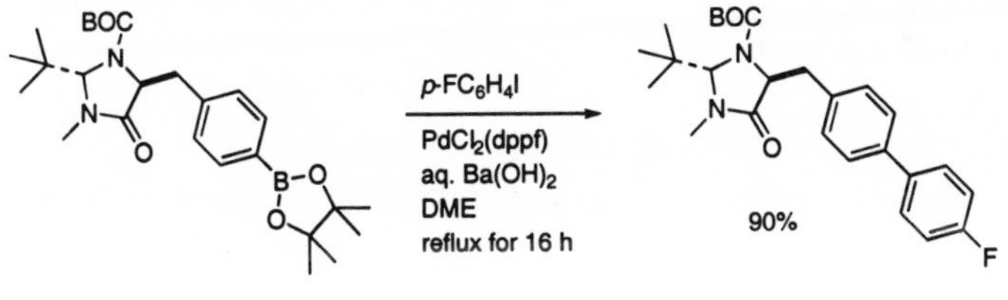

(70)

are known to be deprotected to the corresponding free amino acids which are potent inhibitors of zinc metalloprotease.

Solid-Phase Synthesis

Newly emerging solid-phase synthetic techniques for forming non-peptidic C–C bonds provide a highly efficient methodology for automated syntheses and multiple simultaneous syntheses in generating new compound libraries. Many pharmacologically interesting biaryls have been prepared on resins linking various halobenzenes or arylboronic acids.[121,129–132] The coupling reaction in the solid phase occurs even at room temperature when using a phosphine-free palladium and an activated iodobenzene (Eq. 71),[129] whereas the palladium-phosphine complex gives the best results for the corresponding bromo derivatives (Eq. 72).[130] The heterogeneous reaction in the solid phase is greatly acceler-

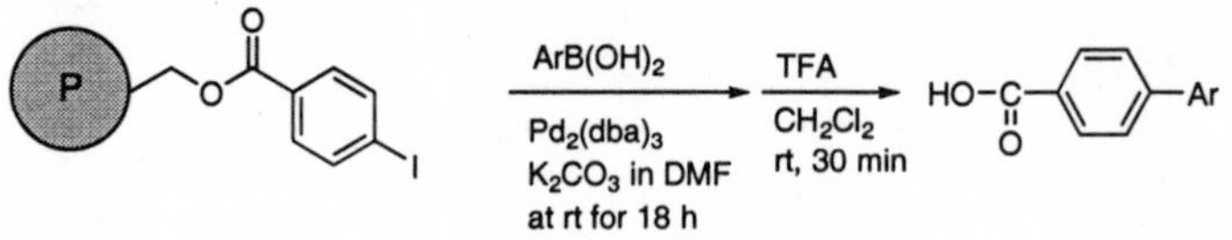

(71)

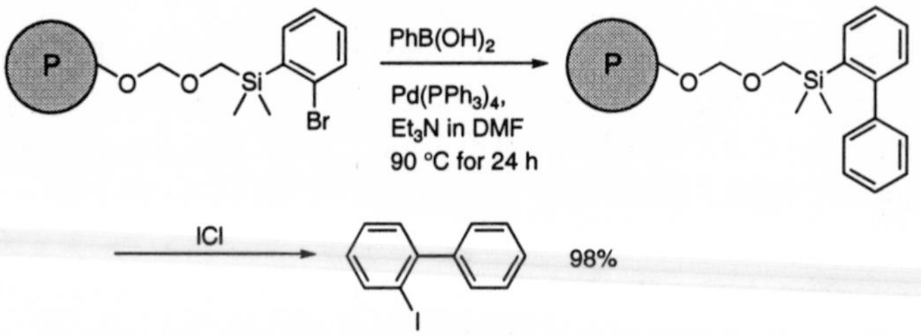

(72)

ated by microwave irradiation.[131] The reaction of arylboronic acids with 4-iodo- and 4-bromobenzoic acid linked to Rink amide (RAM) TentaGel results in almost quantitative yields within 3.8 min by irradiation with 45 W of microwave.

Super Molecules and Metal Ligands

Various molecular recognition molecules have been designed to create new ligands for metals because the semirigid structure of polyaryls allows the rational design of unnatural functional molecules by computational modeling.[133] Novel polydentate hemispherands having nine donor heteroatoms, which form overall neutral complexes with rare earth ions, were prepared with quantitative yields (Eq. 73).[133(a)] The semirigid,

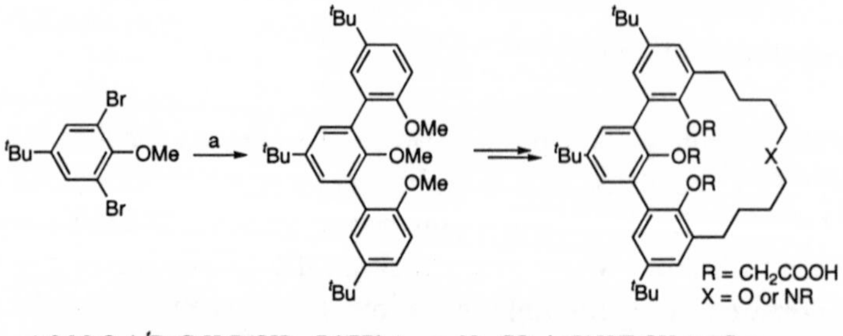

(73)

a) 2-MeO-4-^{t}BuC$_6$H$_3$B(OH)$_2$, Pd(PPh$_3$)$_4$, aq. Na$_2$CO$_3$ in PhH/EtOH at reflux.

saddle-shaped structure based on incorporating three dibenzofuran units into a macroring was expected to have potential binding or catalytic sites on the floor of the cleft. The *ortho* metallation-boronation approach again provides a straightforward method to add two (dihydroxy)boryl groups to dibenzofurane (Eq. 74).[134]

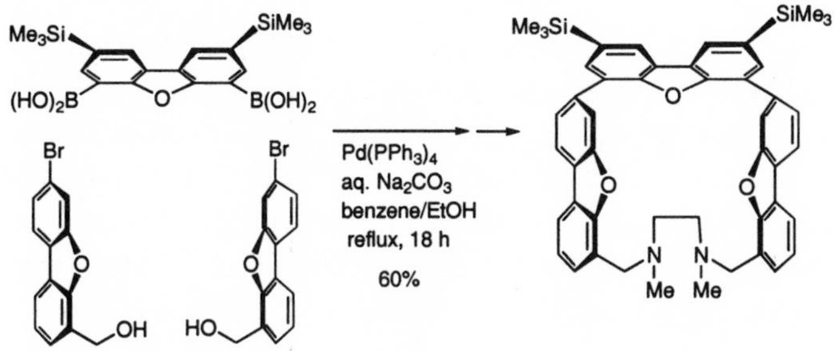

(74)

Chiral Biaryls

Biaryls with axial chirality are potentially important as chiral ligands for asymmetrical reactions and also as intermediates for synthesizing biologically active natural biaryl compounds, for example, michellamine in Eq. 66. (η^6-Disubstituted arene)chromium tricarbonyl complexes exist in two enantiomeric forms based on planar chirality. Biaryl coupling provides a new approach to synthesizing both optically pure atropisomers starting from a single chiral arene chromium complex (Eq. 75).[135]

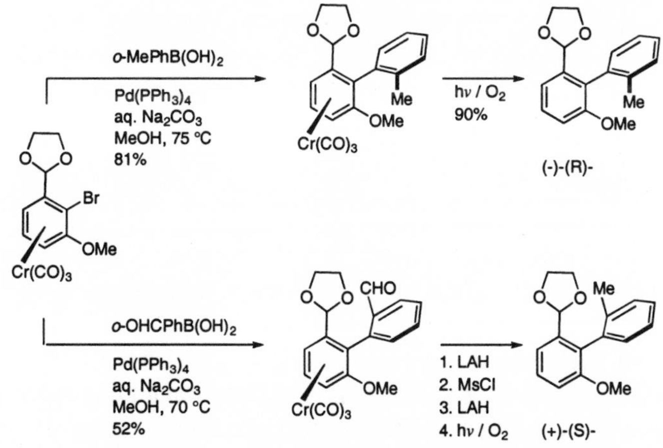

(75)

Coupling with *o*-tolylboronic acid diastereoselectively produces a kinetically controlled product in which the 2-methyl group is in *syn*-orientation to the tricarbonylchromium fragment. On the other hand, a less hindered 2-formylphenyl group allows the axial isomerization of the (R, R)-chromium complex to the thermodynamically more stable (R, S)-isomer during the cross-coupling reaction.

Liquid Crystals

The log, lath like molecular structure of most liquid crystalline compounds and the ever increasing complexity of the more advanced materials demanded by display device engineers make cross-coupling reactions very important in syntheses.[80(b),136] The method based on arylboronic acids simplifies the procedure for making liquid crystal materials with more complex substitution patterns (Eq. 76).[136(a)] The use of Pd/C

(76)

(a) $C_6H_3F_2B(OH)_2$, $Pd(PPh_3)_4$, aq. Na_2CO_3 at 80 °C. (b) (i) BuLi in THF at -78 °C, (ii) $B(OMe)_3$, (iii) 10% HCl. (c) $C_8H_{17}OC_6H_4Br$, $Pd(PPh_3)_4$, aq. Na_2CO_3 at 80 °C

catalysts provides a simple and economical alternative which avoids contaminating the phosphine ligand in the product and allows recycling the catalyst (Eq. 29).[69] This procedure is now used in an industrial process by Merck in Germany for several liquid crystals (1–3 tons/year).[137]

Dendrimers

A fully aromatic, water soluble, hyperbranched polymer, which complexes with small organic molecules in water, was prepared by homo-coupling (3,5-dibromophenyl)boronic acid with 80–95% yields (Eq. 77).[138] The molecular weight of the polymer depends on the organic solvents and temperature used for the coupling, but adding more monomer at the end of polymerization does not increase the molecular weight, though there is no definite steric saturation point at the surface. The polymer is converted into a water-soluble carboxylic acid derivative by sequential treatment with BuLi and CO_2.

$$(77)$$

The stepwise biaryl coupling of arylboronic acid to 3,5-dibromo-1-arylbenzene affords a series of monodisperse aromatic dendrimers having molecular diameters of 15–31 Å.[139]

Poly-(p-phenylene)s

The synthesis of poly(p-phenylene) via the homocoupling of p-bromophenylboronic acids was first reported by the Max Planck Institut.[140] After this discovery, various new poly(p-phenylenes) were synthesized.[141–146] A rigid-rod poly(p-phenylene) was prepared by cross-coupling a biphenyl diboronic ester[147] and a dibromide in aqueous DMF in the presence of a water soluble palladium catalyst (Eq. 78).[148] The

$$(78)$$

boronic ester is insoluble in water, but it is readily hydrolyzed to the acid to participate in cross-coupling.

Poly(p-phenylene), a highly insoluble polymer that has been studied extensively for its possible electronic and photonic applications, has a 23° twist between its consecutive aryl units caused by *ortho* hydrogen interactions. Attempts to enhance the solubility by substitution in the rings force the consecutive aryl units even further out of plane, resulting in plummeting of the extended conjugation. An *ortho* metallation-boronation and biaryl coupling, discussed in Section VI A, provides an excellent route to aromatic ladder polymers, keeping the consecutive aryl units planar while maximizing extended π-conjugation through the poly(p-phenylene) backbone (Eq. 79).[149] The dodecyl groups exert a

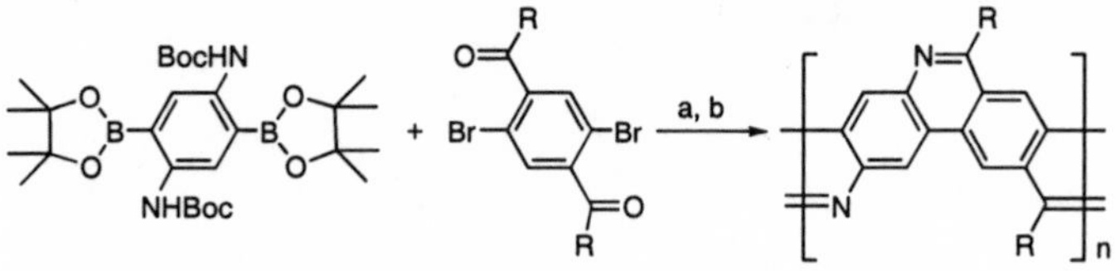

a) Pd(dba)$_2$ + 4.5 PPh$_3$, aq. NaHCO$_3$ in DME at 85 °C
b) CF$_3$COOH

$R = C_4H_{19}, C_{12}H_{25}$ (79)

plasticizing effect so that even this planar, rigid-rod polymer possesses good film-forming properties.

Solid polymer electrolytes have been actively pursued as a major contribution to the development of high-energy density batteries, particularly lithium secondary batteries. The poly(p-phenylene)s substituted with oligo(oxymethylene) side chains and blends of these novel polymers with lithium salts are achieved by biaryl coupling of benzene-1,4-diboronic acids[147] (Fig. 4).[150] Octapoly(p-phenylene) functions as the artificial ion channel which specially recognizes (bio)membranes by their thickness (Fig. 5).[151]

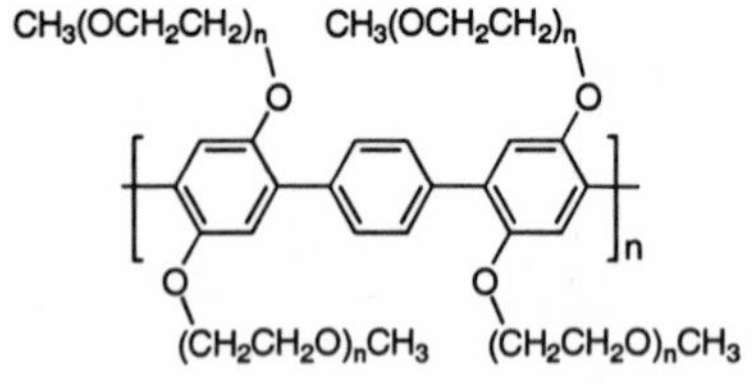

Figure 4. Polymer Electrolyte

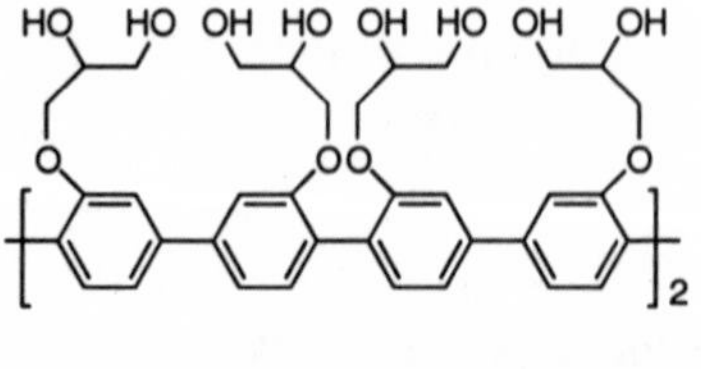

Figure 5. Artificial Ion Channels

Another example of π-conjugated materials is graphite ribbons (Eq. 80).[152] Polymerization by the biaryl-coupling strategy is followed by

(80)

(a) $ROC_6H_4C \equiv CH$, $PdCl_2(PPh_3)_2$, CuI, Bu_4NBr in toluene/Pr_2NH (1/1)
(b) $Pd(dba)_2 \cdot 8$ PPh_3, KOH in $PhNO_2/H_2O$ at 85 °C;
(c) CF_3COOH in CH_2Cl_2 at 25 °C

electrophile-induced cyclization to give a planar polyaromatic material which can be used in light-emitting diodes, nonlinear optical devices, lightweight batteries, and sensory devices. In biaryl-coupling polymerization, nitrobenzene was recognized as the best solvent for the palladium-phosphine catalyst, thus providing a highly efficient coupling method with a small amount of catalyst (0.3 mol%).[152(b)]

The synthesis of highly π-conjugated polymers has been recently reviewed.[153]

D. Tabular Survey

Representative results reported in biaryl coupling are summarized in Table 5. The reaction conditions are not optimized in most cases, and the procedures are still being improved. However, finding arylboronic acid derivatives, aryl electrophiles, and their coupling procedures can be useful.

Table 5. Synthesis of Biaryls via Cross-Coupling Reaction of Arylboronic Acids

Entry	Arylboronic Acid	Aryl Halide	Conditions[a]	Yield (%)	Ref.
1	C_6H_5–B(OH)$_2$	o-BrC$_6$H$_4$Me	PdL$_4$/aq. Na$_2$CO$_3$/benzene	94	15
2		o-BrC$_6$H$_4$OMe -EtOH/reflux/6 h	PdL$_4$/aq. Na$_2$CO$_3$/benzene	99	15
3		p-BrC$_6$H$_4$Cl	PdL$_4$/aq. Na$_2$CO$_3$/benzene -EtOH/reflux/6 h	89	15
4		p-BrC$_6$H$_4$CO$_2$Me	PdL$_4$/aq. Na$_2$CO$_3$/benzene -EtOH/reflux/6 h	94	15
5		(aryl bromide structure, CO$_2$Et)	PdL$_4$/aq. Na$_2$CO$_3$/toluene -EtOH/reflux/17 h	80	15
6		(aryl bromide structure, CN)	PdL$_4$/aq. NaHCO$_3$/benzene /reflux/4 h	72	154
7		(aryl bromide structure, OMOM, CO$_2$Me)	PdL$_4$/aq. Na$_2$CO$_3$/toluene -EtOH/reflux/23 h	90	155
8		(aryl bromide structure, CO$_2$Me)	Pd(OAc)$_2$ + P(o-tolyl)$_3$/Et$_3$N /DMF/100 °C/3.5 h	66	84
9		p-BrC$_6$H$_4$CH$_2$CH$_2$CO$_2$Et	PdL$_4$/CsF/DME-MeOH /reflux/8 h	95	49
10		p-BrC$_6$H$_4$CH$_2$C≡N	PdL$_4$/CsF/DME /reflux/2 h	92	49

#	Substrate	Conditions	Yield	Ref.
11	p-BrC$_6$H$_4$NHCOCF$_3$	PdL$_4$/CsF/DME /reflux/2 h	92	49
12	p-BrC$_6$H$_4$CH$_2$COCH$_3$	PdL$_4$/CsF/DME /reflux/3 h	85	49
13	(HO, Br-pyridine)	PdL$_4$/aq. Na$_2$CO$_3$/benzene -EtOH/reflux/1 h	57	156
14	(CH$_3$, NO$_2$, Br-pyridine)	PdL$_4$/aq. Na$_2$CO$_3$/benzene -EtOH/reflux/4 h	61	156
15	(pyrazole, I)	PdL$_4$/aq. Na$_2$CO$_3$/benzene -EtOH/reflux/22 h	0	156
16	(Br-pyrimidine)	PdL$_4$/aq. Na$_2$CO$_3$/benzene -EtOH/reflux/1 h	87	156
17	p-BrC$_6$H$_4$OMe	Pd(OAc)$_2$/K$_2$CO$_3$/NH$_4$NBr H$_2$O/70 °C/1 h	95	68
18	p-BrC$_6$H$_4$NHAc	Pd(OAc)$_2$/K$_2$CO$_3$/NH$_4$NBr H$_2$O/70 °C/2 h	98	68
19	(OMe, Br-naphthalene)	Pd(OAc)$_2$/K$_2$CO$_3$/NH$_4$NBr H$_2$O/70 °C/1 h	99	68
20	(Cl-pyridine)	PdCl$_2$(dppb)/aq. Na$_2$CO$_3$ /toluene/reflux/24 h	71	64

PhB(OH)$_2$

(continued)

Table 5. Continued

Entry	Arylboronic Acid	Aryl Halide	Conditions[a]	Yield (%)	Ref.
21		(HO, Cl-pyridine structure)	PdCl$_2$(dppb)/aq. Na$_2$CO$_3$ /toluene/reflux/24 h	15	64
22		(Cl-pyrimidine structure)	PdCl$_2$(dppb)/aq. Na$_2$CO$_3$ /toluene/reflux/24 h	65	64
23		(Cl-pyrazine structure)	PdCl$_2$(dppb)/aq. Na$_2$CO$_3$ /toluene/reflux/24 h	78	64
24		(Cl-carboline structure)	PdL$_4$/aq.Na$_2$CO$_3$/toluene /reflux/17 h	87	157
25		(OPh, Cl-triazine structure)	PdL$_4$/aq. Na$_2$CO$_3$/toluene /reflux/48 h	86	158
26		p-ClC$_6$H$_4$CHO	NiCl$_2$(dppf)–BuLi/K$_3$PO$_4$ /dioxane/80 °C/16 h	93	42
27		p-ClC$_6$H$_4$CO$_2$Me	NiCl$_2$(dppf)–BuLi/K$_3$PO$_4$ /dioxane/80 °C/16 h	89	42
28		p-ClC$_6$H$_4$OMe	NiCl$_2$(dppf)–BuLi/K$_3$PO$_4$ /dioxane/80 °C/16 h	83	42
29		p-ClC$_6$H$_4$NHAc	NiCl$_2$(dppf)–BuLi/K$_3$PO$_4$ /dioxane/80 °C/16 h	81	42
30		p-ClC$_6$H$_4$NH$_2$	NiCl$_2$(dppf)–BuLi/K$_3$PO$_4$ /dioxane/80 °C/16 h	89	42
31	CH$_3$–(C$_6$H$_4$)–B(OH)$_2$	(dichlorobiphenyl structure)	PdL$_4$/aq. Na$_2$CO$_3$/benzene -EtOH/reflux/overnight	56	159

Entry	Boronic acid	Aryl halide	Conditions	Yield (%)	Product
32	2,4,6-trimethylphenyl, $B(OH)_2$ (CH$_3$ ×3)	O_2N–, Br	PdL_4/aq. Na_2CO_3/benzene -EtOH/reflux/48 h	88	160
33	4-fluorophenyl, $B(OH)_2$	MeO–, Br	PdL_4/aq. $Ba(OH)_2$/DME /80 °C/16 h	84	161
34	biphenyl-3-yl, $B(OH)_2$	BnO–, Br, $O(CH_2)_3Cl$, C_2H_5	PdL_4/aq. Na_2CO_3/benzene -EtOH/reflux	95	162
35	naphthalen-1-yl, $B(OH)_2$	AcHN–pyridine, Br, CH_2Ph	PdL_4/aq. Na_2CO_3/toluene -EtOH/90 °C/17 h	83	163
36	4-fluorophenyl, $B(OH)_2$	pyridine, Br, ^{i}Bu, HO	PdL_4/aq. Na_2CO_3/toluene -MeOH/85 °C/14 h	77	164
37	2-methoxynaphthalen-1-yl, OMe, $B(OH)_2$	Ph, HO, Br	PdL_4/aq. Na_2CO_3/toluene -EtOH/reflux/10 h	96	165
38	dimethoxyphenyl, OMe, $B(OH)_2$, MeO	isoquinoline N, Cl	PdL_4/aq. Na_2CO_3/DME -EtOH/reflux/16	93	166
39	dimethoxyphenyl, OMe, MeO, $B(OH)_2$	Me_3Si–furan, I	PdL_4/aq. Na_2CO_3/toluene -MeOH/reflux/4 h	96	167
40	3,4-dimethoxyphenyl, $B(OH)_2$, OMe, MeO	chromone, O, I	PdL_4/aq. Na_2CO_3/benzene -EtOH/reflux/15 h		

(continued)

Table 5. Continued

Entry	Arylboronic Acid	Aryl Halide	Conditions[a]	Yield (%)	Ref.
41			PdL$_4$/aq. Na$_2$CO$_3$/benzene -EtOH/reflux/18 h	67	168
42		R= (CH$_2$)$_2$CH=CH(CH$_2$)$_3$CH$_3$	PdL$_4$/aq. Na$_2$CO$_3$/THF /reflux/2 h	60	169
43			PdL$_4$/aq. Na$_2$CO$_3$/benzene -EtOH/reflux	52	170
44			PdL$_4$/aq. NaHCO$_3$/toluene /70 °C/17 h	60	171
45			PdL$_4$/aq. NaHCO$_3$/DME /reflux/1 h	35	172
46			PdL$_4$/aq. Na$_2$CO$_3$/DME /reflux/24 h	40–49	115
47			PdL$_4$/aq. Na$_2$CO$_3$/toluene -EtOH/reflux/14 h; two bromo groups in the boronic acid remain intact	83	155
48			PdL$_4$/Et$_3$N/DME/100 °C 2–3 h	92	86
49			PdL$_4$/aq.Na$_2$CO$_3$/benzene -EtOH/reflux/48 h	99	159

	Boronic acid	Aryl halide	Conditions	Yield (%)	Ref.
50	![structure: 4-(dihydroxyboryl)benzonitrile, N≡C–C6H4–B(OH)2]	![structure: 5-bromo-2-ethoxy-1,3-dimethylbenzene with OC2H5, Me, Me, Br]	Pd(OAc)2 + 2 P(o-tol)3/aq. Na2CO3/toluene-MeOH/70–75 °C/3 h	96	173
51	![structure: 2-(NHCOtBu)phenylboronic acid, B(OH)2, NHCOtBu]	![structure: 2-chloro-3-fluoro-4-iodopyridine, Cl, F, I]	PdL4/aq. Na2CO3/toluene-EtOH/reflux/48 h	94	174
52	![structure: 2-(NHCOtBu)phenylboronic acid, B(OH)2, NHCOtBu]	![structure: 2-chloro-3-amino-4-iodopyridine, Cl, H2N, I]	PdL4/aq. K2CO3/toluene-EtOH/reflux/24 h	91	114
53	![structure: 2-(NHCOtBu)phenylboronic acid, B(OH)2, NHCOtBu]	![structure: 2-chloro-4-iodopyridine with HN= and pyridyl, Cl, I, HN]	PdL4/aq. Ba(OH)2/DME/80 °C/3 days	82	110
54	![structure: 3-methoxy-2-(NHCOtBu)-6-(dihydroxyboryl)phenol, B(OH)2, MeO, NHCOtBu]	![structure: 2-methoxy-3-(tBuCOHN)-4-iodopyridine, N, OMe, tBuCOHN, I]	PdL4/aq. N2CO3/toluene	88	90
55	![structure: 9,10-bis(OCONEt2)phenanthrene with B(OH)2, OCONEt2, B(OH)2]	![structure: 4-bromonitrobenzene, NO2, Br]	PdL4/aq. Na2CO3/DME/reflux	80	19
56	![structure: 2-(CONiPr2)phenylboronic acid, B(OH)2, CONiPr2]	![structure: 3-bromo-4-(Et2NCO2)pyridine, N, Br, Et2NCO2]	PdL4/aq. Na2CO3/toluene-EtOH/reflux/18 h	81	18
57	![structure: 2-(CONiPr2)phenylboronic acid, B(OH)2, CONiPr2]	![structure: 2-bromothiazole, N, S, Br]	PdL4/aq. Na2CO3/toluene/reflux	87	175
58	![structure: 3-methoxy-2-(CONiPr2)-6-(dihydroxyboryl)phenol, B(OH)2, MeO, CONiPr2]	![structure: bromo-trimethoxy-benzaldehyde, CHO, OMe, OMe, Br, MeO]	PdL4/aq. Na2CO3/DME/reflux	93	108

(continued)

235

Table 5. Continued

Entry	Arylboronic Acid	Aryl Halide	Conditions[a]	Yield (%)	Ref.
59			PdL$_4$/aq. Na$_2$CO$_3$/toluene /reflux/8 h	84	176
60			PdL$_4$/aq. Na$_2$CO$_3$/DME /reflux	56	177
61			PdL$_4$/aq. Na$_2$CO$_3$/MeOH reflux	60	178
62			reference 86	83	179
63			PdL$_4$/anhyd. KOH Bu$_4$NBr/benzene/reflux	60	79
64			PdL4/aq. KOH-Bu$_4$NBr /THF/reflux/8 h	77	67
65			PdL$_4$/aq. KOH-Bu$_4$NBr /THF/reflux/8 h	77	67
66			PdL$_4$/aq. KOH-Bu$_4$NBr /THF/reflux/8 h	62	67
67			PdL$_4$/aq. KOH-NH$_4$NBr /THF/reflux/3 h	75	180

236

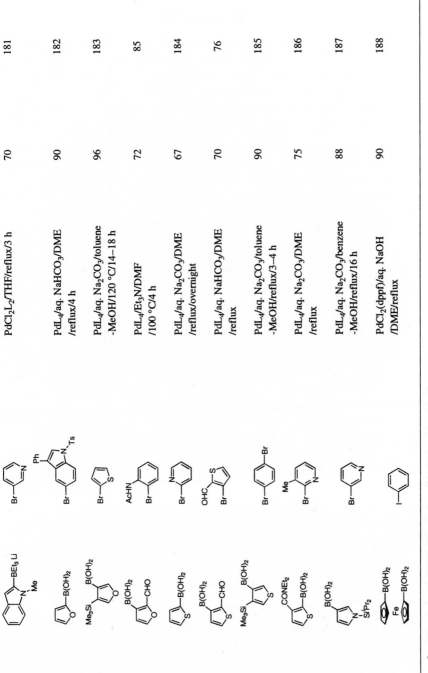

	Conditions	Yield (%)	Ref.
68	PdCl$_2$L$_2$/THF/reflux/3 h	70	181
69	PdL$_4$/aq. NaHCO$_3$/DME /reflux/4 h	90	182
70	PdL$_4$/aq. Na$_2$CO$_3$/toluene -MeOH/120 °C/14–18 h	96	183
71	PdL$_4$/Et$_3$N/DMF /100 °C/4 h	72	85
72	PdL$_4$/aq. Na$_2$CO$_3$/DME /reflux/overnight	67	184
73	PdL$_4$/aq. NaHCO$_3$/DME /reflux	70	76
74	PdL$_4$/aq. Na$_2$CO$_3$/toluene -MeOH/reflux/3–4 h	90	185
75	PdL$_4$/aq. Na$_2$CO$_3$/DME /reflux	75	186
76	PdL$_4$/aq. Na$_2$CO$_3$/benzene -MeOH/reflux/16 h	88	187
77	PdCl$_2$(dppf)/aq. NaOH /DME/reflux	90	188

Note: ^{a}L = PPh$_3$, dppb = Ph$_2$P(CH$_2$)$_4$PPh$_2$, dppf = 1,1′-bis(diphenylphosphino)ferrocene.

237

REFERENCES

1. For reviews on the synthesis of biaryls, see Sainsbury, M. *Tetrahedron* **1980**, *36*, 3327; Bringmann, G.; Walter, R.; Weirich, R. *Angew. Chem., Int. Ed. Engl.* **1990**, *29*, 977; Knight, D. W. *Comprehensive Organic Synthesis*; Trost, B. M.; Fleming, I.; Pattenden, G., Eds.; Pergamon: New York, 1991; Vol. 3, p. 481.

2. Semmelhack, M. F.; Helquist, P. M.; Jones, L. D. *J. Am. Chem. Soc.* **1971**, *93*, 5908. Zembayashi, M.; Tamao, K.; Yoshida, J.; Kumada, M. *Tetrahedron Lett.* **1977**, 4089; Takagi, K.; Hayama, N.; Inokawa, S. *Chem. Lett.* **1979**, 917.

3. (a) Farina, V. *Comprehensive Organometallic Chemistry*; Hegedus, L. S., Ed.; Pergamon: New York, 1995; Vol. 12, pp. 161–240; (b) Larock, R. C. *Metal-Organic Chemistry*, Liebeskind, L. S., Ed; JAI Press: London, 1994; Vol. 3, p. 97.

4. Hegedus, L. S. *Organometallics in Organic Synthesis*; Schlosser, M., Ed.; Wiley: New York, 1994; p. 383.

5. Tsuji, J. *Palladium Reagents and Catalysts*; Wiley: New York, 1995.

6. Murahashi, S.; Yamamura, M.; Yanagisawa, K.; Mita, N.; Kondo, K. *J. Org. Chem.* **1979**, *44*, 2408.

7. Tamao, K. *Comprehensive Organic Synthesis*; Trost, B. M.; Fleming, I.; Pattenden, G., Eds.; Pergamon: New York, 1991; Vol. 3, p. 435.

8. Negishi, E.; King, A. O.; Okukado, N. *J. Org. Chem.* **1977**, *42*, 1821. For a review, Negishi, E. *Acc. Chem. Res.* **1982**, *15*, 340.

9. Gouda, K.; Hagiwara, E.; Hatanaka, Y.; Hiyama, T. *J. Org. Chem.* **1996**, *61*, 7232 and references cited. For a review, Hatanaka, Y.; Hiyama, T. *Synlett.* **1991**, 845.

10. For reviews, Stille, B. J. *Angew. Chem., Int. Ed. Engl.* **1986**, *25*, 508; Farina, V.; Roth, G. P. *Metal-Organic Chemistry*, Liebeskind, L. S., Ed.; JAI Press: London, 1996; Vol. 6, pp. 1–53.

11. Miyaura, N.; Suzuki, A. *Chem. Rev.* **1995**, *95*, 2457.

12. Suzuki, A. *Acc. Chem. Res.* **1982**, *15*, 178; Miyaura, N.; Suzuki, A. *J. Synth. Org. Chem. Jpn.* **1993**, *51*, 1043.

13. Snieckus, V. *Chem. Rev.* **1990**, *90*, 879.

14. Martin, A. R.; Yang, Y. *Acta Chem. Scand.* **1993**, *47*, 221.

15. Miyaura, N.; Yanagi, T.; Suzuki, A. *Synth. Commun.* **1981**, *11*, 513.

16. (a) Gerrard, W. *The Chemistry of Boron*; Academic: New York, 1961. (b) Nesmeyanov, A. N.; Sokolik, R. A. *Methods of Elemento-Organic Chemistry*; North-Holland: Amsterdam, 1967; Vol. 1. (c) Onak, T. *Organoborane Chemistry*; Academic: New York, 1975.

17. Brown, H. C.; Cole, T. E. *Organometallics* **1983**, *2*, 1316. Brown, H. C.; Srebnik, M.; Cole, T. E. *Organometallics* **1986**, *5*, 2300.

18. Kandil, B. I. A.; Patil, P. A.; Sharp, M. J.; Siddiqui, M. A.; Snieckus, V. *J. Org. Chem.* **1991**, *56*, 3763.

19. Fu, J.; Sharp, M. J.; Snieckus, V. *Tetrahedron Lett.* **1988**, *29*, 5459.

20. James, C. A.; Snieckus, V. *Tetrahedron Lett.* **1997**, *38*, 8149. Wang, W.; Snieckus, V. *J. Org. Chem.* **1992**, *57*, 424.

21. Rocca, P.; Marsais, F.; Godard, A.; Queguiner, G. *Tetrahedron* **1993**, *49*, 49.

22. Myers, A. G.; Tom, N. J.; Fraley, M. E.; Cohen, S. B.; Madar, D. J. *J. Am. Chem. Soc.* **1997**, *119*, 6072.

23. Hensel, V.; Schluter, A.-D. *Liebigs Ann.* **1997**, 303.

24. Kaufmann, D. *Chem. Ber.* **1987**, *120*, 901.
25. Ye, X-S.; Wong, H. N. C. *J. Org. Chem.* **1997**, *62*, 1940.
26. Breuer, S. W.; Thorpe, F. G. *Tetrahedron Lett.* **1974**, 3719.
27. Gerrard, W.; Howarth, M.; Mooney, E. F.; Pratt, D. E. *J. Chem. Soc.* **1963**, 1582.
28. Zheng, Q.; Yang, Y.; Martin, A. R. *Tetrahedron Lett.* **1993**, *34*, 2235.
29. (a) Terashima, M.; Kakimi, H.; Ishikura, M.; Kamata, K. *Chem. Pharm. Bull.* **1983**, *31*, 4573. (b) Ishikura, M.; Oda, I.; Terashima, M. *Heterocycles* **1985**, *23*, 2375.
30. Muetterties, E. L. *J. Am. Chem. Soc.* **1960**, *82*, 4163.
31. Maddaford, S. P.; Keay, B. A. *J. Org. Chem.* **1994**, *59*, 6501; Cristofoli, W. A.; Keay, B. A. *Tetrahedron Lett.* **1991**, *32*, 5881.
32. Snyder, H. R.; Reedy, A. J.; Lennarz, W. J. *J. Am. Chem. Soc.* **1958**, *80*, 835.
33. Bean, F. R.; Johnson, J. R. *J. Am. Chem. Soc.* **1932**, *54*, 4415; Torssell, K. *Arkiv. Kemi.* **1956**, *10*, 473.
34. Michaelis, A.; Richter, E. *Liebigs Ann.* **1901**, *315*, 33.
35. Yamamoto, Y.; Seki, T.; Nemoto, H. *J. Org. Chem.* **1989**, *54*, 4734.
36. Malan, C.; Morin, C. *Synlett* **1996**, 167.
37. Ishiyama, T.; Murata, M.; Miyaura, N. *J. Org. Chem.* **1995**, *60*, 7508.
38. Ishiyama, T.; Ito, Y.; Kitano, T.; Miyaura, N. *Tetrahedron Lett.* **1997**, *38*, 3447.
39. Murata, M.; Watanabe, S.; Masuda, Y. *J. Org. Chem.* **1997**, *62*, 6458.
40. Aliprantis, A. O.; Canary, J. W. *J. Am. Chem. Soc.* **1994**, *116*, 6985.
41. Saito, S.; Sakai, M.; Miyaura, N. *Tetrahedron Lett.* **1996**, *37*, 2993.
42. Saito, S.; Oh-tani, K.; Miyaura, N. *J. Org. Chem.* **1997**, *62*, 8024.
43. Indolese, A. F. *Tetrahedron Lett.* **1997**, *38*, 3513.
44. Percec, V.; Bae, J-Y.; Hill, D. H. *J. Org. Chem.* **1995**, *60*, 1060.
45. Unpublished results.
46. Portony, M.; Milstein, D. *Organometallics* **1993**, *12*, 1665.
47. Foa, M.; Cassar, L. *J. Chem. Soc., Dalton Trans.* **1975**, 2572.
48. Kobayashi, Y.; Mizojiri, R. *Tetrahedron Lett.* **1996**, *37*, 8531.
49. Wright, S. W.; Hageman, D. L.; McClure, L. D. *J. Org. Chem.* **1994**, *59*, 6095.
50. Yoshida, T.; Okano, T.; Otsuka, S. *J. Chem. Soc., Dalton Trans.* **1976**, 993.
51. Grushin, V. V.; Alper, H. *Organometallics* **1993**, *12*, 1890.
52. Amatore, C.; Carré, Jutand, A.; M'Barki, M. A. *Organometallics* **1992**, *11*, 3009; Amatore, C.; Carré, Jutand, A.; M'Barki, M. A. *Organometallics* **1995**, *14*, 1818.
53. Moriya, T.; Miyaura, N.; Suzuki, A. *Synlett* **1994**, 149.
54. Otsuka, S. *J. Organomet. Chem.* **1980**, *200*, 191.
55. Kang, S-K.; Lee, H-W.; Jang, S-B.; Ho, P-S. *J. Org. Chem.* **1996**, *61*, 4720.
56. (a) Darses, S.; Jeffery, T.; Genet, J-P. *Tetrahedron Lett.* **1996**, *37*, 3857; (b) Sengupta, S.; Bhattacharyya, *J. Org. Chem.* **1997**, *62*, 3405.
57. Siegmann, K.; Pregosin, P. S.; Venanzi, L. M. *Organometallics* **1989**, *8*, 2659.
58. Ozawa, F.; Kubo, A.; Hayashi, T. *Chem. Lett.* **1992**, 2177.
59. Amatore, C.; Carré, E.; Jutand, A.; M'Barki, M. A. *Organometallics* **1995**, *14*, 1818.
60. Amatore, C.; Jutand, A.; Suarez, A. *J. Am. Chem. Soc.* **1993**, *115*, 9531.
61. Amatore, C.; Broeker, G.; Jutand, A.; Khalil, F. *J. Am. Chem. Soc.* **1997**, *119*, 5176.
62. Farina, V.; Krishnan, B. *J. Am. Chem. Soc.* **1991**, *113*, 9585.
63. Shen, W. *Tetrahedron Lett.* **1997**, *38*, 5575.

64. Ali, N. M.; McKillop, A.; Mitchell, M. B.; Rebelo, R. A.; Wallbank, P. J. *Tetrahedron* **1992**, *48*, 8117.

65. Beller, M.; Fischer, H.; Herrmann, W. A.; Ofele, K.; Brossmer, C. *Angew. Chem., Int. Ed. Engl.* **1995**, *34*, 1848.

66. Wallow, T.; Novak, B. M. *J. Org. Chem.* **1994**, *59*, 5034.

67. Bumagin, N. A.; Bykov, V. V.; Beletskaya, I. P. *Dan. SSSR.* **1990**, *315*, 1133.

68. Badone, D.; Baroni, M.; Cardamone, R.; Ielmini, A.; Guzzi, U. *J. Org. Chem.* **1997**, *62*, 7170.

69. Marck, G.; Villiger, A.; Buchecker, R. *Tetrahedron Lett.* **1994**, *35*, 3277.

70. Casalnuovo, A. L.; Calabrese, J. C. *J. Am. Chem. Soc.* **1990**, *112*, 4324.

71. Hoechst, WO97/5104, WO97/5151.

72. Jang, S-B. *Tetrahedron Lett.* **1997**, *38*, 1793.

73. Ishikura, M.; Kamada, M.; Terashima, M. *Synthesis* **1984**, 936.

74. Watanabe, T.; Miyaura, N.; Suzuki, A. *Synlett* **1992**, 207.

75. (a) Saito, S.; Kano, T.; Hatanaka, K.; Yamamoto, H. *J. Org. Chem.* **1997**, *62*, 5651. (b) Rocca, P.; Marsais, F.; Godard, A.; Queguiner, G. *Tetrahedron Lett.* **1993**, *34*, 2937.

76. Gronowitz, S.; Bobosik, V.; Lawitz, K. *Chim. Scripta* **1984**, *23*, 120.

77. Muller, D.; Fleury, J-P. *Tetrahedron Lett.* **1991**, *32*, 2229.

78. Kuvilla, H. G.; Nahabedian, K. V. *J. Am. Chem. Soc.* **1961**, *83*, 2159, 2164, and 2167; Kuvila, H. G.; Reuwer, J. F.; Mangravite, J. A. *J. Am. Chem. Soc.* **1964**, *86*, 2666.

79. Fernando, S. R. L.; Maharoof, U. S. M.; Deshayes, K. D.; Kinstle, T. H.; Ogawa, M. Y. *J. Am. Chem. Soc.* **1996**, *118*, 5783.

80. (a) Kallitsis, J. K.; Gravalos, K. G.; Hilberer, A.; Hadziioannou, G. *Macromolecules* **1997**, *30*, 2989; (b) Muller, H.; Tschierske, C. *J. Chem. Soc., Chem. Commun.* **1995**, 645.

81. Oh-e, T.; Miyaura, N.; Suzuki, A. *J. Org. Chem.* **1993**, *58*, 2201.

82. Shieh, W-C.; Carlson, J. A. *J. Org. Chem.* **1992**, *57*, 379.

83. Kumar, S. *Tetrahedron Lett.* **1996**, *37*, 6271.

84. Muller, W.; Lowe, D. A.; Neijt, H.; Urwyler, S.; Herrling, P. L. *Helv. Chim. Acta* **1992**, *75*, 855.

85. Yang, Y. *Synth. Commun.* **1989**, *19*, 1001.

86. Thompson, W. J.; Gaudino, J. *J. Org. Chem.* **1984**, *49*, 5237.

87. Ritter, K. *Synthesis* **1993**, 735.

88. Huth, A.; Beetz, I.; Schumann, I. *Tetrahedron Lett.* **1989**, *45*, 6679.

89. Kelly, T. R.; Garcia, A.; Lang, F.; Walsh, J. J.; Bhaskar, K. V.; Boyd, M. R.; Gotz, R.; Keller, P. A.; Walter, R.; Bringmann, G. *Tetrahedron Lett.* **1994**, *35*, 7621.

90. Godard, A.; Rovera, J-C.; Marsais, F.; Ple, N.; Queguiner, G. *Tetrahedron* **1992**, *48*, 4123.

91. Oh-e, T.; Miyaura, N.; Suzuki, A. *Synlett* **1990**, 221.

92. Kamikawa, T.; Hayashi, T. *Tetrahedron Lett.* **1997**, *38*, 7087.

93. Kowalski, M. H.; Hinkle, R. J.; Stang, P. J. *J. Org. Chem.* **1989**, *54*, 2783.

94. Darses, S.; Genet, J-P. *Tetrahedron Lett.* **1997**, *38*, 4393.

95. Ozawa, F.; Hidaka, T.; Yamamoto, T.; Yamamoto, A. *J. Organomet. Chem.* **1987**, *330*, 253.

96. Kochi, J. K. *Organometallic Mechanisms and Catalysis*; Academic: New York, 1978; pp. 419–421.

97. Tsou, T. T.; Kochi, J. K. *J. Am. Chem. Soc.* **1979**, *101*, 6319.

98. Kong, K-C.; Cheng, C-H. *J. Am. Chem. Soc.* **1991**, *113*, 6313.

99. Morita, D. K.; Stille, J. K.; Norton, J. R. *J. Am. Chem. Soc.* **1995**, *117*, 8576.

100. O'Keefe, D. F.; Dannock, M. C.; Marcuccio, S. M. *Tetrahedron Lett.* **1992**, *33*, 6679.

101. Ishiyama, T.; Miyaura, N.; Suzuki, A. *Bull. Chem. Soc. Jpn.* **1991**, *64*, 1999; Ishiyama, T.; Kizaki, H.; Miyaura, N.; Suzuki, A. *Tetrahedron Lett.* **1993**, *34*, 7595.

102. Moreno-Manas, M.; Perez, M.; Pleixats, R. *J. Org. Chem.* **1996**, *61*, 2346.

103. Yatagai, H. *Bull. Chem. Soc. Jpn.* **1980**, *53*, 1670.

104. Cho, C. S.; Uemura, S. *J. Organomet. Chem.* **1994**, *465*, 85; Cho, C. S.; Motofusa, S.; Ohe, K.; Uemura, S. *J. Org. Chem.* **1995**, *60*, 883.

105. Sheldon, R. A.; Kochi, J. K. *Metal-Catalyzed Oxidations of Organic Compounds*; Academic Press; New York, **1981**; pp. 79–83.

106. Smith, K. A.; Campi, E. M.; Jackson, W. R.; Marcuccio, S.; Naeslund, C. G. M.; Deacon, G. B. *Synlett* **1997**, 131.

107. Miyaura, N.; Suzuki, A. *Main Group Met. Chem.* **1987**, 295.

108. Zhao, B.; Snieckus, V. *Tetrahedron Lett.* **1991**, *32*, 5277.

109. Godard, A.; Rocca, P.; Pomel, V.; Dumont, T.; Rovera, J. C.; Thaburet, J. F.; Marsais, F.; Queguiner, G. *J. Organomet. Chem.* **1996**, *517*, 25.

110. Guillier, F.; Nivoliers, F.; Godard, A.; Marsais, F.; Queguiner, G. *Tetrahedron Lett.* **1994**, *35*, 6489.

111. Siddiqui, M. A.; Snieckus, V. *Tetrahedron Lett.* **1988**, *29*, 5463.

112. Cochennec, C.; Rocca, P.; Marsais, F.; Godard, A.; Queguiner, G. *J. Chem. Soc., Perkin Trans.* **1995**, 979.

113. Cochennec, C.; Rocca, P.; Marsais, F.; Godard, A.; Queguiner, G. *Synthesis* **1995**, 321.

114. Yang, Y.; Hornfeldt, A.-B.; Gronowitz, S. *Chim. Scripta* **1988**, *28*, 275.

115. (a) Siddiqui, M. A.; Snieckus, V. *Tetrahedron Lett.* **1990**, *31*, 1523; (b) The synthesis of 2-formyl arylboronic acids: see Gronowitz, S.; Hornfeldt, A. B.; Yang, Y. H. *Chim. Scripta* **1986**, *26*, 311. The synthesis of *o*-acyl arylboronic acid: Demeter, A.; Timari, G.; Kotschy, A.; Berces, T. *Tetrahedron Lett.* **1997**, *38*, 5219.

116. Rocca, P.; Cochennec, C.; Marsais, F.; Dumont, T.; Mallet, M.; Godard, A.; Queguiner, G. *J. Org. Chem.* **1993**, *58*, 7832.

117. Cheng, W.; Snieckus, V. *Tetrahedron Lett.* **1987**, *28*, 5097; Bahl, A.; Grahn, W.; Stadler, S.; Feiner, F.; Bourhill, G.; Brauchle, C.; Reisner, A.; Jones, P. G. *Angew. Chem., Int. Ed. Engl.* **1995**, *34*, 1485.

118. Unrau, C. M.; Campbell, M. G.; Snieckus, V. *Tetrahedron Lett.* **1992**, *33*, 2773.

119. Liess, P.; Hensel, V.; Schluter, A.-D. *Liebigs Ann.* **1996**, 1037.

120. Giroux, A.; Han, Y.; Prasit, P. *Tetrahedron Lett.* **1997**, *38*, 3841.

121. Piettre, S. R.; Baltzer, S. *Tetrahedron Lett.* **1997**, *38*, 1197.

122. Hoye, T. R.; Mi, L. *Tetrahedron Lett.* **1996**, *37*, 3097.

123. Hobbs, P. D.; Upender, V.; Liu, J.; Pollart, D. J.; Thomas, D. W.; Dawson, M. I. *J. Chem. Soc., Chem. Commun.* **1996**, 923.

124. Hoye, T. R.; Chen, M. *J. Org. Chem.* **1996**, *61*, 7940.

125. Larson, R. D.; King, A. O.; Chen, C. Y.; Corley, E. G.; Foster, B. S.; Roberts, F. E.; Yang, C. Y.; Lieberman, D. R.; Reamer, R. A.; Tschaen, D. M.; Verhoeven, T. R.; Reamer, R. A.; Arnett, J. F. *J. Org. Chem.* **1994**, *59*, 6391.

126. Kogan, T. P.; Dupre, B.; Keller, K. M.; Scott, I. L.; Bui, H.; Market, R. V.; Beck, P. J.; Voytus, J. A.; Revelle, B. M.; Scott, D. *J. Med. Chem.* **1995**, *38*, 4976.

127. A private communication; see also, Yasuda, N.; Xavier, L.; Rieger, D. L.; Li, Y.; DeCamp, A. E.; Dolling, U-H. *Tetrahedron Lett.* **1993**, *34*, 3211.

128. Satoh, Y.; Gude, C.; Chan, K.; Firooznia, F. *Tetrahedron Lett.* **1997**, *38*, 7645.

129. Guiles, J. W.; Johnson, S. G.; Murray, W. V. *J. Org. Chem.* **1996**, *61*, 5169.

130. Han, Y.; Walker, S. D.; Young, R. N. *Tetrahedron Lett.* **1996**, *37*, 2703.

131. Larhed, M.; Lindeberg, G.; Hallberg, A. *Tetrahedron Lett.* **1996**, *37*, 8219.

132. Yoo, S.; Seo, J.; Yi, K.; Gong, Y. *Tetrahedron Lett.* **1997**, *38*, 1203.

133. (a) Wolbers, M. P. O.; van Veggel, F. C. J. M.; Snellink-Ruel, B. H. M.; Hofstraat, J. W.; Geurts, F. A. J.; Reinhoudt, D. N. *J. Am. Chem. Soc.* **1997**, *119*, 138. (b) Helically chiral ligand: Judice, J. K.; Keipert, S. J.; Cram, D. J. *J. Chem. Soc., Chem. Commun.* **1993**, 1323 and 1325. (c) N-C-N Hexadentate ligand for Ru(II): Beley, M.; Chodorowski, S.; Collin, J-P.; Sauvage, J.-P. *Tetrahedron Lett.* **1993**, *34*, 2933. (d) Tetrakis(2-methoxy-1-naphthyl)porphyrin: Hayashi, T.; Miyahara, T.; Koide, N.; Kato, Y.; Masuda, H.; Ogoshi, H. *J. Am. Chem. Soc.* **1997**, *119*, 7281. (e) 3,3′-Diaryl binaphthols: Cox, P. J.; Wang, W.; Snieckus, V. *Tetrahedron Lett.* **1992**, *33*, 2253.

134. Schwartz, E. B.; Knobler, C. B.; Cram, D. J. *J. Am. Chem. Soc.* **1992**, *114*, 10775.

135. Kamikawa, K.; Watanabe, T.; Uemura, M. *J. Org. Chem.* **1996**, *61*, 1375.

136. (a) Hird, M.; Gray, G. W.; Toyne, K. J. *Mol. Cryst. Liq. Cryst.* **1991**, *206*, 187. (b) Trollsas, M.; Ihre, H.; Geddle, U. W.; Hult, A. *Macromol. Chem. Phys.* **1996**, *197*, 767. (c) Wulff, G.; Schmidt, H.; Witt, H.; Zentel, R. *Angew. Chem., Int. Ed. Engl.* **1994**, *33*, 188.

137. Poetsch, E.; Meyer, V. DE 4 241 747 C2. Poetsch, E.; Meyer, V.; Kompter, H. M.; Krause, J. DE 4 340 490 A1. Poetsch, E.; Meyer, V. DE 4 326 169 A1.

138. Kim, Y. H.; Webster, O. W. *J. Am. Chem. Soc.* **1990**, *112*, 4592.

139. (a) Miller, T. M.; Neenan, T. X.; Zayas, R.; Bair, H. E. *J. Am. Chem. Soc.* **1992**, *114*, 1018. (b) Percec, V.; Chu, P.; Ungar, G.; Zhou, J. *J. Am. Chem. Soc.* **1995**, *117*, 11441. (c) Wallimann, P.; Seiler, P.; Diederich, F. *Helv. Chim. Acta* **1996**, *79*, 779.

140. Rehahn, M.; Schluter, A.-D.; Wegner, G.; Feast, W. *J. Polymer* **1989**, *30*, 1054 and 1060.

141. Chmil, K.; Scherf, U. *Makromol. Chem.* **1993**, *194*, 1377.

142. Rau, I. U.; Rehahn, M. *Makromol. Chem.* **1993**, *194*, 2225.

143. Dendrimers based on poly(*p*-phenylene): Karakaya, B.; Claussen, W.; Gessler, K.; Saenger, W.; Schluter, A.-D. *J. Am. Chem. Soc.* **1997**, *119*, 3296.

144. Tanigaki, N.; Masuda, H.; Kaeriyama, K. *Polymer* **1997**, *38*, 1221.

145. Koch, F.; Heitz, W. *Makromol. Chem. Phys.* **1997**, *198*, 1531.

146. Propeller-like chiral conjugated polymer based on binaphthol: Ma, L.; Hu, Q-S.; Vitharana, D.; Wu, C.; Kwan, C. M. S.; Pu, L. *Macromolecules* **1997**, *30*, 204.

147. An improved procedure for benzene diboronic acid and ester (50–60%): Todd, M. H.; Balasubramanian, S.; Abell, C. *Tetrahedron Lett.* **1997**, *38*, 6781.

148. Wallow, T. I.; Novak, B. M. *J. Am. Chem. Soc.* **1991**, *113*, 7411.

149. Tour, J. M.; Lamba, J. J. S. *J. Am. Chem. Soc.* **1993**, *115*, 4935.
150. Lauter, U.; Meyer, W. H.; Wegner, G. *Macromolecules* **1997**, *30*, 2092.
151. Sakai, N.; Brennan, K. C.; Weiss, L. A.; Matile, S. *J. Am. Chem. Soc.* **1997**, *119*, 8726.
152. (a) Goldfinger, M. B.; Swager, T. M. *J. Am. Chem. Soc.* **1994**, *116*, 7895; (b) Goldfinger, Crawford, K. B.; M. B.; Swager, T. M. *J. Am. Chem. Soc.* **1997**, *119*, 4578.
153. Scherf, U.; Mullen, K. *Synthesis* **1992**, 23.
154. Wrobel, J. et al. *J. Med. Chem.* **1992**, *35*, 4613.
155. Manabe, K.; Okamura, K.; Date, T.; Koga, K. *J. Org. Chem.* **1993**, *58*, 6692.
156. Stavenuiter, J.; Hamzink, M.; Hulst, R. *Heterocycles* **1987**, *26*, 2711.
157. Bracher, F.; Hildebrand, D. *Liebigs Ann.* **1992**, 1315.
158. Janietz, D.; Bauer, M. *Synthesis* **1993**, 33.
159. Helms, A.; Heiler, D.; McLendon, G. *J. Am. Chem. Soc.* **1992**, *114*, 6227.
160. Miller, R. B.; Dugar, S. *Organometallics* **1984**, *3*, 1261.
161. Sawyer, J. S. et al. *J. Med. Chem.* **1993**, *36*, 3982.
162. Kelly, T. R.; Bridger, G. J.; Zhao, C. *J. Am. Chem. Soc.* **1990**, *112*, 8024.
163. Bolm, C.; Ewald, M.; Felder, M.; Schlingloff, G. *Chem. Ber.* **1992**, *125*, 1169.
164. Yang, H.; Hay, A. S. *J. Polym. Sci. Part A: Polym. Chem.* **1993**, *31*, 2015.
165. Alcock, N. W.; Brown, J. M.; Hulmes, D. I. *Tetrahedron: Asymmetry* **1993**, *4*, 743.
166. Song, Z. Z.; Wong, H. N. C. *Liebigs Ann. Chem.* **1994**, 29.
167. Yokoe, I.; Sugita, Y.; Shirataki, Y. *Chem. Pharm. Bull.* **1989**, *37*, 529.
168. Nair, V.; Powell, D. W.; Suri, S. C. *Synth. Commun.* **1987**, *17*, 1897.
169. Fukuyama, Y.; Kiriyama, Y.; Kodama, M. *Tetrahedron Lett.* **1993**, *34*, 7637.
170. Feldman, K. S.; Campbell, R. F. *J. Org. Chem.* **1995**, *60*, 1924.
171. Getahun, Z.; Jurd, L.; Chu, P. S.; Lin, C. M.; Hamel, E. *J. Med. Chem.* **1992**, *35*, 1058.
172. Keseru, G. M.; Mezey-Vandor, G.; Nogradi, M.; Vermes, B.; Kajtar-Peredy, M. *Tetrahedron* **1992**, *48*, 913.
173. Wong, M. S.; Nicoud, J.-F. *Tetrahedron Lett.* **1993**, *34*, 8237.
174. Rocca, P.; Marsais, F.; Godard, A.; Queguiner, G. *Tetrahedron* **1993**, *49*, 3325.
175. Sharp, M. J.; Snieckus, V. *Tetrahedron Lett.* **1985**, *26*, 5997.
176. Sharp, M. J.; Cheng, W.; Snieckus, V. *Tetrahedron Lett.* **1987**, *28*, 5093.
177. Iwao, M.; Iihama, T.; Mahalanabis, K. K.; Perrier, H.; Snieckus, V. *J. Org. Chem.* **1989**, *54*, 24.
178. Uemura, M.; Nishimura, H.; Kamikawa, K.; Nakayama, K.; Hayashi, Y. *Tetrahedron Lett.* **1994**, *35*, 1909.
179. Achab, S.; Guyot, M.; Potier, P. *Tetrahedron Lett.* **1993**, *34*, 2127.
180. Ishikura, M.; Kamada, M.; Terashima, M. *Heterocycles* **1984**, *22*, 265.
181. Ishikura, M.; Terashima, M. *J. Chem. Soc., Chem. Commun.* **1989**, 135.
182. Yang, Y.; Martin, A. R. *Synth. Commun.* **1992**, *22*, 1757.
183. Zhong, Z.; Zhou, Z. Y.; Mak, T. C. W.; Wong, H. N. C. *Angew. Chem., Int. Ed. Engl.* **1993**, *32*, 432.
184. Gronowitz, S.; Lawitz, K. *Chim. Scripta* **1984**, *24*, 5.
185. Song, Z. Z.; Ho, M. S.; Wong, H. N. C. *J. Org. Chem.* **1994**, *59*, 3917.
186. Brandao, M. A. F.; Oliveira, A. B.; Snieckus, V. *Tetrahedron Lett.* **1993**, *34*, 2437.
187. Alvarez, A.; Duzman, A.; Ruiz, A.; Velarde, E. *J. Org. Chem.* **1992**, *57*, 1653.
188. Knapp, R.; Rehahn, M. *J. Organomet. Chem.* **1993**, *452*, 235.

INDEX

245

Advances in Metal-Organic Chemistry

Edited by **Lanny S. Liebeskind,**
Department of Chemistry, Emory University

Organometallic chemistry is having a major impact on modern day organic chemistry in industry and academia. Within the last ten years, the use of transition metal based chemistry to perform reactions of significant potential in organic synthesis has come of age. *Advances in Metal-Organic Chemistry* contains in-depth accounts of newly emerging synthetic organic methods and of important concepts that highlight the unique attributes of organometallic chemistry applied to problems in organic synthesis. Particular emphasis will be given to transition metal organometallics. Each issue contains six to eight articles written by leading investigators in the field. Emphasis is placed on giving the reader a true feeling of the particular strengths and weaknesses of the new chemistry with ample experimental details for typical procedures. Contributors have been urged to write in an informal style in order to make the material accessible to interested readers who are not experts in the field.

Volume 1, 1989, 393 pp. $109.50/£69.50
ISBN 0-89232-863-0

CONTENTS: Introduction to the Series: An Editor's Foreword, *Albert Padwa.* Preface. *Lanny S. Liebeskind.* Recent Developments in the Synthetic Applications of Organoiron and Organomolybdenum Chemistry, *Anthony J. Pearson.* New Carbonylations by Means of Transition Metal Catalysts, *Iwao Ojima.* Chiral Arene Chromium Carbonyl Complexes in Asymmetric Synthesis, *Arlette Solladie-Cavallo.* Metal Mediated Additions to Conjugated Dienes, *Jan-E. Backvall.* Metal-Organic Approach to Stereoselective Synthesis of Exocyclic Alkenes, *Ei-ichi Negishi.* Transition Metal Carbene Complexes in Organic Synthesis, *William D. Wulff.* Index.

Volume 2, 1991, 300 pp. $109.50/£69.50
ISBN 0-89232-948-3

CONTENTS: Introduction, *Lanny Liebeskind.* Synthetic Applications of Chromium Tricarbonyl Stabilized Benzylic Carbanions, *Steven J. Coote, Stephen G. Davies and Craig L. Goodfellow.* Palladium-Mediated Arylation of Enol Ethers, *G. Doyle Daves, Jr.* Transition-Metal Catalyzed Silymetallation of

JAI PRESS INC.
100 Prospect Street, P. O. Box 811
Stamford, Connecticut 06904-0811
Tel: (203) 323-9606 Fax: (203) 357-8446

J
A
I

P
R
E
S
S

J A I   P R E S S

Acetylenes and Et_3B Induced Radical Addition of Ph_3SnH to Acetylenes: Selective Synthesis of Vinylsilanes and Vinlystannanes, *Koichiro Oshima*. Development of Carbene Complexes of Iron as New Reagents for Synthetic Organic Chemistry, *Paul Helquist*. Tricarbonyl (n^6-Arene) Chromium Complexes in Organic Synthesis, *Motokazu Uemura*. p-Bond Hybridization in Transition Metal Complexes: A Stereoelectronic Model for Conformational Analysis, *William E. Crowe and Stuart L. Schreiber*. Palladium Mediated Methylenecyclopropane Ring Opening: Applications to Organic Synthesis, *William A. Donaldson*. Index.

Volume 3, 1994, 321 pp. $109.50/£69.50
ISBN 1-55938-406-9

CONTENTS: Introduction, *Lanny Liebeskind*. Orthomanganated Aryl Ketones and Related Compounds in Organic Synthesis, *Lindsay Main and Brian K. Nicholson*. Cyclopropylcarbene-Chromium Complexes: Versatile Reagents for the Synthesis of Five-Membered Rings, *James W. Herndon*. Organic Synthesis via Vinylpalladium Compounds, *Richard C. Larock*. Ruthenium Catalyzed Oxidative Transformation of Alcohols, *Shun-Ichi Murahashi and T. Naota*. Palladium-Catalyzed Carbonyl Allylation via Allylpalladium Complexes, *Yoshiro Masuyama*. Index.

Volume 4, 1995, 317 pp.
ISBN 1-55938-709-2 $109.50/£69.50

CONTENTS: Preface, *Lanny Liebeskind*. Recent Progress in Higher Order Cyanocuprate Chemistry, *Bruce H. Lipshutz*. The Evolution of a Commercially Feasible Prostaglandin Synthesis, *James R. Behling, John S. Ng and Paul W. Collins*. Transition Metal Promoted Higher Order Cycloaddition Reactions, *James H. Ridgy*. Acyclic Diene Tricarbonyiron Complexes in Organic Synthesis, *Rene Gree and J.P. Lellouche*. Novel Carbonylation Reactions Catalyzed by Transitions Metal Complexes, *Masanobu Hidai and Youichi Ishii*. Index.

Volume 5, 1996, 267 pp. $109.50/£69.50
ISBN 1-55938-789-0

CONTENTS: Preface, *Lanny S. Liebeskind*. Recent Advances in the Stille Reaction, *Vittorio Farina and Gregory P. Roth*. Seven-Membered Ring Synthesis via Iron-Mediated Carbonylative Ring Expansion and s-Alkyl-p-Allyl Complexes, *Peter Eilbracht*. New Catalytic Asymmetric Carbon-Carbon Bond-Forming Reactions, *Masakatsu Shibasaki*. Recent Improvements and Developments in Heck-Type Reactions and Their Potential in Organic Synthesis, *Tuyêt Jeffery*. Index.

JAI PRESS INC.

100 Prospect Street, P. O. Box 811
Stamford, Connecticut 06904-0811
Tel: (203) 323-9606 Fax: (203) 357-8446